BACTERIOLOGY RESEARCH DEVELOPMENTS

THE MANY BENEFITS OF LACTIC ACID BACTERIA

BACTERIOLOGY RESEARCH DEVELOPMENTS

Additional books and e-books in this series can be found on Nova's website under the Series tab.

PUBLIC HEALTH IN THE 21ST CENTURY

Additional books and e-books in this series can be found on Nova's website under the Series tab.

BACTERIOLOGY RESEARCH DEVELOPMENTS

THE MANY BENEFITS OF LACTIC ACID BACTERIA

JEAN GUY LEBLANC
AND
ALEJANDRA DE MORENO DE LEBLANC
EDITORS

nova
science publishers
New York

NOTICE TO THE READER

Library of Congress Cataloging-in-Publication Data

ISBN: 978-1-53615-388-0
Library of Congress Control Number: 2019938763

Published by Nova Science Publishers, Inc. † New York

CONTENTS

PREFACE

Lactic Acid Bacteria (LAB) are a heterologous group of microorganisms that have been isolated from numerous ecological niches, including fermented foods, plants, and the gastrointestinal tract of animals. Because of their generally regarded as safe status (GRAS), there has been great interest in using these microorganisms in food production, as probiotic microorganisms or as biotechnological tools. This book will describe some of the many benefits of LAB including: i) their use in foods where advances in the fight against spoilage and pathogenic microorganisms in foods, their thermotolerance, their microencapsulation and responses to osmotic challenges will be discussed; ii) their capacity to produce beneficial compounds including bioactive peptides, biosurfactants, gamma-aminobutyric acid and antimicrobial products such as organic acids, hydrogen peroxide, bacteriocins and peptidoglycan hydrolases; and iii) their effect on health and other applications such as their use as DNA vaccine delivery system, bile-salt hydrolase and exopolysaccharides production as well as the use of spore forming LAB. This new book is a compilation of topics that have been written by experts from all over the world (Argentina, Brazil, Greece, Mexico, and Thailand) who work in different research settings offering varying viewpoints on the most up-to-date information currently available on the uses and many benefits of Lactic Acid Bacteria.

This book contains 13 chapters that have been divided into 3 main sections: i) the use of LAB in foods; ii) beneficial compounds produced by LAB; and iii) the effects of LAB on health and other applications.

In the first section the use of LAB in the fight against spoilage and pathogenic microorganisms in foods, the themotolerance of probiotic LAB in cooked meat products, the microencapsulation of LAB in meat products and LAB responses to osmotic challenges will be discussed by an Argentinean, a Mexican, a Brazilian, and a Greek group.

In the second section, LAB as a source of bioactive peptides and surfactants, antimicrobial products such as bacteriocins and peptidoglycan hydrolases, and of gamma-aminobutyric acid will be discussed by Argentinean, Mexican, and Thai groups.

In the final section, the effects of LAB on health and other applications will be reviewed. Here, the use of LAB as a DNA vaccine delivery system, the potential of bile salt hydrolase activities of LAB on health, exopolysaccharide-related aspects of LAB and the utilization of spore-forming LAB will be discussed by Brazilian, Thai and Argentinean researchers.

ACKNOWLEDGMENTS

We would like to express our recognition to effort of all the authors that have contributed to this book. We are also grateful to staff of Nova Publishers for their professionalism and their confidence in allow us to put together this important work. On a personal note we would like to acknowledge the Centro de Referencia para Lactobacilos (CERELA) and the Universidad Nacional de Tucumán for their academic support and the CONICET and FonCyT for their constant financial support.

USE OF LACTIC ACID BACTERIA IN FOODS

In: The Many Benefits of Lactic Acid Bacteria
Editors: J. G. LeBlanc and A. de Moreno

ISBN: 978-1-53615-388-0
© 2019 Nova Science Publishers, Inc.

Chapter 1

LACTIC ACID BACTERIA AGAINST SPOILAGE AND PATHOGENIC MICROORGANISMS IN FOOD

Marcela P. Castro[1,2], Carmen A. Campos[3,4,*], María E. Cayré[1],
Laura I. Schelegueda[3,4], Nadia Galante[1,2] and Sofía B. Delcarlo[3,4]

[1]Laboratorio de Microbiología de Alimentos, Departamento de Ciencias Básicas
y Aplicadas, Universidad Nacional del Chaco Austral, Sáenz Peña, Chaco, Argentina
[2]Consejo Nacional de Investigaciones Científicas y Técnicas de la República
(CONICET), Buenos Aires, Argentina
[3]Departamento de industrias, Facultad de Ciencia Exactas y Naturales,
Universidad de Buenos Aires, Buenos Aires, Argentina
[4]Instituto de Tecnología de Alimentos y Procesos Químicos (ITAPOQ), CONICET –
Universidad de Buenos Aires, Buenos Aires, Argentina

ABSTRACT

Advances in the microbiological safety of foods have been largely driven by public demand in response to disease outbreaks. A vast array of possibilities has emerged straightforward, being biopreservation the most suitable alternative to please consumers claims for healthier and "greener" foods. Biopreservation uses microorganisms, generally lactic acid bacteria (LAB) as protective cultures (PC). LAB can be added not only to prolong the shelf-life but also to preserve the microbial and/or sensorial quality of food products. The microbial antagonism of LAB against undesirable microorganisms has been attributed in many cases to the effect of one or synergism between several mechanisms, such as competition for nutrients, and production of different metabolites, including bacteriocins. Having GRAS status, LAB-PC can be safely added to foods. Advantages of PC include additive-free preservation, natural image, and food safety

---

* Corresponding Author's Email: carmen@di.fcen.uba.ar.

improvements. Hundreds of strains from many LAB species have been extensively studied for the preservation of foods showing promising results. The objective of this chapter is to describe current methods and developing technologies for the application of LAB-PC in food products throughout the world. The benefits of new technologies and their industrial limitations are also presented and discussed. Information given will provide an integrated approach on this topic.

INTRODUCTION

Food matrices are prone to microbial colonization due to their nutritional compounds. Some of these microbes are responsible for food spoilage, and some others are food-borne pathogens which can cause serious health problems once consumed (Varsha & Nampoothiri, 2016). Hence, it is necessary the implementation of adequate technologies to preserve and prolong the shelf-life of food products as well as the health of consumers. Moreover, there is a rising consumer demand for chemical preservative-free products, which results in a perfect opportunity for the implementation of natural additives with antibacterial and antifungal properties.

Among the new approaches, the so-called biopreservation stands out as a promising alternative, specifically the protective cultures (PC) incorporation into food matrices. Microorganisms that produce inhibitory substances, are able to compete against spoilage bacteria and fungi, and extend food shelf life, are consider as PC (Alexandraki et al., 2013; Fernández et al., 2017). The practice of PC inoculation into food matrices is the conception of the biopreservation approach.

Lactic acid bacteria (LAB) have been used traditionally as starter cultures in fermentation processes, being natural inhabitants of fermented food products. Besides, their implementation provides several technological advantages (Bourdichon et al., 2012). LAB incorporation also results in properties changes of the matrix that are non-beneficial for pathogenic and spoilage microorganisms. Furthermore, considering their GRAS (Generally Regarded as Safe) status (EFSA, 2011), LAB become a potential perfect group to be regarded as PC contributing to biopreservation (Divya et al., 2012 a,b).

Lactic acid is the main metabolic end-product from LAB, as a result of carbohydrates degradation. Its production causes a pH decrease, preventing undesirable microorganism proliferation. Lactic acid together with carbon dioxide, hydrogen peroxide, acetoin and other organic acids, all products from LAB metabolism, induce physicochemical changes on food matrices which create a disadvantageous environment for the development of spoilage and pathogenic fungi and bacteria (Deegan et al., 2006). Despite LAB metabolism products have been studied throughout, being their beneficial effects widely known, other LAB molecules have gained newly interest. Among them, bacteriocins are bioactive peptides ribosomally synthetized that have antimicrobial activity against other bacteria (Field, Ross & Hill, 2018). Bacteriocins are produced by many Gram-positive and Gram-negative species (Riley & Wertz, 2002). In spite of this, the ones produced by LAB are of particular interest to the food industry. LAB bacteriocins demonstrated antagonistic action against many food spoilage and pathogenic microbes, in consequence they are an important alternative to chemical additives, prolonging food shelf life, improving safety and preserving quality of food products (Gálvez et al., 2007; Field, Ross & Hill, 2018). Likewise, as mentioned before, many bacteriocinogenic LAB have GRAS status, which make them potentially applicable to food products being non-toxic to consumers and even enhancing health effects. Beside small metabolites, also peptides with antifungal activity have been reported as resulting from LAB bioactivity. These antifungal peptides are not produced *de novo* by LAB; however, they are produced during proteolysis of proteins present in their surroundings (Siedler et al., 2019).

The objective of this chapter is to describe current methods and developing technologies for the application of LAB-PC in food products throughout the world. The benefits of new technologies and their industrial limitations are also presented and discussed. Information given will provide an integrated approach on this topic. Table 1 is presented herein as a brief sample of the magnitude of the diversity of LAB species, target microorganisms and food matrices comprised within this biopreservation approach.

SCREENING METHODOLOGIES AND STRAIN SELECTION FOR FOOD BIOPRESERVATION

LAB bacteriocins play an important role in the future of food biopreservation. Their narrow killing spectrums and their antimicrobial mechanisms distinguish them from traditional antibiotics. These particular characteristics infer an alternative to fight against drug-resistant bacteria, preserving and prolonging food shelf-life and also decreasing antibiotics ingestion by the consumers, fulfilling consumer demands for more natural and chemical preservative-free products (Cavera et al., 2015; Kaskoniene et al., 2017; Zou et al.,2018).

Bacteriocin-producing LAB strains are isolated from diverse sources, but it is considered that those isolated from a certain matrix constitute the best candidates for biopreservation in that particular matrix, since they have better adaptability to the source environment than LAB isolated from other sources (Castro et al., 2011). In recent years bacteriocins have been discovered from various sources, particularly fermented food products (Castro et al., 2011; Abdel-Haliem et al., 2013; Gao et al., 2015; Gautam & Sharma, 2015) but also fish (An et al., 2015), chicken intestines (Rani et al., 2016), soil (Smitha & Bhat, 2013; Lim and Balolong, 2016), human body and its excrement (Lakshminarayanan et al., 2013; Wannun et al., 2014; Maldonado-Barragán et al., 2016) and citrus (Noda et al., 2015), among others. However, in the USA and Canada, the only bacteriocins commercially available are pediocin, produced by *Pediococcus* (*Ped.*) *acidilactici*, Micocin®, a blend of three bacteriocins (carnocyclin A, carnobacteriocin BM1, and piscicolin 126) produced by *Carnobacterium* (*Cb.*) *maltaromaticum* UAL307, and nisin, produced by *Lactococcus* (*Lc.*) *lactis*. Considering these commercial bacteriocins, nisin is the only one approved by the FDA as a food preservative (Martin-Visscher et al., 2011; Huang et al., 2016; Zou et al., 2018).

Hence, as a result of the multiples sources in which bacteriocins can be found, the processes implied in their discovery are very complex and significant resources and time must be invested in the screening, identification, purification, and characterization stages (Kaskoniene et al., 2017). Then, it has become a necessity the development of novel and faster methods to search new bacteriocins as an alternative to the traditional slower and laborious ones.

Traditional screening methodologies for antagonistic activity of bacteriocins, like disc-diffusion, agar well diffusion and spot-on-lawn assays, uses strains in the exponential phase of *Escherichia* (*E.*) *coli, Staphylococcus* (*S.*) *aureus* or *Listeria* (*L.*) *monocytogenes* as indicators to inoculate a soft agar culture media (Pingitore et al., 2007; Song et al., 2014a; Pasteris et al., 2014; Gao et al., 2015; An et al., 2015; Maldonado-Barragán et al., 2016). Samples are centrifuged to obtain free-cells supernatant possibly containing the bacteriocin if the evaluated bacteria are bacteriocinogenic. With a plastic straw, wells are created on the inoculated soft agar media and they are filled with the supernatant or with a purified extract for the spot-on-the-lawn assay (Zou et al., 2018). After incubation, the presence of a clear zone around the wells is considered as a positive inhibitory effect, and its activity is related to the measure of the area of these inhibitory zones. Even though the positive results are easy to identify and discriminate from negative ones, these colonies based assays turn out to be time-consuming since the samples screened are often quite numerous.

One of the novel methods is the PCR array assay developed to identify the presence of genes that codify for bacteriocin production on LAB isolated from food samples. This screening methodology is based on the comparison with known bacteriocin-related genes from the NCBI GenBank database (Macwana & Muriana, 2012; Juturu & Wu, 2108). However, PCR-based methods require the design of primers for each bacteriocin-producing strain since different strains contain different gene sequences that codify for bacteriocin production (Zou et al., 2018). Therefore, PCR-based methods are user-friendlier but also time-consuming. Another new screening methodology is the multi-well plate protocol for bacteriocin detection and it is based on an *in vivo* screening platform that uses 96-well plates and fluorescence imaging analysis to identify bacteriocins from natural sources (Son et al., 2016). The simplicity, fastness, reliability and the high cost-effective ratio of this method are the main advantages against traditional ones.

Table 1. Brief compilation of several LAB PC species, target microorganisms and food matrices comprised within the scope of bioprotective cultures.

LAB strain	Food product	Target microornganism	Reference
Lb. curvatus 54M16	Italian Fermented sausage	*L. monocytogenes* Spoilge microorganisms	Gilleo et al., 2018
Lb. curvatus *Lb. sakei*	Vacuum-packaged sliced raw beef	Spoilage microorganisms	Zhang et al., 2018b
Lb. plantarum subsp. *plantarum* CICC 6257	CO_2-MAP Ground Pork	*L. monocytogenes*	Zhang et al., 2018a
Lb. plantarum PSC20	*Chorizo* sausage	*L. monocytogenes*	Nikodinoska et al., 2019
Lb. plantarum EC52	Meat sausage	*L. monocytogenes* – *E. coli* O157: H7	Díaz –Ruiz et al., 2012
Lb. plantarum LMG P-26358 *Lc. lactis* CSK2775	Cheese	*L. innocua*	Mills et al., 2017
Lb. plantarum B2	Melon	*L. monocytogenes*	Russso et al., 2015
Lb. plantarum CUK-501	Cucumber	*Aspergillus flavus, Fusarium graminearum, Rhizopus stolonifer* and *Botrytis cinerea*	Sathe et al., 2007
Lb. PlantarumLc. lactis	Sea bass	*Listeria* spp,	Boulares et al., 2016
Lb. rhamnosus GG	Fresh-cut pear	*Salmonella*and*L.monocytogenes*	Iglesias et al., 2018
Lb. rhamnosus 131RZFAUU	Domiati soft cheese	*St. aureus*	Ibrahim et al., 2017
Lb. harbinensis K.V9.3.1Np	Yogurt Fermented milk	*Yarrowia lipolytica* *Penicillum expansum* Spoilage fungi	Mieszin et al., 2017 Mosbah et al., 2018 Delavenne et al., 2013, 2015
Lb. fermentum R6	Chicken breast meat	*C. perfringens*	Li et al., 2017
Lc. lactis CBM21	Apple Lamb´s lettuce	*L. monocytogenes* Yeasts Native microflora	Siroli et al., 2016 Siroli et al., 2017
Lc. lactis PSY 2	Reef cod	Total mesophilic, psychrophilic LAB, *Staphylococcus* sp., Enterobacteriaceae and *Pseudomonas* spp.	Sarika et al., 2012
Lc lactis CH-HP15	Model ham liquid medium	Pathogenic or spoilage spore-forming species	Ramaroson et al.,2018
Lc. piscium	Shrimps	*Br. thermosphacta*	Fall et al., 2007
Ec. lactis 4CP3	Chicken breast	*L. monocytogenes*	Ben Braiek et al., 2019
Ec. mundtii CRL 35	Beaker sausage	*L. monocytogenes*	Orihuel et al., 2018
Ec. faecium DM224 and DM270	Whey cheese	*L. monocytogenes*	Aspri et al., 2016 (a, b), 2017
Leuc. mesenteroides CM135, CM160 and PM249	Golden Delicious apples and Iceberg lettuce	*L. monocytogenes*	Trias et al., 2008
Carnobacterium spp.	Ricotta	*Pseudomonas* spp.	Spanu et al., 2017, 2018
Cb. maltaromaticum	Fillets of surubim	*L. monocytogenes*	dos Reis et al., 2011
Cb. maltaromaticum	Lean beef	*Salmonella* spp.	Hu et al., 2019
Ped. acidilactici CRL 1753	Bread	Spoilage fungi	Bustos et al., 2018

Once the screening process is complete, the identification of bacteriocin-producing strains is the next step. Normally, the identification of bacteriocin-producing species and strains is based on physiological and biochemical features and 16S rDNA gene sequences (Barbosa et al., 2014; Zhu et al., 2014; Gao et al., 2015). Subsequently after the identification comes the most critical and complex step, purification. This process implies an intensive labor and it is also quite time wasting. Purification of bacteriocins is directly related with applications, purity and price of the end product. The methodology applied to purify bacteriocins depends on the features of each particular bacteriocin, so there is not a single method suitable for all of them and each applied technique has its advantages and disadvantages (Sabo et al., 2018; Zou et al., 2018; Juturu & Wu, 2108). Afterward the screening, identification and purification phases, the characterization of bacteriocins is the last step. There are novel methods, capillary electrophoresis-mass spectrometry designed by Catala-Clariana et al., (2010) for the identification of bioactive peptides in hypoallergenic infant milk formula and, the micellar electrokinetic chromatography, developed by Soliman & Donkor (2010) to quantify nisin on food products. These methods possess the advantage of being faster, more sensitive and efficient than traditional ones.

As mentioned before, the best candidates as biopreservatives on a given matrix are the strains isolated specifically from that matrix, although not always the bacteriocin-producing LAB discovered from a given source is applied on the same source. Hence, an additional and necessary step is to prove the strains on different model mediums, changing certain conditions to determinate the adaptability of the strains to the new environment and confirm the production of the bacteriocin on those conditions, the so-called selection phase.

Selection consists on examination of the strains that exhibited inhibitory activity on the screening stage, and determinate their survival in simulated conditions, such as starvation, presence of NaCl and bile salts, low pH and high temperatures (Matamoros et al., 2009). Another important aspect is to confirm the maintenance or improvement of the sensory, chemical and physical properties of the target matrix since LAB may have negative effects in different types of food and contribute to food spoilage (Gram et al., 2002; Maragkoudakis et al., 2009). Then, innocuousness is also a factor to take into account, specifically no biogenic amine production and antibiotics resistance. As an example, in Wessels & Huss (1996) research work, some selected strains presented good efficiency in model medium but cannot be applied in seafood because their growth was inhibited by this low-sugar and low-temperature environment, and even could spoil the product.

Summarizing, there are many methods to select, identify, purify and characterize bacteriocin-producing LAB strains in search for the best candidates as biopreservatives. Even though there are no severe rules on the order of the steps to follow or the methodologies to implement, the chosen method must be in accordance to the future application together with the enhancement of food safety and quality.

LIMITATIONS AND FURTHER CONSIDERATIONS

Despite the fact that LAB can produce a variety of compounds with antimicrobial activity, the amounts of individual compounds produced by LAB are often insufficient to sustain the inhibitory phenotype, which is most likely to arise from synergistic or additive bioactivity of the blend of compounds. Thus, identification of new antimicrobial compounds is hampered by their occurrence in low amounts in complex matrices. Moreover, they may not exert inhibition *per se*, which hinders the identification by fractionation techniques. According to Siedler et al., (2019), new approaches are needed to identify and study these complex relationships. Furthermore, knowledge on the physiology of the spoilage organisms is required to understand the mechanisms behind the bioactive compounds.

The main drawback in LAB-PC application into food is the inherent antimicrobial activity loss due to interactions with ingredients and/or processing driven physicochemical instability. Peptide function can be affected by several factors, such as fat content, proteolytic degradation, polar or non-polar food components, pH (which influences the solubility of the bacteriocin) and sodium chloride concentrations, among others. To overcome these obstacles, many bacteriocins have undergone several bioengineering strategies to improve bacteriocin efficacy in the food environment. These strategies comprised a wide array of alternatives, from solubility enhancement (Rouse et al., 2012) to new bacteriocin biosynthetic gene clusters which could be used to guide the bioengineering of new and existing peptide structures (Song et al., 2014b; Sun et al., 2015). The driving force behind these scientific facts is that bioengineered strains which have been tailored through food-grade approaches are not considered genetically modified microorganisms (GMMs) and can be directly added to food (Field et al., 2018). To expand these concepts, the reader would find interesting the reviews from Field at al. (2015), and Ramu et al., (2015).

Among the emerging potential roles for bacteriocin-producing LAB, one of the most outstanding is in functional foods where they will be playing a health-promoting task beyond the biopreservative one. The use of multipurpose bacteria will enable to market food products with an over-the-counter formula that could provide, for instance, a positive modulation of the gastrointestinal (GI) microbiome. As an illustration, a study showed that the strain *Lactobacillus (Lb.) salivarius* UCC118 was able to modify the GI microbiota in diet-induced obese mice whereas its bacteriocin-free derivative did not exert this effect (Murphy et al., 2013). Conversely, many LAB strains with proved probiotic activity, intended at a beginning to be used as health enhancers, are being under study in order to expand their utility to PC. *Lb. sakei* CRL 1862 comprises a good example of this fact, being a well-known antihypertensive biopeptide producer, it had been proved to control food pathogens (*L. monocytogenes*) colonization on food-industry surfaces (Pérez Ibarreche et al., 2014, 2016).

APPLICATIONS OF PROTECTIVE CULTURES IN MILK AND DAIRY PRODUCTS

Milk and dairy products owe their excellent growth medium conditions to their highly nutritious nature. As a matter of fact, microbial contamination of milk and dairy products is a worldwide problem. Currently, foodborne illnesses from consumption of dairy products have been mainly related to *Campylobacter* (*C.*) *jejuni, E. coli* 0157:H7, *L. monocytogenes, Salmonella* spp., *Yersinia* (*Ye.*) *enterocolitica*, and *Bacillus* (*B.*) *cereus*. Moulds, mainly species of *Aspergillus, Fusarium*, and *Penicillium,* can grow in milk and dairy products. Under the appropriate conditions, these moulds may produce mycotoxins which can be a health hazard. Contamination of raw milk comprises the main microbiological hazard associated with dairy products; pathogens like *S. aureus* may be part of the resident microflora of the living animal, whereas other pathogens such as *E. coli* 0157:H7 or *Salmonella* spp. may originate from faecal contamination during initial milk collection. Despite the many methods applied to control it, contamination post processing can also happen in every stage of the product shelf-life, such as transport, storage and handling. LAB occur naturally in foods and have a long history of safe use in dairy industry. Lack of health risk concerns make them the best candidates for biopreservation and a great alternative to the use of chemical preservatives in dairy products, where they can be introduced as part of the fermentation process or as adjuvant cultures. A number of applications of bacteriocin-producing LAB have been reported to successful control pathogens in milk, yogurt, and cheeses (Silva et al., 2018). Applying bacteriocin-producing LAB strains as antibacterial starter cultures and PC may confer an advantage over the use of semi-purified/purified bacteriocins as much as they are adsorbed into food matrices and are easily degraded, which results in a loss of antibacterial activity.

Anti-listerial activity of many LAB strains has been exploited to control Listeria growth in different ripened and fresh cheeses (Coelho et al., 2014; Ibrahim et al., 2017; Silva et al., 2018), while their ability to exert antifungal activity has been optimally used to preserve fermented milk, cheeses and yogurt (Mieszkin et al., 2017; Mosbah et al., 2018; Leyva Salas et al., 2018). Studies described herein have been selected from a vast amount of scientific reports taking into consideration their relatively novelty and groundbreaking practices.

Milk and dairy products are recognized as a good source for the isolation of bacteriocin-producing LAB. Accordingly, donkeys' milk has been a subject of research in recent years, even though its nutritional and therapeutic properties have been recognized since ancient times. Resembling human milk in many ways, donkey milk is recommended as a substitute for individuals presenting cows' milk protein allergy. Aspri et al., (2016 a,b; 2017) had isolated LAB from raw donkey milk and characterized the strains by phenotypical and molecular methods. Besides, they evaluated their technological properties and safety characteristics as potential starter or adjunct cultures in the production of fermented dairy products. Based on these features, 79 strains were selected and analysed for inhibitory activity against spoilage bacteria and foodborne pathogens. Amongst them, 3 strains belonging to *Enterococcus* (*Ec.*) *faecium* displayed antimicrobial activity against *L. monocytogenes*, *S. aureus* and *B. cereus*. Bacteriocins produced by these strains were tested for their capability to control post-processing contamination and growth of *L. monocytogenes* during refrigerated storage of artificially contaminated fresh whey cheese; two of them (*Ec. faecium* DM 224 and 270) were bactericidal while one was bacteriostatic (*Ec. faecium* DM 33). These strains were tested in a food model and showed a good performance within the cheese matrix, thus they have a promising perspective for biopreservation.

Cheese is considered to be one of the foods most frequently contaminated with *L. monocytogenes* (Aspri et al., 2017). In particular, blue-veined and mould cheeses such as Brie, Camembert, Danish Blue, Stilton and Gorgonzola represent a suitable substrate for the growth of this pathogen (Ramaswamy et al., 2007). The famous Italian PDO Gorgonzola cheese is made using pasteurized cow's milk inoculated with thermophilic cultures (*Streptococcus thermophilus* and *Lb. delbrueckii*), yeasts and, finally, *Penicillium* spp. which confers the typical blue-green veined. Albeit different physical treatments had been studied to control Listeria in Gorgonzola rinds, irregular and rough surfaces hampered treatment efficacy and imparted visible changes in the aspect of the rinds. As a consequence, the use of LAB-PC resulted in a promising alternative to limit the growth of *L. monocytogenes* and *L. innocua* on Gorgonzola surface during ripening process (Morandi et al., 2019). The anti-Listeria activity differed among the LAB strains and it was Listeria biotype-dependent. Amongst 27 strains, *Carnobacterium* SCA, SCB, *Lb. plantarum* NA18, *Lb. sakei* SCC, *Lc. lactis* FT27, N16 and SV77 were selected for their marked anti-Listeria activity. When Listeria was inoculated on the cheese surface at the end of ripening process (after 50 days; pH: 6.7), only *Lc. lactis* FT27 exerted a significant inhibition on the growth of the two Listeria species if the cheese was strictly maintained at 4°C. Hence, this research highlighted that the susceptibility of Listeria biotypes to LAB antimicrobial activity is strain-dependent, thus a blend of different LAB strains could represent an effective tool to develop PC for blue-veined cheeses. Seemingly, Mills et al., (2017) proposed a multi-strain approach to tackle this problem. They generated single- and double-bacteriocin-producing cultures of *Lc. lactis* CSK2775 with the capacity to produce the class I bacteriocins lacticin 3147 (lacticin), nisin A (nisin), or lacticin and nisin. This multi-bacteriocin producer was used in combination with plantaricin from *Lb. plantarum* LMG P-26358 to create a cheese starter system.

When PC are meant to be incorporated into a cheese making process, one of the main aspects to be aware of is their contribution to the sensory profile of the final product. The application of new strains to the cheese making process requires a positive sensory acceptance by the consumers. Consequently, regardless of the antimicrobial efficiency of the culture, the generation of volatile compounds (possibly off-odors), and/or high acidification (potential off-flavors) have to be checked and control throughout the cheese manufacture and ripening. In a study by Coelho et al., (2014), 8 bacteriocin producer strains identified as *Lc. lactis* (1) and *Ec. faecalis* (7) were assessed in order to identify those with the greatest potential in reducing *L. monocytogenes* in fresh cheese. A non-trained panel evaluated the attributes acidity, salty taste, firmness, flavor and general acceptability of cheeses made with the bacteriocin-producer strains. All cheeses made with LAB cultures were found to have flavor, salty taste and general acceptability comparable to the control cheese made without any PC, i.e., no significant differences from control cheeses where found among the sensory attributes tested. According to Pereira et al., (2011), microorganisms inoculated in cheeses made with cows' milk do not exert any influence on the flavor of the cheese in the early days of ripening, which confirms the results obtained in the aforementioned study since LAB-PC were intended to be applied to fresh cheeses with short ripening periods. Regarding bacteriocin activity, its availability comprised the determining factor for an *in situ* antibacterial action. Hence, bacteriocin-like activity should be produced at sufficient levels and be active enough during cheese manufacture and storage. Results derived from this study showed that *in situ* bacteriocin producers clearly showed a high antilisterial activity in fresh cheese made with the *Enterococcus* strains. The *Lc. lactis* strain tested produced lacticin 481, which proved to be less effective in reducing *L. monocytogenes* (Coelho et al., 2014). In accordance to the latter study aims, *Lb. rhamnosus* 131RZFAUU was tested in Domiati soft cheese as a PC against *S. aureus* without affecting flavor acceptability of final product. This LAB-PC was selected among three potential candidates which were analyzed separately and all together. Interestingly, no negative sensory properties were observed by the panelists in fresh cheeses made with LAB strains as a single culture. Conversely, unacceptable flavor was observed in Domiati cheese made with a combination of three strains (Ibrahim et al., 2017).

Ricotta fresca is a whey cheese susceptible of secondary contamination mainly due to its naturally poor competitive microflora and composition, inherent physical and chemical properties and the absence of preservatives. *Pseudomonas* spp, yeasts, molds and *Enterobacteriaceae* comprise the main psychotropic microorganisms usually found in refrigerated whey cheeses (Pala et al., 2016). The use of biopreservatives against these psychotropic spoilage microorganisms was preliminary assessed by Spanu et al., (2017,

2018) on the surface of MAP *ricotta fresca* during refrigerated storage. A commercial biopreservative, Lyofast CNBAL, comprising *Carnobacterium* spp. was tested against *Pseudomonas* spp. This was the first successful attempt to use biopreservatives against psychotropic spoilage microorganism in sheep ricotta cheese.

Selected LAB strains (as in fermentation) or the bioactive compounds purified from the culture medium can be exploited as efficient alternatives for food preservation. Eleven antifungal *Lactobacillus* strains, isolated from cow and goat milk, were fully characterized by Delavenne et al., (2013)using molecular and phenotypic methods. Their antifungal activities were tested in milk and yogurt, against fungal species (*Debaryomyces hansenii, Kluyveromyces lactis, K. marxianus, Penicillium brevicompactum, Rhodotorula mucilaginosa* and *Yarrowia (Y.) lipolytica*) commonly involved in the spoilage of dairy products. The antifungal strains belonged to *Lb. paracasei, Lb. rhamnosus, Lb. zeae* and *Lb. harbinensis* species and showed different acidifying and growth capacities in milk. The antifungal activity in milk developed by lactobacilli had a strain-dependent activity spectrum. A very strong antifungal effect in yogurt was displayed by *Lb. harbinensis* against all tested fungi as compared to control. Accordingly, the strain *Lb. harbinensis* K.V9.3.1Np showed a remarkable antifungal activity for the bioprotection of fermented milk without modifying their organoleptic properties (Delavenne et al., 2015). In keeping with these findings, Mieszin et al., (2017) studied the action mechanism of this bioprotective strain against the spoilage yeast *Y. lipolytica*. The tested supernatants were obtained after milk fermentation with yogurt starter cultures either in co-culture with *L. harbinensis* K.V9.3.1Np (active supernatant) or not (control supernatant). Morphological changes (membrane collapsing and cell lysis) were observed in the target yeast. The authors attributed these effects to the synergistic action of organic acids since 5 of the 8 organic acids quantified in the active supernatant (acetic, lactic, 2-pyrrolidone-5-carboxylic, hexanoic and 2-hydroxybenzoic acids) were at significantly higher concentrations than in the control one. These physiological mechanisms induced by an antifungal LAB could be used as part of the hurdle technology to prevent fungal spoilage in dairy products. On the other hand, Mosbah et al., (2018) studied the antifungal molecules produced by *Lb. harbinensis* K.V9.3.1Np when used in yogurt fermentation. Activity tests against *Penicillum (Pe.) expansum* and *Y. lipolytica* showed that the active compounds from this LAB-PC were benzoic acid and a polyamine identified as a spermine analog, which had not been reported earlier. Nevertheless, the highest activity was shown by a mixture of short polycyclic lactates.

Leyva Salas et al., (2017) found that, for a same antifungal LAB culture, a variable antifungal activity occurred between different semi-hard cheeses, suggesting a role of biotic and abiotic factors resulting from interactions between the food, the antifungal cultures and the fungal contaminants. The potential influencing factors could be the dry matter content in cheeses, ripening chamber humidity, interactions between the smear and the fungi, and non-sterility of the ripening chambers.

APPLICATIONS OF PROTECTIVE CULTURES IN MEAT PRODUCTS

Meat and meat products are perishable foods and they should be properly handled along the productive chain in order to prevent microbial growth. If spoilage microorganisms such as *Pseudomonas*, *Acinetobacter*, *Brochothrix thermosphacta*, *Moraxella*, *Enterobacter*, *Lactobacillus*, *Leuconostoc*, and *Proteus* are present and highly proliferated, the meat will spoil and will be unfit for human consumption (Jayasena & Jo, 2013). Additionally, pathogenic bacteria such as *Salmonella* spp., thermophilic *C. jejuni*, enterohemorrhagic *E. coli* O157:H7, *Clostridium* (*Cl.*) *perfringens*, anaerobic *Cl. botulinum*, *L. monocytogenes*, *S. aureus*, *B. cereus*, and *Ye. enterocolitica*, can be present in meat and meat products turning them hazardous for consumers (Hui, 2012). Therefore, assurance of meat safety and quality is of utmost importance. In this sense, LAB have been used as PC alone or in combination with several physicochemical treatments, modified atmosphere packaging (MAP), high hydrostatic pressure (HHP), and chemical preservatives, as an additional hurdle to control the proliferation of spoilage and pathogens microorganisms in meat and meat products. An endless variety of studies regarding LAB-PC application against target microorganisms on several meat products can be profoundly found in the literature (Castro et al., 2016, Castellano et al., 2017, Oliveira
et al., 2018). Nevertheless, in this section the studies have been compiled with the selection criteria of up-to-date alternatives for meat biopreservation.

Several LAB species, including *Lactobacillus*, *Lactococcus*, *Pediococcus* and *Enterococcus* spp., have been evaluated to enhance microbial safety of meat and meat products due to their ability to produce bacteriocins active against *L. monocytogenes* (Oliveira et al., 2018, Woraprayote et al., 2016). As demonstrated by Giello et al., (2018), the use of the bacteriocin-producing *Lb. curvatus* 54M16 in fermented sausages could be an important contribution to improve the quality of final product. This strain produced several bacteriocins: sak X, sak Tα, sak Tβ and sak P which are active against *L. monocytogenes*. The anti-listerial effect during the manufacture of fermented sausages naturally contaminated and co-inoculated with a mixture of *L. monocytogenes* OH and Scott A (10^4 CFU/g) was investigated. Although this LAB-PC was partially effective in the control system against *L. monocytogenes* OH and Scott A, it was able to inhibit native *L. monocytogenes* over the course of the fermentation. Besides, *Lb. curvatus* 54M16 dominated and affected the bacterial ecosystem as it was determined by 16S rRNA-based analysis. It could be assumed that inoculated samples with PC helped on the dominance of *Lactobacillus* since un-inoculated sausages showed more diverse microbiota in which spoilage-associated bacteria, such as *Brochothrix*, *Psychrobacter*, *Pseudomonas* and some *Enterobacteriaceae*, were metabolically active until the end of the ripening.

Application of CO_2 modified atmosphere packaging in combination with antagonistic strain *Lb. plantarum* subsp. *plantarum* CICC 6257 as PC resulted in an effective strategy for enhancing microbiological safety of ground pork (Zhang et al., 2018a). The solely use of *Lb. plantarum* CICC 6257 had no significant effect on the lag-phase (λ) of *L. monocytogenes*, whereas its combination with CO_2 – MAP prolonged the λ and reduced the maximum specific growth rate (μ_{max}) and N_{max} of the pathogen in meat. The selected optimal treatment was the combined effect between *Lb. plantarum* CICC 6257 and 60% CO_2, by means of which the λ of *L. monocytogenes* increased from 3.43h to 8.33h, and the N_{max} and μ_{max} were decreased by 46% and 66%, respectively.

Even though *Enterococcus* spp. are able to produce bacteriocins with strong anti-listerial activity (Khan et al., 2010; Gaaloul et al., 2015), safety aspects should be carefully assessed before their use as PC in meat and meat products since enterococci are known to be opportunistic pathogens causing several nosocomial infections (Rehaiem et al., 2014). In this sense, Ben Braïek et al., (2017, 2018) confirmed the innocuity of two *Ec. lactis* strains named Q1 (producer of enterocin P) and 4CP3 (producer of the enterocins A, B and P) to determine the effect of separate addition of these two strains in the control of *L. monocytogenes* in refrigerated chicken breast meat. These meat models were artificially contaminated with 10^5 CFU/g of *L. monocytogenes* and inoculated with each of the enterocin-producing *Ec. lactis* strains at a concentration of 10^7 CFU/g. In the presence of *Ec. lactis* Q1, viable populations of *L. monocytogenes* were reduced to 1.90 log CFU/g after 14 days, and then remained constant up to the end of the challenge experiment. While the addition of *Ec. lactis* 4CP3 caused a significant decrease of Listeria counts below the detection limit after 7 days and totally inhibited pathogen growth after 14 days of refrigerated storage (Ben Braiek et al., 2019). On the other hand, *Ec. mundtii* CRL 35 was found a promising strain to be used as a safe adjunct culture in sausage fermentation (Orihuel et al., 2018). This strain produces Enterocin CRL35, a pediocin like bacteriocin with anti-listerial and anti-viral activity (Salvucci et al., 2010; Wachsman et al., 2003). The effect of this PC on *L. monocytogenes* growth was evaluated in a meat model (beaker sausage) in the presence or absence of curing additives (NaCl, $NaNO_2$ and ascorbic acid). *L. monocytogenes* inhibition was enhanced when beaker sausage was supplemented with the curing mixture reaching a decrease greater than 2 log CFU/g. These results demonstrate that *Ec. mundtii* CRL 35 could be an efficient PC able to enhance hygienic quality of fermented sausages.

The effect of PC on other pathogenic microorganisms associated to meat and meat products has been also informed. *Cl. perfringens* strains have been implicated in outbreaks of food poisoning, particularly of meat and meat products (Grass et al., 2013). *Cl. perfringens* and its spores are supposed to grow or germinate under temperature abuse conditions, which commonly occur during the transportation, storage, or retail display of the meat chill chain. In order to present a viable alternative for the control of *Cl. perfringens* on chicken breast meat, Li et al., (2017) investigated the effects of five

LAB strains on *Cl. perfringens* growth, germination, and outgrowth of its spores under temperature abuse conditions. Among strain tested, *Lb. fermentum* R6 and *P. pentosaceus* P1, effectively inhibited the growth of vegetative cells and repressed germination and outgrowth of *Cl. perfringens* spores. However, the growth of *P. pentosaceus* P1 caused a deep decrease in pH values to 5.3 which made it not suitable as PC for this food matrix.

Ramaroson et al., (2018) selected LAB strains to be used as PC in cooked ham based on their antagonistic activity against pathogenic or spoilage spore-forming species as *B. cereus*, *Clostridium* sp. like *botulinum*, *Cl. frigidicarnis*, and *Cl. algidicarnisand*. Besides, the authors tested LAB-PC ability to resist High Pressure Processing (HPP). According to the results in model ham liquid medium, one *Lc. lactis* strain was selected because of its widest inhibitory spectrum, potential production of bacteriocin and ability to regrow following HPP. This PC could bring the opportunity of combining biopreservation and HPP to control spore-forming pathogens in cooked ham.

On the other hand, the combination of PC and chemical additives represents an alternative for the control of contaminants in meat environments, mainly enteric pathogens. Regarding this, Casquete et al., (2016) reported the efficacy of chitosan in combination with *P. acidilactici* bacteriocinogenic and non-bacteriocinogenic strains, in the control of *L. innocua*, *S. aureus*, *E. coli* and *Salmonella* Typhimurium in the fermented paste meat product Alheira (Portugal). Similarly, Hu et al., (2019) demonstrated that the combined use of chitosan and a bacteriocin-producing *Cb. maltaromaticum* strain reduced cell counts of *Salmonella* in lean beef.

The use of LAB as PC to inhibit growth of spoilage bacteria has been suggested by many authors (Hu et al., 2008; Jones et al., 2009: Biscola et al., 2014), and it is being considered as a strategy to improve the shelf life of meat and meat products. For instance, Zhang et al., (2018b) studied the bacterial diversity of vacuum-packaged sliced raw beef inoculated with the commercial cultures Bactoferm B-2 (*Lb. sakei*) and SafePro B-LC-48 (*Lb. curvatus*) as PC by means of polymerase chain reaction-denaturing gradient gel electrophoresis (PCR-DGGE). The DGGE profiles of DNA extracted from non-inoculated samples showed that *Enterbacteriaceae* and *Pseudomonas fragi* emerged early, then *Brochothrix thermosphacta* shared the dominant position, and finally, *Ps. putida* showed up to became predominant. The addition of either *Lb. sakei* or *Lb. curvatus* reduced the complexity of microbial diversity and inhibited the growth of spoilage bacteria with *Lb. sakei* exerting greater inhibitory effect. In other study, Comi et al., (2016) examined the use of the strains *Lb. sakei* (B-2 Safe Pro® e CHR HANSEN) and *Lc. lactis* subsp. *lactis* (Rubis e CHR HANSEN) as PC to reduce the risk of *Leuconostoc* (*Leuc*) *mesenteroides* spoilage in artisanal cooked bacon. LAB strains were inoculated on pieces of bacon together with a mix of three different *Leuc. mesenteroides* strains isolated from spoiled cooked bacon and responsible for the greening color and spoilage. After 90 days of storage, sticky-white slime and a greening color were observed in the cooked bacon inoculated only with *Leuc. mesenteroides*. While cooked bacon with

PC did not present greening, slime or inflated packaging. At this stage of the storage, the ratios *Leuc. mesenteroides/Lc. lactis* subsp. *lactis* and *Leuc. mesenteroides/Lb. sakei* were about 1/100 and 1/1000, respectively. Based on these results, the authors assumed that substrate competition by PC was responsible for *Leuc. mesenteroides* growth inhibition, thus suppressing its spoilage effects. Moreover, the acceptability of the bacon added with PC was confirmed with sensorial analysis. These results illustrate the benefits of inoculation with LAB, especially *Lb. sakei* as a potential way to inhibit growth of spoilage microorganisms and to prolong the shelf life of meat and meat products.

APPLICATIONS OF PROTECTIVE CULTURES IN FRESH SEAFOOD AND SEAFOOD PRODUCTS

Fish is a very delicate product since its shelf life is remarkably short due to autolytic reactions and microbial proliferation. Autolytic reactions involve the action of proteases that generate an increase of free amino acids and free volatile bases (Huss, 1995), which allow microbial growth to be faster than in other types of meat. Main deteriorative strains present in fish include Gram negative microorganisms such as *Shewanella* sp., *Aeromonas* sp., *Pseudomonas* sp., *Photobacterium* sp., Enterobacteriaceae and some Gram positive bacteria such as LAB, including *Carnobacterium* sp., and *Lactobacillus* sp. (Cifuentes Bachmann & Leroy, 2015). Main pathogenic bacteria responsible for fish poisoning are *Aeromonas hydrophila, Cl. botulinum,* pathogenic *Vibrio* spp. and *L. monocytogenes* (Løvdal, 2015). Additionally, fatty fish lipid composition is characterized for a high proportion of polyunsaturated fatty acids, which are highly oxidable, and thus, nutritional value is lost at very fast rates. Furthermore, after fish is captured in coastal areas, it is firstly stored at fishing boats, and then distributed to non-coastal areas, so a long period of time passes before fish products reach consumers (Wiernasz et al., 2017). All these characteristics make fish an important vector for food-borne diseases, yielding between a 10% and a 20% of total food borne diseases worldly (Pilet & Leloir, 2011).

In the last years LAB have been widely studied to be used on fish preservation. Even though their main action is usually against Gram positive microorganisms, -as an extra hurdle, they contributed to control native fish flora and specifically to control *L. monocytogenes* growth. Additional hurdles in fish preservation include modified atmosphere package, vacuum package, addition of essential oils, plant extracts, refrigeration, addition of active films, marinade and diminishment of water activity (Boulares et al., 2016; Concha-Meyer et al., 2010; López de Lacey, López-Caballero, & Montero, 2013; Sarika et al., 2012).

LAB have been added to fish products in different manners: as PC, as bacteriocins purified or semi purified, and either LAB or bacteriocins immobilized in different films or coatings. Regarding the application as protective culture, LAB may produce bacteriocins *in situ* and compete with native fish bacteria, generating a bacteriostatic effect as well as a bactericidal one. Limitations to these applications may include the generation of off flavors (Saraoui et al., 2017) and the difficulty of LAB to grow and produce bacteriocins *in situ*. To partially overcome LAB *in situ* lack of activity, antagonist activity may be tested in fish juice (Wiernasz et al., 2017), to get a more accurate prediction of LAB activity when applied to fish products. When purified or semi-purified bacteriocins are applied directly to fish, issues related to cell viability are bypassed, but it is necessary to foresee a bacteriocin concentration capable to maintain a threshold concentration at which native flora would stay inhibited, since bacteriocin concentration may be affected by lytic action of proteolytic bacteria (Giribaldi et al., 2018). Finally, when immobilizing bacteriocins, their structure may be protected, and their release modulated (Settier-Ramírez et al., 2018). When LAB are immobilized, they may be released in a controlled manner to food, thus preventing them from overgrowing and acidifying. Anyhow, immobilization must ensure cell viability and bacteriocin bonding in a manner that would not affect their antagonist action.

Comparing some of the aforementioned preservation techniques, Gómez-Sala et al., (2016) evaluated the use of *Lb. curvatus* BCS35 and *Ec. faecium* BNM58 on young hake and megrim, both applied as PC, liquid food ingredient or lyophilized food ingredient. In this case, when strains were used as PC the best results were achieved: coliform population was diminished throughout 14 days of storage, as well as the total mesophilic count, and fish appearance was improved, compared to control system. Boulares et al., (2016), added *Lb. plantarum* and *Cb. piscicola* and *Lc. lactis* as PC to sea bass, and found an improvement on fatty acid composition, and protein profile, indicating that the use of both LAB could control both proteolytic and lipolytic activities of deteriorative native bacteria. Their study also showed that the best combinations to control the growth of *Listeria* spp. population were mixtures of *Lc. lactis-Lb. plantarum* and *Lc. lactis-Cb. piscicola,* and a mixture of *Lb. plantarum-Cb. piscicola, which* was the best to control total basic volatile nitrogen and trimethylamine. Co-cultures had also been capable of diminishing undesired off-flavors generated by very active bacteriocinogenic strains, which by itself could not be applied to fish products. Such is the example of a co-culture of *Lc. piscium* and *Cb. divergens* since the addition of *Lc. piscium* inhibited the production of off flavors produced by *Cb. divergens* in shrimp preservation, extending their shelf life for 10 extra days in comparison to shrimps only inoculated with *Cb. divergens* (Saraoui et al., 2016). From these studies it may be concluded that the use of LAB strains may control different aspects of fish deterioration, and that there is a strong dependence on strain mixture.

The application of LAB cell free supernatant (CFS), purified or semi purified bacteriocins has also been assessed on fish preservation. The most used purified bacteriocin has been nisin in combination with extra hurdles. Novel applications of nisin include the addition of the bacteriocin in alginate films. Concha-Meyer et al., (2010), applied nisin together with two different LAB strains of *Cb. maltaromaticum*, to prolong shelf life of smoked salmon, and analyzed the anti-Listeria effect on samples. Systems containing nisin alone, or together with both LAB strains, showed a bacteriostatic effect n *L. monocytogenes*; in addition, in systems inoculated with both LAB, there was a decrease of the pathogens on day 28,due to the release of the bacteriocinogenic strains into the salmon filets. Lu et al., (2009), managed to increase the shelf life of northern sea snake filets by adding nisin together with EDTA, mixed or not on a calcium alginate film. Their results showed that the coating did not improve overall quality indicators (total mesophilic and psychrophilic bacterial counts, TBV-N, TMA-N, pH, or TBA) compared to samples only sprayed with the antimicrobials. Still, all treated samples showed values below maximum allowed for fish products throughout the storage, indicating that the mixture of nisin and EDTA had a positive effect on fish shelf-life. Nevertheless, sensory characteristics were improved by the application of the preservatives together with the alginate coating. These results showed that coatings are a promising technique to use on fish preservation, but factors such as release speed of antimicrobials (related to calcium-alginate concentration) should be further studied (Fang & Tsai, 2003).

Biopreservation of fish, has also been used together with other naturally occurring additives such as plant extracts, since they have been informed to have antagonist activity against Gram negative microorganisms (Djeussi et al., 2013), and their richness in polyphenols and components with antioxidant activity may favor fish preservation. López de Lacey, López-Caballero, & Montero (2013) used a bioactive film prepared with agar in addition with green tea and *Lb. paracasei* L26 and *Bifidobacterium lactis* B94, obtaining a shelf life extension of hake fillets for 7 extra days (14 days in total). This study also showed that the probiotic strains could migrate from the film to the fillets increasing their population. It is important to highlight that the green tea extract did not affect the viability of inoculated bacteria, but diminished the presence of native LAB on samples. Dos Reis et al., (2011) investigated the synergic action of two different strains of *Cb. maltaromaticum* cultures and hydroalcoholic extract of *Lippia sidoides* (LSE) on fish food models: surubim fish peptone broth and surubim homogenate. In this case, *Cb. maltaromaticum* by itself did not control the growth of *L. monocytogenes*, neither did the LSE, but a synergic activity of both antimicrobial factors was found, and the pathogen population was controlled for 28 days on fish homogenate.

LAB can be used as an extra hurdle together with other preservation techniques which require LAB to grow under refrigeration, without the presence of oxygen as well as under low water activity conditions. Giribaldi et al., (2018) obtained an improvement on swordfish fillets fatty acid profile, for 60 days, on ready to eat fillets treated with marinade water solution containing 3% NaCl for 2 days together with an immersion on a probiotic culture of *Lb. paracasei* IMPC 2.1. Tomé, Gibbs, & Teixeira (2007) used another preservation technique (dry salted for 6 hours; drying for 6 hours; smoking for 2 hours), and compared the biopreservation capacity of 5 LAB (*Enterococcus faecium* ET05, *Lb. curvatus* ET06, *Lb. curvatus* ET30, *Lb. deldrueckii* ET32 and *Ped. acidilactici* ET34) on salmon. After 21 days of storage at 5°C, the growth of *L. innocua* 2030c was best controlled by the strain *E. faecium* ET05, achieving a reduction of 5 log UFC at the end of storage in comparison to control samples.

PROTECTIVE CULTURES IN THE BIOPRESERVATION OF FRUITS, VEGETABLES AND MISCELLANEOUS FOOD PRODUCTS

Fruits and Vegetables

Fruits and vegetables are highly perishable foods. Their deterioration is the result of biochemical changes, physiological ageing and microbial spoilage. Most of the bacteria causing the spoilage of fruits and vegetables are Gram-negative and belong to the genera *Pseudomonas* and *Enterobacteriaceae*. Within the present Gram-positive bacteria, it is usual to detect LAB. Regarding pathogenic microorganisms, fruits and vegetables may harbor *L. monocytogenes*, *Salmonella* spp. and *Escherichia coli* O157:H7. Finally, fungi and yeasts could also be found, which cause the deterioration of these foods although to a minor proportion than bacteria (Rico et al., 2007; Oliveira et al., 2015).

Taking into account the risks to consumer health and the economic losses that microorganisms' proliferation in fruits and vegetables may cause, the application of some preserving methods is necessary. However, in this type of food it is extremely important to maintain its condition of "fresh" or "minimally processed," being inconvenient to use severe methodologies. In this context, the combination of various emerging preservation technologies is presented as a possible solution. Among them, the use of PC or the bacteriocins produced by them is outstanding. While the biopreservation of fruits and vegetables has not been studied as well as the biopreservation of other types of foods, there are numerous reports regarding its effect on these matrices and some of them are summarized in Table 1.

Regarding minimally processed fruits, a large amount of investigations has focused on isolating native LAB showing inhibitory activity against the deteriorative or pathogenic microorganisms usually found in these products. For example, Siroli et al., (2015) isolated native LAB from sliced apples and reported that, among the isolates with inhibitory activity against *L. monocytogenes*, *Salmonellaenteritidis* E5 and *E. coli* 555, the strains identified as *Lb. plantarum* CIT3 and *Lb. paracasei* M3B6 were the most notable. The latter significantly decreased the population of *L. monocytogenes* and *E. coli* in minimally processed apples. In addition, when *Lb. plantarum* CIT3 was combined with natural antimicrobials the observed effect was greater. In a similarly targeted study, Siroli et al., (2016) evaluated the application of *Lc. lactis* CBM21, alone or in combination with natural antimicrobials, as PC in sliced apples. The authors reported that the use of *Lc. lactis* limited the growth of yeast and inhibited the development of *L. monocytogenes*. The effectiveness of the PC in controlling the growth of *L. monocytogenes* was attributed to the ability of the former to produce nisin. It should be noted that the inhibitory effect was greater when the PC was combined with natural antimicrobials.

Probiotic fortification has been extensively studied to produce foods with functional properties. Regularly, dairy products are more prone to be labelled as functional foods; however, probiotic fruits and vegetables could provide a good alternative for vegetarian or vegan consumers. In addition to beneficial effects on the health of consumers, some probiotic microorganisms can act as PC. Russo et al., (2015) studied the capability of *Lb. plantarum* B2 and *Lb. fermentum* PBCC11.5 to control the growth of *L. monocytogenes* in refrigerated melon and found that the 2 PC showed antagonistic activity against the pathogen. Furthermore, Iglesias et al., (2018) reported that the probiotic *Lb. rhamnosus* GG inhibited the growth of *L. monocytogenes* on fresh-cut pear but did not show antagonistic action against *Salmonella*.

In products derived from fruits, such as juices, the application of PC has not been widely studied. However, the application of bacteriocins has proved to be very effective. *Alicyclobacillus acidoterrestris* is a spoilage-causing bacterium in fruit juices. The addition of different bacteriocins, for example enterocin AS-48, nisin or bificin C6165, produced the inhibition of this microorganism and in some cases inactivated its endospores (Grande et al., 2005; Walker & Phillips, 2008; Pei et al., 2014).

The use of native flora to control microbial growth was also studied in vegetables. Siroli et al., (2015) isolated LAB from lamb lettuce and selected *Lb. plantarum* V7B3 and *Lb. casei* V4B4 as potential biocontrol agents for minimal processed lamb lettuce. Both PC were able to inhibit the growth of *L. monocytogenes* and *E. coli,* and also reduced the population of aerobic mesophilic microorganisms. Seemingly to the aforementioned application on apples, the combination of the PC with natural antimicrobials improved the biopreservation results. In another study using the same raw matrix, Siroli et al., (2017) compared the effectiveness of different washing solutions for lettuce. For this purpose, the samples were washed with: i) chlorine 120 mg/L, ii) thyme essential oil 250 mg/L, iii) *Lc. lactis* CBM21 10^6 CFU/ml, iv) thyme essential oil and *Lc. lactis* CBM21. The results showed no differences between the populations of microorganisms in the different systems. In addition, the sensory characteristics of the samples washed with thyme essential oil and *Lc. lactis* CBM21 were superior.

Moreover, Trias et al., (2008) studied the effect of 3 strains of *Leuc. mesenteriodes* (CM135, CM160 and PM249), previously isolated from fresh fruits and vegetables, on the development of *L. monocytogenes* in Iceberg lettuce. Before being applied to vegetables, the 3 strains were characterized and it was found that 2 of them (CM135 and CM160) produced bacteriocins with antagonistic activity against *L. monocytogenes* and against other LAB. When inoculating the lettuce with 10^9 CFU/g of *Leuc. mesenteriodes*, no development of *L. monocytogenes* was detected. However, as the concentration of PC decreased, the inhibitory effect also declined. The application of these 3 strains to fresh fruits and vegetables did not cause sensory changes.

As mentioned above, the deterioration of vegetables may be a consequence of the development of fungi and yeasts. In this context, Sathe et al., (2007) isolated LAB from different vegetables and studied their antifungal activity. Among the isolated strains, *Lb. plantarum* CUK-501 showed antifungal activity against all the indicators studied *in vitro*. For *in vivo* studies, *Lb. plantarum* was inoculated into cucumber. Results showed that *Aspergillus flavus*, *Fusarium graminearum*, *Rhizopus stolonifer* and *Botrytis cinerea* growths were delayed. Similar results were found when the PC was replaced by the CFS.

As well as fruit juices, the use of PC has not been studied in canned vegetables. However, studies were carried out applying bacteriocins. Lucas et al., (2006) studied the antimicrobial activity of the enterocin AS-48 against *B. coagulans* in canned vegetables. After 24 hours of storage at different temperatures, the enterocin AS-48 reduced *B. coagulans* population, and after 15 days of storage, no viable cells were detected. In addition, the enterocin AS-48 increased the thermal sensitivity of *B. coagulans* spores.

Finally, lactic acid fermentation of vegetables should not be omitted. This technology has historically been used in many cultures as an effective method of food preparation and preservation. Lactic acid fermentation is a simple and valuable biotechnological tool to maintain and/or improve the safety, nutritional, sensory and preservation properties of fruits and vegetables (Di Cagno et al., 2013). In general, the starters cultures are selected because of their technological potential, such as the production of acid or exopolysaccharides. However, their antimicrobial activity should not be ignored (Settanni & Corsetti, 2008). For example, Di Cagno (2008) isolated autochthonous microorganisms (*Ln. mesenteroides* C1, *Lb. plantarum* M1 and *Ped. pentosaceus* C4) from carrots, French beans and marrows, and applied them as fermentation starters for the same vegetables. In addition to the good sensory characteristics, the authors found that the starters were able to inhibit the growth of *Enterobacteriaceae* and yeasts.

Considering the topics discussed, it can be concluded that the use of PC and bacteriocins plays a determining role in the control of deteriorative and pathogenic flora of fruits and vegetables. However, in both cases, they should always be applied within the context of hurdle technology, since their spectrum of action is reduced. In addition, it is important to note that each strain used as PC must be tested on the food to be applied and against a broad spectrum of indicator microorganisms.

Miscellaneous Food Products

Bioprotective cultures are also used to preserve a diverse kind of food products such as different beverages and bakery products. Bakery products are prone to post-treatment contamination by mold spores. Fungal development conducts to mycotoxin production and sensory spoilage which limit shelf life and causes significant health and economic losses. To overcome this issue, the search for bioprotective cultures exhibiting antifungal activity has gained interest in the last years as an alternative to chemical preservatives. Furthermore, some LAB also exhibited antimycotoxin properties (Guimarães et al., 2018; Juodeikiene et al., 2018). Le Lay et al., (2016) isolated several strains for the species *Leuc. citreum, Lb. sakei, Lb. plantarum, Lb. spicheri, Lb. reuteri and Lb. brevis* which were able to delay fungal growth by surface spraying to bakery products. Furthermore, some of these protective cultures led to delay fungal growths after incorporation in milk bread rolls preparation. Cizeikiene et al., (2013) used a spray containing a mixture of *Ped. acidilactici* KTU05-7, *Ped. pentosaceus* KTU05-8 and KTU05-10 strains that produce organic acid and bacteriocin like inhibitory substances to suppress superficial mold development on bread for 8 days of storage at 15°C in polyethylene bags. Trends commented suggest that bioprotective cultures can be used for controlling molds in bakery products.

Guimarães et al., (2018) selected the strain *Lb. plantarum* UM55 among different LAB due to its ability for inhibiting the growth of *Aspergillus* (*A.*) *flavus*. Futhermore, the cell-free supernatant (CFS) of this strain was able to inhibit by 91% the production of aflatoxins (AFLs). It had also inhibited the growth and AFLs production of *A. parasiticus, A. arachidicola, A. nomius* and *A. minisclerotigenes*. This antimycotoxyn activity was due to the presence of phenyllactic acid, hydroxyphenyllactic acid, and indole lactic acid, being phenyllactic acid the one that exerted the stronger effect. Similar results were found by Schmidt et al., (2018) who reported that *Lb. reuteri* R29 exhibited a broad spectrum of antifungal activity due to the presence of phenyllactic acid. Based on mentioned trends, Bustos et al., (2018) using a factorial design optimized a low-cost medium to obtain biomass and phenlyllactic acid production by *Ped. acidilactici* CRL 1753 strain. Furthermore, this strain was proposed to be used as a biopreservative in bread making to increase the shelf life of wheat bread.

Although is it known that nisin producer strains could prevent malolactic fermentation in wines and also they may serve to reduce the amount of sulfur dioxide, their use in winemaking to control bacterial spoilage (Daeschel et al., 1991) was evaluated only recently. Fernández-Perez et al., (2018) assessed the use of *Lc. lactis* LM29 under enological conditions to decrease sulfur dioxide addition. Results showed that this lactococci strain was able to produce nisin in the presence of 2 and 4% ethanol (v/v), while higher concentrations of ethanol fully inhibited bacteriocin production.

Furthermore, the use of 50 mg/L nisin as additive decreased fourfold the concentration of sulfur dioxide required to prevent LAB growth during wine ageing.

CONCLUSION

Taking into consideration the abundant advantages of LAB strains, their potentiality as PC or biopreservatives is undeniably, providing therapeutic value to fermented food products, contributing to public health and reducing economic losses related to spoilage. Since fermented foods are consumed daily by people, the proper selection of protective cultures with different beneficial properties to enhance safety and promote health is a need within novel and reliable methods for food preservation.

REFERENCES

Abdel-Haliem M.E.F., Tartour E., Enan G. (2013). Characterization, production and partial purification of a bacteriocin produced by *Lactobacillus plantarum* LPS10 isolated from pickled olives. *Res. J. Pharm., Biol. Chem. Sci.,* 7 (5), 2362–2371.

Abriouel H.. Lucas R., Ben Omar N.. Valdivia E.. Gálvez A. (2010). Potential applications of the cyclic peptide enterocin AS-48 in the preservation of vegetable foods and beverages. *Probiot. Antimicrob. Prot.,* 2, 77–89.

Ahmadi S., Soleimanian-Zad S, Sheikh-Zeinoddin S. (2016). Effect of heat, nisin and ethylene diamine tetra-acetatetreatments on shelf life extension of liquid whole egg. *International Journal of Food Science and Technology*, 51, 396–402.

Alexandraki V., Tsakalidou E., Papadimitriou K. Holzapfel, W. (2013). *Status and trends of the conservation and sustainable use of micro-organisms in food processes.* Background Study Paper (FAO).

An J., Zhu W., Liu Y., Zhang X., Sun L., Hong P., Wang Y., Xu C., Xu D., Liu H. (2015). Purification and characterization of a novel bacteriocin CAMT2 produced by *Bacillus amyloliquefaciens* isolated from marine fish *Epinephelus areolatus*. *Food Control*, 51, 278–282.

Ananou S., Rivera S., Madrid M. I., Maqueda M., Martínez-Bueno M., Valdivia E. (2018). Application of enterocin AS-48 as biopreservative in eggs and egg fractions: Synergism through lysozyme. *LWT - Food Science and Technology*, 89, 409–417.

Aspri M., O'Connor P.M., Field D., Cotter P.D., Ross P., Hill C., Papademas P. (2017). Application of bacteriocin-producing *Enterococcus faecium* isolated from donkey milk, in the bio-control of *Listeria monocytogenes* in fresh whey cheese. *International Dairy Journal*, 73, 1-9.

Aspri M., Bozoudi D., Tsalta, D., Hil, C., Papadema, P. (2016, a). Raw donkey milk as a source of *Enterococcus* diversity: Assessment of their technological properties and safety characteristics. *Food Control*, 73, 81-90.

Aspri M., Economou,N., Papademas P. (2016, b). Donkey milk: An overview on functionality, technology and future prospects. *Food Reviews International*, 33, 316-333.

Barbosa M. S., Todorov S. D., Belguesmia Y., Choiset Y., Rabesona H., Ivanova I. V., Chobert J.M., Haertlé T., Franco B. D.G. M. (2014). Purification and characterization of the bacteriocin produced by *Lactobacillus sakei* MBSa1 isolated from Brazilian salami. *J. Appl. Microbiol,* 116 (5), 1195–1208.

Ben Braïek O., Cremonesi P., Morandi, S. Smaoui, S. Hani K., Ghrairi,T. (2018). Safety characterisation and inhibition of fungi and bacteria by a novel multiple enterocin-producing *Enterococcus lactis* 4CP3 strain. *Microbial Pathogenesis*, 118, 32 –38.

Ben Braïek O., Ghomrassi H., Cremonesi P., Morandi S., Fleury Y., Le Chevalier P., et al.,(2017). *Isolation and characterization of an enterocin P-producing Enterococcus lactis strain from a fresh shrimp (Penaeus vannamei).* Antonie Van Leeuwenhoek, 110(6), 771–786.

Ben Braïeka O., Smaouic S., Ennouric K., Morandid S., Cremonesi P., Hanif K., Ghrairia T. (2019) RAPD-PCR characterisation of two *Enterococcus lactis* strains and their potential on *Listeria monocytogenes* growth behaviour in stored chicken breast meats: Generalised linear mixed-effects approaches. *LWT - Food Science and Technology,* 99, 244–253.

BiscolaV., Abriouel H., Todorov S. D., Capuano V. S. C., Gálvez A., de Melo Franco B. D. G. (2014). Effect of autochthonous bacteriocin-producing *Lactococcus lactis* on bacterial population dynamics and growth of halotolerant bacteria in Brazilian charqui. *Food Microbiology*, 44, 296 –301.

Boulares M., Mankai M., Sadok S., Hassouna M. (2016). Anti-Listerial inhibitory lactic acid bacteria in fresh farmed sea bass (*Dicentrarchus labrax*) fillets during storage at 4°C under vacuum-packed conditions. *Journal of Food Safety*, 37(3). https:// doi.org/10.1111/jfs.12323.

Bourdichon F., Casaregola S., Farrokh C., Frisvad J. C., Gerd, M. L., Hammes W. P., Harnet, J., Huys G., Laulund S., Ouwehand A., Powell I. B., Prajapati J. B., Seto Y., Schure E. T., Van Boven A., Van Kerckhoven V., Zgoda A., Tuijtelaars S., Hansen E. B. (2012). Food fermentations: Microorganisms with technological beneficial use. *International Journal of Food Microbiology*, 154(3), 87–97 (Erratum in 156(3):301).

Bustos A. N., Font de Valdez G., Gerez C. L. (2018). Optimization of phenyllactic acid production by *Pediococcus acidilactici* CRL 1753. Application of the formulated bio-preserver culture in bread. *Biological Control*, 123, 137–143.

Casquete R. Castro S. M, Teeixeira P. (2016) Evaluation of the Combined Effect of Chitosan and Lactic Acid Bacteria in Alheira (Fermented Meat Sausage) Paste. *Journal of Food Processing and Preservation*, 41, 1-8.

Castellano P, Pérez Ibarreche M., Blanco Massani M., Fontana C., Vignolo G. (2017) Strategies for Pathogen Biocontrol Using Lactic Acid Bacteria and Their Metabolites: A Focus on Meat Ecosystems and Industrial Environments. *Microorganisms* 5, 38.

Castro M. P., Palavecino N. Z., Herman C., Garro O.A., Campos C.A. (2011). Lactic acid bacteria isolated from artisanal dry sausages: Characterization of antibacterial compounds and study of the factors affecting bacteriocin production. *Meat Science*, 87, 321–329.

Castro M., Cayré M. E., Herman C., Alvarez O. (2016) Bacteriocin Application in Meats: The Protective Culture Approach. In *Bacteriocins: Production, Applications and Safety*. Troy Padilla. Nova Science Publishers, Inc. Nueva York, USA, 99-136.

Catala-Clariana S., Benavente F., Gimenez E., Barbosa J., Sanz-Nebot V. (2010). Identification of bioactive peptides in hypoallergenic infant milk formulas by capillary electrophoresis-mass spectrometry. *Anal. Chim. Acta*, 683 (1), 119–125.

Cavera V.L., Arthur T.D., Kashtanov D., Chikindas M.L. (2015). Bacteriocins and their position in the next wave of conventional antibiotics. *Int. J. Antimicrob. Agents*, 46 (5), 494–501.

Cifuentes Bachmann, D. E., & Leroy, F. (2015). Use of bioprotective cultures in fish products. *Current Opinion in Food Science*, 6, 19–23. https://doi.org/10.1016/j.cofs.2015.11.009.

Cizeikiene D., Juodeikiene G., Paskevicius A, Bartkiene, E. (2013). Antimicrobial activity of lactic acid bacteria against pathogenic and spoilage microorganism isolated from food and their control in wheatbread. *Food Control* 31, 539-545.

Coelho M.C., Silva C. C. G., Ribeiro S. C., Dapkevicius M. L. N. E., Rosa H. J. D. (2014). Control of *Listeria monocytogenes* in fresh cheese using protective lactic acid bacteria. *International Journal of Food Microbiology*, 191, 53–59.

Comi G., Andyanto D., Manzano M., Iacumin L. (2016) *Lactococcus lactis* and *Lactobacillus sakei* as bio-protective culture to eliminate *Leuconostoc mesenteroides* spoilage and improve the shelf life and sensorial characteristics of commercial cooked bacon. *Food Microbiology*, 58, 16-22.

Concha-Meyer A., Schöbitz R., Brito C., Fuentes R. (2010). Lactic acid bacteria in an alginate film inhibit *Listeria monocytogenes* growth on smoked salmon. *Food Control*, 22(3–4), 485–489. https://doi.org/10.1016/j.foodcont.2010.09.032.

Daeschel M. A., Dong-Sun Jung Watson B. T. (1991). Controlling Wine Malolactic Fermentation with Nisin and Nisin-Resistant Strains of *Leuconostoc oenos*. *Applied and Environmental Microbiology*, 57, 601-603.

Deegan L. H., Cotter P. D., Hill C., Ross P. (2006). Bacteriocins: Biological tools for bio-preservation and shelf-life extension. *International Dairy Journal*, 16, 1058–1071.

Delavenne E., Cliquet S., Trunet C., Barbier G., Mounier J., Le Blay G. (2015). Characterization of the antifungal activity of *Lactobacillus harbinensis* K. V9. 3.1 Np and *Lactobacillus rhamnosus* K. C8. 3.1 I in yogurt. *Food Microbiol.*, 45, 10–17.

Delavenne E., Ismail R., Pawtowski A., Mounier J., Barbier G., Le Blay G. (2013). Assessment of lactobacilli strains as yogurt bioprotective cultures. *Food Control*, 30, 206-213.

Delves-Broughton J. (2005). Nisin as a food preservative. *Food Australia*, 57, 525–532.

Di Cagno R., Cosa,R., De Angelis M. Gobbetti M. (2013). Exploitation of vegetables and fruits through lactic acid fermentation. *Food Microbiology*, 33, 1-10.

Di Cagno R., Surico R.F., Siragusa S., De Angelis M., Paradiso A., Minervini F., De Gara L. Gobbetti M. (2008). Selection and use of autochtonous mixed starter for lactic acis fermentation of carrots, French beans or marrows. *International Journal ofFood Microbiology*, 127, 220-228.

Diaz-Ruiz, G.; Omar N.B.; Abriouel H.; Cantilde M. M.; Galvez A. (2012). Inhibition of *Listeria monocytogenes* and *Escherichia coli* by bacteriocin-producing *Lactobacillus plantarum* EC52 in a meat sausage model system. *Afr. J. Microbiol. Res.*, 6, 1103–1108.

Divya J. B., Varsha K. K., Nampoothir, K. M. (2012a). Newly isolated lactic acid bacteria with probiotic features for potential application in food industry. *Applied Biochemistry and Biotechnology,* 167, 1314-1324.

Divya J. B., Varsha K. K., Nampoothiri K. M., Ismail B., Pandey A. (2012b). Probiotic fermented foods for health benefits. *Engineering Life Sciences*, 12, 377-390.

Djeussi D. E., Noumedem J. A. K., Seukep J. A., Fankam A. G., Voukeng I. K., Tankeo S. B., Kuete V. (2013). Antibacterial activities of selected edible plants extracts against multidrug-resistant Gram-negative bacteria. *BMC Complementary and Alternative Medicine,* 13. https://doi.org/10.1186/1472-6882-13-164.

dos Reis F. B., de Souza V. M., Thomaz M. R. S., Fernandes L. P., de Oliveira W. P., De Martinis E. C. P. (2011). Use of *Carnobacterium maltaromaticum* cultures and hydroalcoholic extract of *Lippia sidoidesCham.* against *Listeria monocytogenes* in fish model systems. *International Journal of Food Microbiology*, 146(3), 228–234. https://doi.org/10.1016/j.ijfoodmicro.2011.02.012.

EFSA on Biological Hazards (BIOHAZ) (2011). Scientific opinion on the maintenance of the list of QPS biological agents intentionally added to food and feed (2011 update). *The European Food Safety Agency Journal*, 9, 2497.

Fall P. A., Leroi F., Cardinal M., Chevalier F., Pilet M. F. (2007). Inhibition of *Brochothrix thermosphacta* and sensory improvement of tropical peeled cooked shrimp by *Lactococcus piscium* CNCM I-4031. *Letters in Applied Microbiology*, 50(4), 357–361. https://doi.org/10.1111/j.1472-765X.2010.02801.x.

Fang T. J., Tsai H. C. (2003). Growth patterns of *Escherichia coli* O157:H7 in ground beef treated with nisin, chelators, organic acids and their combinations immobilized in calcium alginate gels. *Food Microbiology*, 20(2), 243–253. https://doi.org/10.1016/S0740-0020(02)00081-3.

FAO/WHO (2002). *Report of a Joint FAO/WHO expert consultation on guidelines for the evaluation of probiotics in food.* World Health Organization and Food and Agriculture Organization of the United Nations, London Ontario, Canada.

Fernandez B., Vimont A., Desfossés-Foucault É. Daga M., Arora G., Fliss I. (2017). Antifungal activity of lactic and propionic acid bacteria and their potential as protective culture in cottage cheese. *Food Control*, 78, 350-356.

Fernandez Perez R., Saenz Y., Rojo-Bezares B., Zarazaga M., Rodriguez J. M., Torres C., Tenorio C., Ruiz-Larrea F. (2018). Production and Antimicrobial Activity of Nisin Under Enological Conditions. *Frontiers in Microbiology*, 9 | Article 1918. doi: 10.3389/fmicb,2018.01918.

Field D., Cotter P.D., Ross R.P., Hill C. (2015). Bioengineering of the model lantibiotic nisin. *Bioengineered*, 6, 187-192.

Field D., Ross R. P., Hill C., (2018). Developing bacteriocins of lactic acid bacteria into next generation biopreservatives. *Current Opinion in Food Science*, 20, 1-6.

Gaaloul N., ben Braiek O., Han, K., Volski A., Chikindas M. L., Ghrairi T. (2015). Isolation and characterization of large spectrum and multiple bacteriocin-producing Enterococcus faecium strain from raw bovine milk. *Journal of Applied Microbiology*, 118, 343–355.

Galvez A. (2006) Inhibition of toxicogenic Bacillus cereus in rice-based foods by enterocin AS-48. *Int. J. Food Microbiol.*, 106, 185–194.

Gálvez A., Abriouel H., López R.L., Omar N.B., (2007). Bacteriocin-based strategies for food biopreservation. *International Journal of Food Microbiology*, 120, 51-70.

Gao Y., Li D., Liu S., Zhang L. (2015). Garviecin LG34, a novel bacteriocin produced by *Lactococcus garvieae* isolated from traditional Chinese fermented cucumber. *Food Control*, 50, 896–900.

Gautam N., Sharma N. (2015). A study on characterization of new bacteriocin produced from a novel strain of *Lactobacillus spicheri* G2 isolated from Gundruk - a fermented vegetable product of North East India: a novel bacteriocin production from *Lactobacillus spicheri* G2. *J. Food Sci. Technol.*, 52 (9), 5808–5816.

Giello M., La Storia A., De Filippis F., Ercolini D., Villani F. (2018). Impact *of Lactobacillus curvatus* 54M16 on microbiota composition and growth of *Listeria monocytogenes* in fermented sausages. *Food Microbiology*, 72, 1–15.

Giribaldi M, Gai F, Peiretti P. G., Ortoffi M. F., P. Lavermicocca, Lonigro S. L., Valerio F., Cavallarin L. (2018). Quality of ready-to-eat swordfish fillets inoculated by *Lactobacillus paracasei IMPC 2.1. Journal of the Science of Food and Agriculture,* doi: 10.1002/jsfa.9161.

Gómez-Sala B., Herranz C., Díaz-Freitas B., Hernández P. E., Sala A., Cintas L. M. (2016). Strategies to increase the hygienic and economic value of fresh fish: Biopreservation using lactic acid bacteria of marine origin. International Journal of *Food Microbiology*, 223, 41–49. https://doi.org/10.1016/j.ijfoodmicro.2016.02.005

Grande Burgos M r, Lopez Aguayo C., Perez Pulid, R., Galvez A., Lucas Lopez R. (2015). Inactivation of *Staphylococcus aureus* in Oat and Soya Drinks by Enterocin AS-48 in Combination with Other Antimicrobials. *Journal of Food Science*, 80, M2030-2034.

Grande M.J., Lucas R., Abriouel H., Valdivia E., Ben Omar N., Maqueda M., Martinez-Cañamero M, Galvez A. (2007).Treatment of vegetable sauces with enterocin AS-48 alone or in combination with phenolic compounds to inhibit proliferation of *Staphylococcus aureus*. *J Food Prot*, 70, 405–411.

Grande M.J., Lucas R., Abriouel H., Ben Omar N., Maqueda M., Martínez-Bueno M., Mertínez-Cañamero M., Valdivia E. Gálvez A. (2005). Control of *Alicyclobacillus acidoterrestris* in fruit juices by enterocin AS-48. *International Journal of Food Microbiology,* 104, 289-297.

Grass J. E., Gould L. H., Mahon B. E. (2013). Epidemiology of foodborne disease outbreaks caused by *Clostridium perfringens*, United States, 1998–2010. *Foodborne Pathogens and Disease,* 10(2), 131–136.

Guimarães A., Santiago A., Teixeira J. A., Venâncio A., Abrunhosa L. (2018). Anti-aflatoxigenic effect of organic acids produced by *Lactobacillus plantarum*. *International Journal of Food Microbiology*, 264, 31–38.

Hu P Xu XL, Zhou GH, et al., (2008) Study of the *Lactobacillus sakei* protective effect towards spoilage bacteria in vacuum packed cooked ham analyzed by PCR-DGGE. *Meat Science*, 80, 462-469.

Hu Z.Y., Balay D., H, Y., McMullen L. M., Gänzle M. G. (2019). Effect of chitosan, and bacteriocin – Producing *Carnobacterium maltaromaticum* on survival of *Escherichia coli* and *Salmonella Typhimurium* on beef. *International Journal of Food Microbiology* 290, 68–75.

Huang T., Zhang X., Pan J., Su X., Jin X., Guan X. (2016). Purification and characterization of a novel cold shock protein-like bacteriocin synthesized by *Bacillus thuringiensis*. *Sci. Rep.,* 6, 35560.

Hui Y. H. (2012). Hazard analysis and critical control point system. In Y. H. Hui (Ed.), *Handbook of meat and meat processing*. BocaRaton, FL:Taylor & Francis Group., 741–767.

Huss H. H. (1995). *Quality and quality Changes in fresh fish. Rome: Food and Agriculture organization of the United States*. http://www.fao.org/docrep/V7180E/V7180e06.htm.

Ibrahim A., Awad S. (2017). Selection and identification of protective culture for controlling *Staphylococcus aureus* in fresh Domiati like cheese. *Journal of Food Safety,* 38, e12418. https://doi.org/10.1111/jfs.12418.

Iglesias M.B., Echeverría G., Viñas I., López M. L. Abadias, M. (2018). Biopreservation of fresh-cut pear using *Lactobacillus rhamnosus* GG and effect on quality and volatile compounds. *Food Science and Technology*, 87, 581-588.

Jayasena D. D., Jo C. (2013). Essential oils as potential antimicrobial agents in meats and meat products: A review. *Trends in Food Science & Technology*, 34, 96 –108.

Jones R.. J., Zagorec M., Brightwell G., Tagg J. R. (2009) Inhibition by *Lactobacillus sakei* of other species in the flora of vacuum packaged raw meats during prolonged storage. *Food Microbiol* 26, 876-881.

Juodeikiene G., Bartkiene E., Cernauskas D., Cizeikiene D., Zadeike D., Krungleviciute V., Bartkevics V. (2017). Antifungal activity of lactic acid bacteria and their application for *Fusarium* mycotoxin reduction in malting wheat grains. *LWT – Food Science and Technology*, doi:10.1016/j.lwt.2017.10.061.

Juturu V., Wu J. C. (2018). Microbial production of bacteriocins: Latest research development and applications. *Biotechnol Adv.*, 36 (8), 2187-2200.

Kaskoniene V., Stankevicius M., Bimbiraite-Surviliene K., Naujokaityte G., Serniene L., Mulkyte K., Malakauskas M., Maruska A. (2017). Current state of purification, isolation and analysis of bacteriocins produced by lactic acid bacteria. *Appl. Microbiol. Biotechnol.*, 101 (4), 1323–1335.

Khan H., Flint S., Yu P-K (2010) Enterocins in food preservation. *International Journal of Food Microbiology*, 141, 1–10.

Lakshminarayanan B., Guinane C.M., O'Connor P.M., Coakley M., Hill C., Stanton C., O'Toole P. W., Ross R. P. (2013). Isolation and characterization of bacteriocin-producing bacteria from the intestinal microbiota of elderly Irish subjects. *J. Appl. Microbiol,* 114 (3), 886–898.

Le Lay, C., Mounier, J., Vasseur, V. Weill, A., Le Blay, G., Barbier, G., Coton, E. (2016). *In vitro* and *in situ* screening of lactic acid bacteria and propionibacteria antifungal activities against bakery product spoilage molds. *Food Control*, 60, 247-255.

Leyva Salas M., Thierry A., Lemaître M., Garric G., Harel-Oger M., Chatel M., Lê S., Mounier J., Valence F., Coton E. (2018). Antifungal Activity of Lactic Acid Bacteria Combinations in Dairy Mimicking Models and Their Potential as Bioprotective Cultures in Pilot Scale Applications. *Front Microbiol.*, 9, 1787.

Leyva Salas M., Mounier J., Valence F., Coton,M., Thierry A., Coton E. (2017). Antifungal microbial agents for food biopreservation—a review. *Microorganisms,* 5, 37. doi: 10.3390/microorganisms5030037.

Li P., Jia S., Zhou C., Fang H., Chen C. (2017) Protective role of *Lactobacillus fermentum* R6 against *Clostridium perfringensin vitro* and in chicken breast meat under temperature abuse conditions. Innovative *Food Science and Emerging Technologies* 41, 117–123.

Li P., Jia S., Zhou C., Fang H.,Che, C. (2017) Protective role of *Lactobacillus fermentum* R6 against *Clostridium perfringensin vitro* and in chicken breast meat under temperature abuse conditions. *Innovative Food Science & Emerging Technologies* 41, 117-123.

Lim K. B., Balolong M. P. (2016). Isolation and characterization of a broad spectrum bacteriocin from *Bacillus amyloliquefaciens* RX7. *Biomed. Res. Int.* 2016 (2) 8521476.

López de Lacey A. M., López-Caballero M. E., Montero P. (2013). Agar films containing green tea extract and probiotic bacteria for extending fish shelf-life. *LWT - Food Science and Technology*, 55(2), 559–564. https://doi.org/10.1016/j.lwt.2013.09.028.

Løvdal T. (2015). The microbiology of cold smoked salmon. *Food Control*, 54, 360–373. https://doi.org/10.1016/j.foodcont.2015.02.025.

Lu F., Liu D., Ye X., Wei Y., Liu F. (2009). Alginate-calcium coating incorporating nisin and EDTA maintains the quality of fresh northern snakehead (*Channa argus*) fillets stored at 4°C. *Journal of the Science of Food and Agriculture*, 89(5), 848–854. https://doi.org/10.1002/jsfa.3523.

Lucas R., Grande M.J., Abriouel H., Maqueda M., Ben Omar N., Valdivia E., Martínez-Cañamero M. Gálvez A. (2006). Application of the broad-spectrum bacteriocinenterocin AS-48 to inhibit *Bacillus coagulans* in canned fruit and vegetable foods. *Food and Chemical Toxicology*, 44, 1774-1781.

Macwana S. J., Muriana P. M. (2012). A 'bacteriocin PCR array' for identification of bacteriocin-related structural genes in lactic acid bacteria. *J. Microbiol. Methods*, 88 (2), 197–204.

Maldonado-Barragán A., Caballero-Guerrero B., Martín V., Ruiz-Barba J. L., Rodríguez J. M. (2016). Purification and genetic characterization of gassericin E, a novel coculture inducible bacteriocin from *Lactobacillus gasseri* EV1461 isolated from the vagina of a healthy woman. *BMC Microbiol*, 16 (1), 37.

Maragkoudakis P. A., Mountzouris K. C., Psyrras D., Cremonese S., Fischer J., Cantor M. D., Tsakalidou E. (2009). Functional properties of novel protective lactic acid bacteria and application in raw chicken meat against *Listeria monocytogenes* and *Salmonella enteritidis*. *International Journal of Food Microbiology*, 130, 219–226.

Martinez-Viedma P, Abriouel H, Ben Omar N, Lucas López R, Valdivia E, Galvez A (2009,c) Antibacterial protection by enterocin AS-48 in sport and energy drinks with less acidic pH values. *J Food Prot*, 72, 881–884.

Martinez-Viedma P, Abriouel H, Ben Omar N, Lucas R, Gálvez A (2009, b) Anti-staphylococcal effect of enterocin AS-48 in bakery ingredients of vegetable origin, alone and in combination with selected antimicrobials. *J. Food Sci.*, 74 (7), M384-389.

Martínez-Viedma P, Abriouel H, Ben Omar N, Lucas R, Valdivia E, Gálvez A (2009, a). Assay of enterocin AS-48 for inhibition of foodborne pathogens in desserts. *J. Food Prot.*, 72, 1654–1659.

Martinez-Viedma P., Abriouel H., Ben Omar N., Lucas R., Gálvez A. (2011). Inhibition of spoilage and toxinogenic *Bacillus* species in dough from wheat flour by the cyclic peptide enterocin AS-48. *Food Control*, 22, 756–761.

Martin-Visscher L. A., Yoganathan S., Sit C. S., Lohans C. T., Vederas J. C. (2011). The activity of bacteriocins from *Carnobacterium maltaromaticum* UAL307 against gram-negative bacteria in combination with EDTA treatment. *FEMS Microbiol. Lett.*, 317 (2), 152–159.H.

Matamoros S., Pilet M. F., Gigout F., Prévost H., Leroi F. (2009). Selection and evaluation of seafood-borne psychrotrophic lactic acid bacteria as inhibitors of pathogenic and spoilage bacteria. *Food Microbiology*, 26, 638–644.

Mieszkin S. Hymery N., Debaets S., Coton E., Le Blay G., Valence F., Mounier J. (2017). Action mechanisms involved in the bioprotective effect of *Lactobacillus harbinensis* K.V9.3.1.Np against *Yarrowia lipolytica* in fermented milk. *Int. J. Food. Microbiol.*, 248, 47-55.

Mieszkin S., Hymery N., Debaets S., Cotona E., Le Blay G. Valence F., Mounie J. (2017). Action mechanisms involved in the bioprotective effect of *Lactobacillus harbinensis* K.V9.3.1.Np against *Yarrowia lipolytica* in fermented milk. International *Journal of Food Microbiology*, 248, 47–55.

Mille, P., Haveroen M.E., Solichova K. et al., (2010). Shelf life extension of liquid whole eggs by heat and bacteriocin treatment. *Czech Journal of Food Sciences*, 28, 280–289.

Mills S., Griffin C., O'Connor P. M., Serrano L. M., Meijer W. C., Hill C., Ross R. P. (2017). A Multibacteriocin Cheese Starter System, Comprising Nisin and Lacticin 3147 in *Lactococcus lactis*, in Combination with Plantaricin from *Lactobacillus plantarum*. *Appl. Environ. Microbiol.*, 83, e00799-17. https://doi.org/10.1128/AEM.00799-17.

Morandi S., Silvetti T., Batelli G., Brasca M. (2019). Can lactic acid bacteria be an efficient tool for controlling *Listeria monocytogenes* contamination on cheese surface? The case of Gorgonzola cheese. *Food Control*, 96, 499-507.

Morsy M. K., Sharoba A. M., Khalaf, H. H., El-Tanahy, H. H., Cutter, C. N. (2015). Efficacy of Antimicrobial Pullulan-Based Coatingt o Improve Internal Quality and Shelf-Life of Chicken Eggs During Storage. *Journal of Food Science*, 0, M1-9.

Mosbah A., Delavenne E., Souissi Y., Mahjoubi M., Jéhan P., Le Yondre N., Cherif A., Bondon A., Mounier J., Baudy-Floc'h M., Le Blay G. (2018) Novel Antifungal Compounds, Spermine-Like and Short Cyclic Polylactates, Produced by *Lactobacillus harbinensis* K.V9.3.1Np in Yogurt. *Front Microbiol.*, 9, 2252.

Murphy E. F., Clarke C. F., Marques T. M., Hill C., Stanton C., Ross R. P., O'Doherty R. M., Shanahan F., Cotter P. D. (2013). Antimicrobials: strategies for targeting obesity and metabolic health?. *Gut Microbes*, 4, 48-53.

Nikodinoska I., Baffoni L., Di Gioia D., Mansob B., García-Sánchez L., Melero B., Rovir J. (2019). Protective cultures against foodborne pathogens in a nitrite reduced fermented meat product *LWT - Food Science and Technology*, 101, 293–299.

Noda M., Miyauchi R., Danshiitsoodol N., Higashikawa F., Kumagai T., Matoba Y., Sugiyama M. (2015). Characterization and mutational analysis of a two-polypeptide bacteriocin produced by Citrus iyo-derived *Lactobacillus brevis* 174A. *Biol. Pharm. Bull.*, 38 (12), 1902–1909.

Oliveira M., Abadias M., Colás-Meda P., Usall J. Viñas I. (2015b). Biopreservative methods to control the growth of foodborne pathogens on fresh-cut lettuce. *International Journal of Food Microbiology*, 2015, 4-11.

Oliveira M., Abadias M., Usall J., Torres R., Teixidó N. Viñas I. (2015a). Application of modified atmosphere packaging as a safety approach to fresh-cut fruits and vegetables – A review. *Trends in Food Science & Technology*, 46, 13-26.

Oliveira M., Ferreira V., Magalhaes R., Teixeira P. (2018). Biocontrol strategies for Mediterranean-style fermented sausages. *Food Research International*, 103, 438-449.

Orihuel A., Bonacina J., Vildoza M.J., Bru E., Vignolo G., Saavedra L., Fadda S. (2018) Biocontrol of *Listeria monocytogenes* in a meat model using a combination of a bacteriocinogenic strain with curing additives. *Food Research International*, 107, 289-296.

Pala C., Scarano C., Venusti M., Sardo D., Casti D., Cossu F., Lamon S., Spanu V., Ibba M., Marras M., Paba A., De Santis E. P. L. (2016). Shelf-life evaluation of sheep's *ricotta fresca* cheese in modified atmosphere packaging. *Italian J. Food Saf.*, 5 (3), 134-139.

Pasteris S. E., Pingitore V. E., Ale C. E., Nader-Macias M. E. (2014). Characterization of a bacteriocin produced by *Lactococcus lactis subsp. lactis* CRL 1584 isolated from a *Lithobates catesbeianus* hatchery. *World J. Microbiol. Biotechnol.*, 30 (3), 1053–1062.

Pei J., Yue T. Yuan Y. (2014). Control of *Alicyclobacillus acidoterrestrisin* fruit juices by a newly discovered bacteriocin. *World Journal of Microbiology and Biotechnology*, 30(3), 855-863.

Pereira C. I., Franco M. I., Gomes A. M. P., Malcata F. X. (2011). Microbiological, rheological and sensory characterization of Portuguese model cheeses manufactured from several milk sources. *LWT Food Sci. Technol.*, 44, 2244–2252.

Pérez Ibarreche M., Castellano P., Leclercq A., Vignolo G. (2016). Control of *Listeria monocytogenes* biofilms on industrial surfaces by the bacteriocin-producing *Lactobacillus sakei* CRL1862. *FEMS Microbiol. Lett.*, 363 (12). http://dx.doi.org/10.1093/femsle/fnw118.

Pérez Ibarreche M., Castellano P., Vignolo G. (2014). Evaluation of anti-Listeria meat borne *Lactobacillus* for biofilm formation on selected abiotic surfaces. *Meat Sci.*, 96, 295–303.

Pilet M.-F., Leroi F. (2011). Applications of protective cultures, bacteriocins and bacteriophages in fresh seafood and seafood products, In: Lacroix, C. (Ed.), "Protective Cultures, Antimicrobial Metabolites and Bacteriophages for Food and Beverage Biopreservation," *Woodhead Publishing Series in Food Science, Technology and Nutrition.* Woodhead Publishing, pp. 324–347.

Pingitore E. V., Salvucci E., Sesma F., Nader-Macías M. (2007). Different strategies for purification of antimicrobial peptides from Lactic Acid Bacteria (LAB). *Communicating Current Research and Educational Topics and Trends in Applied Microbiology* A. Méndez-Vilas (Ed), 2, 557–568.

Ramaroson M., Guillou S., Rossero A., Rezé S., Anthoine V., Moriceau N., Martin J-L, Duranton F., Zagorec M. (2018) Selection procedure of bioprotective cultures for their combined use with High Pressure Processing to control spore-forming bacteria in cooked ham. *International Journal of Food Microbiology* 276, 28–38.

Ramaswamy V., Cresence V. M., Rejitha J.S., Lekshmi M. U., Dharsana K. S., Prasad S. P., Vijila H. M. (2007). Listeria--review of epidemiology and pathogenesis. *J. Microbiol. Immunol. Infect.,* 40(1), 4-13.

Ramu R., Shirahatti P. S., Devi A. T., Prasad A., Kumuda J., Lachana M. S., Zameer F., Dhananjaya B. L., Nagendra Prasad M. N. (2015). Bacteriocins and their applications in food preservation. *Crit. Rev. Food Sci. Nutr.* http://dx.doi.org/10.1080/10408398.2015.1020918.

Rani R. P., Anandharaj M., Hema S., Deepika R., Ravindran A. D. (2016). Purification of antilisterial peptide (Subtilosin A) from novel *Bacillus tequilensis* FR9 and demonstrate their pathogen invasion protection ability using human carcinoma cell line. *Front. Microbiol,* 7, 1910.

Rehaiem A., Ben Belgacem Z., Edalatian M. R., Martínez B., Rodríguez A., Manai M., et al., (2014). Assessment of potential probiotic properties and multiple bacteriocin encoding-genes of the technological performing strain *Enterococcus faecium* MMRA. *Food Control*, 37, 343–350.

Rico D., Martín-Diana A.B., Barat J.M. Barry-Ryan C. (2007). Extending and measuring the quality of fresh-cut fruit and vegetables: a review. *Trends in Food Science & Technology,* 18, 373-386.

Riley M. A., Wertz J. E. (2002). Bacteriocins: evolution, ecology, and application. *Annu Rev Microbiol*, 56, 117-137.

Rouse S., Field D., Daly K. M., O'Connor P. M., Cotter P. D., Hill C., Ross R. P. (2012). Bioengineered nisin derivatives with enhanced activity in complex matrices. *Microb Biotechnol*, 5(4), 501-8. doi: 10.1111/j.1751-7915.2011.00324.

Russo P., Peña N., de Chiara M.L.V., Amodio M.L., Colelli G. Spano G. (2015). Probiotic lactic acid bacteria for the production of multifunctional fresh-cut cantaloupe. *Food Research International*, 77, 762-772.

Sabo SdS., Lopes A. M., Santos-Ebinuma V.., Rangel-Yagui C., Oliveira R P . (2018). Bacteriocin partitioning from a clarified fermentation broth of *Lactobacillus plantarum* ST16Pa in aqueous two-phase systems with sodium sulfate and choline-based salts as additives. *Process. Biochem*, 66, 212-221.

Salvucci,E., Hebert E. M., Sesma F., Saavedra L. (2010). Combined effect of synthetic enterocin CRL35 with cell wall, membrane-acting antibiotics and muranolytic enzymes against *Listeria* cells. *Letters in Applied Microbiology*, 51(2), 191–195.

Saraoui T., Cornet J., Guillouet E., Pile, M. F., Chevalier F., Joffraud,J. J., Leroi F. (2016). Improving simultaneously the quality and safety of cooked and peeled shrimp using a cocktail of bioprotective lactic acid bacteria. *International Journal of Food Microbiology*, 241, 69–77. https://doi.org/10.1016/j.ijfoodmicro.2016.09.024.

Sarika A. R., Lipton A. P., Aishwarya M. S., Dhivya R. S. (2012). Isolation of a bacteriocin-producing *lactococcus lactis* and application of its bacteriocin to manage spoilage bacteria in high-value marine fish under different storage temperatures. *Applied Biochemistry and Biotechnology*, 167(5), 1280–1289. https://doi.org/10.1007/s12010-012-9701-0.

Sathe S.J., Nawani N. N., Dhakephalkar,P. K. Kapadnis B. P. (2007). Antifungical lactic acid bacteria with potential to prolong shelf-life of fresh vegetables. *Journal of Applied Microbiology*, 103, 2622-2628.

Schmidt M., Lynch K. M., Zannini E., Arendt E. K. (2018). Fundamental study on the improvement of the antifungal activity of *Lactobacillus reuteri* R29 through increased production of phenyllactic acid and reuterin. *Food Control*, 88, 139-148.

Settanni L., Corsett, A. (2008). Application of bacteriocins in vegetable food biopreservation. *International Journal of Food Microbiology*, 121, 123-138.

Settier-Ramírez L., López-Carballo G., Gavara R., Hernández-Muñoz P. (2018). Antilisterial properties of PVOH-based films embedded with *Lactococcus lactis* subsp. *lactis*. *Food Hydrocolloids*, 87(July 2018), 214–220. https://doi.org/10.1016/j.foodhyd.2018.08.007.

Shirazinejad,A. R., Noryati I., Rosma A., Darah I. (2010). Inhibitory Effect of Lactic Acid and Nisin on Bacterial Spoilage of Chilled Shrimp. *Engineering and Technology*, 4(5), 163–167. https://doi.org/10.1109/5.726791.

Siedler S., Balti R., Neves A. R. (2019). Bioprotective mechanisms of lactic acid bacteria against fungal spoilage of food. *Current Opinion in Biotechnology*, 56, 138-146.

Silva C. C. G., Silva S. P. M., Ribeiro S. C. (2018). Application of Bacteriocins and Protective Cultures in Dairy Food Preservation. *Front Microbiol.*, 9, 594.

Siroli L., Patrignani,F., Serrazanetti,D. I., Tabanelli G., Montanari C., Gardini, F. Lanciotti R. (2015). Lactic acid bacteria and natural antimicrobials to improve the safety and shelf-life of minimally processed sliced apples and lamb´s lettuce. *Food Microbiology,* 47, 74-84.

Siroli L., Patrignani F., Serrazanetti D. I., Vannini,L., Salvetti E., Torriani S., Gardini F. Lanciotti R. (2016). Use of a nisin-producing *Lactococcus lactis* strain, combined with natural antimicrobials, to improve the safety and shelf-life of minimally processed sliced apples. *Food Microbiology*, 54, 11-19.

Siroli L., Patrignani F., Serrazanetti,D. I., Vernocchi P., Del Chierico F., Russo A., Torriani, S., Putignani, L., Gardini, F. & Lanciotti, R. (2017). Effect of thyme essential oil and *Lactococcus lactis* CBM21 on the microbiota composition and quality of minimally processed lamb´s lettuce. *Food Microbiology*, 68, 61-70.

Smitha S., Bhat S. G. (2013). Thermostable Bacteriocin BL8 from *Bacillus licheniformis* isolated from marine sediment. *J. Appl. Microbiol.*, 114 (3), 688–694.

Soliman L. C., Donkor K. K. (2010). Method development for sensitive determination of nisin in food products by micellar electrokinetic chromatography. *Food Chem.*, 119 (2), 801–805.

Son S. J., Park M. R., Ryu S. D., Maburutse B.E., Oh N.S., Park J., Oh S., Kim Y. (2016). Short communication: in vivo screening platform for bacteriocins using *Caenorhabditis elegans* to control mastitis-causing pathogens. *J. Dairy Sci.*, 99 (11), 8614–8621.

Song D. F., Li X., Zhang Y. H., Zhu M. Y., Gu Q. (2014, b). Mutational analysis of positively charged residues in the N-terminal region of the class IIa bacteriocin pediocin PA-1. *Lett. Appl. Microbiol.*, 58, 356-361.

Song D. F., Zhu M. Y., Gu Q. (2014, a). Purificati on and characterization of Plantaricin ZJ5, a new bacteriocin produced by *Lactobacillus plantarum* ZJ5. *PLoS One*, 9 (8), e105549.

Spanu C., Piras F., Mocci A. M., Nieddu G., De Santis E. P. L., Scarano C. (2018). Use of *Carnobacterium spp* protective culture in MAP packed *Ricotta fresca* cheese to control *Pseudomonas spp*. *Food Microbiology*, 74, 50-56.

Spanu C., Scarano C., Piras F., Spanu V., Pala C., Casti D., Lamon S., Cossu F., Ibba M., Nieddu G., De Santis E. P. L. (2017). Testing commercial biopreservative against spoilage microorganisms in MAP packed *Ricotta fresca* cheese. *Food Microbiol.*, 66, 72-76.

Sun L., Song H., Zheng W. (2015). Improvement of antimicrobial activity of pediocin PA-1 by site-directed mutagenesis in C-terminal domain. *Protein Pept. Lett.*, 22, 1007-1012.

Tomé E., Gibbs P. A., Teixeira P. C. (2007). Growth control of *Listeriainnocua* 2030c on vacuum-packaged cold-smoked salmon by lactic acid bacteria. *International Journal of Food Microbiology*, 121(3), 285–294. https://doi.org/10.1016/j.ijfoodmicro. 2007.11.015.

Trias R., Badosa E., Montesinos E. Bañeras L. (2008). Bioprotective *Leuconostoc* strains against *Listeria monocytogenes* in fresh fruits and vegetables. *International Journal of Food Microbiology*, 127, 91-98.

Varsha K. K., Nampoothiri K. M. (2016). Appraisal of lactic acid bacteria as protective cultures. *Food Control*, 69, 61-64.

Wachsman M. B., Castilla V., de Ruiz Holgad, A. P., de Torre, R. A., Sesma F., Coto C. E. (2003). Enterocin CRL35 inhibits late stages of HSV-1 and HSV-2 replication *in vitro. Antiviral Research*, 58(1), 17–24.

Walker M., Phillips C. (2008). The effect of preservatives on *Alicyclobacillus acidoterrestris* and *Propionibacterium cyclohexanicum* in fruit juice. *Food Control*, 19(10), 974-981.

Wannun P., Piwat S., Teanpaisan R. (2014). Purification and characterization of bacteriocin produced by oral *Lactobacillus paracasei* SD1. *Anaerobe*, 27, 17–21.

Wei H., Wolf G., Hammes W. P. (2006). Indigenous microorganisms from iceberg lettuce with adherence and antagonistic potential for use as protective culture. *Innovative Food Science & Technologies*, 7(4), 294-301.

Wessels S., Huss H. H. (1996). Suitability of *Lactococcus lactis subsp lactis* ATCC 11454 as a protective culture for lightly preserved fish products. *Food Microbiol.*, 13, 323–332.

Wiernasz N., Cornet J., Cardina, M., Pilet M.-F., Passerini,D., Leroi F. (2017). Lactic Acid Bacteria Selection for Biopreservation as a Part of Hurdle Technology Approach Applied on Seafood. *Frontiers in Marine Science*, 4(May), 1–15. https://doi.org/10.3389/fmars.2017.00119.

Woraprayote W., Malila Y., Sorapukdee S, Swetwiwathana A., Visessanguan W. (2016) Bacteriocins from lactic acid bacteria and their applications in meat and meat products. *Meat Science*, 120, 118-132.

Woraprayote W., Pumpuang L., Tosukhowong A., Zendo T., Sonomoto K., Benjakul S., Visessanguan,W. (2017). Antimicrobial biodegradable food packaging impregnated with Bacteriocin 7293 for control of pathogenic bacteria in pangasius fish fillets. *LWT - Food Science and Technology*, 89, 427–433. https://doi.org/10.1016/ j.lwt.2017.10.026.

Zhang W. Wang X., Xu C., Chen Y., Sun W., Liu,Q., Dong Q. (2018a) Modeling inhibition effects of *Lactobacillus plantarum* subsp. *plantarum* CICC 6257 on growth of *Listeria monocytogenes* in ground pork stored at CO_2-rich atmospheres. *LWT - Food Science and Technology*, 97, 811–817.

Zhang Y., Zhu,L., Dong P., Liang R., Mao Y., Qi, S., Luo X. (2018b) Bio-protective potential of lactic acid bacteria: Effect of *Lactobacillus sakei* and *Lactobacillus curvatus* on changes of the microbial community in vacuum-packaged chilled beef. *Asian-Australas J. Anim. Sci.*, 31 (4), 585-594.

Zhu X., Zhao Y., Sun Y., Gu Q. (2014). Purification and characterisation of plantaricin ZJ008, a novel bacteriocin against *Staphylococcus spp.* from *Lactobacillus plantarum* ZJ008. *Food Chem.*, 165, 216–223.

Zou J., Jiang H., Cheng H., Fang J., Huang G. (2018). Strategies for screening, purification and characterization of bacteriocins. *International Journal of Biological Macromolecules*, 117, 781–789.

In: The Many Benefits of Lactic Acid Bacteria
Editors: J. G. LeBlanc and A. de Moreno

ISBN: 978-1-53615-388-0
© 2019 Nova Science Publishers, Inc.

Chapter 2

THERMOTOLERANT PROBIOTIC LACTIC ACID BACTERIA IN COOKED MEAT PRODUCTS

María de Lourdes Pérez-Chabela[*]
Biotechnology Department, Universidad Autónoma Metropolitana,
Iztapalapa, Alcaldía de Iztapalapa, México City, Mexico

ABSTRACT

Lactic acid bacteria are part of the native flora in fresh meat, but certain strains have been employed as starter cultures in fermented meat products, becoming dominant flora due to lactic acid production, and a natural preservative. Nonetheless, although lactic acid bacteria are mesophilic, they cannot be employed as starter cultures in cooked meat products due to their limited survival to thermal processes. However, certain lactic acid bacteria isolated from cooked meats present the characteristics of thermotolerance, i.e., they survive thermal processes above 70°C for at least 30 min. These bacteria can overexpress heat shock proteins that are produced in response to exposure to a stressful environment, such as temperature or pH. These thermotolerant lactic acid bacteria present probiotic characteristics: being able to survive gastric tract conditions to colonize the small intestine and cell adhesion capacity. In addition, inoculation of these lactic acid bacteria in on cooked meat batters has no detrimental effect on texture, yield, color and sensory acceptation. The probiotication of cooked and emulsified meat products can be achieved with thermotolerant probiotic lactic acid bacteria.

[*]Corresponding Author's Email: lpch@xanum.uam.mx

INTRODUCTION

Lactic acid bacteria (LAB) are a group of Gram-positive bacteria, nonsporing cocci or rods, which produce lactic acid as the major end product of the fermentation of carbohydrates (Axelsson, 2004). They have a low content of GC, tolerate acidic conditions, are immobile, nutritionally demanding and must be administered vitamins, amino acids, purines and pyrimidines, lack catalase and facultative anaerobes, but are sometimes classified as aerotolerant anaerobes (Willey et al., 2008). They are found in dairy products, meat products and the human body, and are classified as GRAS (generally recognized as safe).

Lactic acid bacteria are divided into homolactic and heterolactic fermentative. Homolactic fermentation (Glycolysis Embden-Meyerhof-Parnas pathways) results almost exclusively in lactic acid as the end product, and heterolactic fermentation (the 6-phosphogluconate/phosphoketolase pathways) results in significant amounts of other end products, such as acetate, ethanol, CO_2 and lactic acid (Axelsson, 2004).

The LAB considered natural strains in meat and meat products are *Carnobacterium piscicola, Carnobacterium divergens, Lactobacillus sakei, Lactobacillus curvatus, Lactobacillus plantarum, Leuconostoc mesenteroides, Leuconostoc gelidum* and *Leuconostoc carnosum* (Hugas, 1998). LAB are employed as starter cultures in non-thermal processed fermented meat products, such as salami, but can easily grow at refrigeration temperatures, although they are considered mesophilic.

In this chapter, probiotic characteristics of thermotolerant LAB such as the survival in the gastrointestinal tract, the colonization of the intestine and cell adhesion capacity of thermotolerant lactic acid bacteria will be presented. In addition the inoculation of these LAB on cooked meat batters will be discussed.

THERMOTOLERANCE

Thermal tolerance or thermotolerance has been defined as the ability of an organism and of its cell structures to withstand (survive) destructive heat stress exceeding the range of optimal temperature for its development (Parsell and Lindquist, 1993). Mammalian cells can dramatically increase their tolerance to thermal damage after prior heat conditioning. The evidence suggested that heat shock proteins (HSP) provide the protective mechanism (Burdon, 1987).

Heat Shock Proteins. Classification

The first to publish the observation that cells could mount very strong transcriptional activity when exposed to elevated temperatures was Ritossa (1962), being coined the heat shock response. This discovery led to the identification of heat shock proteins.

Heat shock proteins (HSPs), or stress proteins, are present in all organisms and in all of their cells. Selected HSPs, also known as chaperones, play crucial roles in folding/unfolding of proteins, assembly of multiprotein complexes, transport/sorting of proteins into correct subcellular compartments, cell-cycle control and signaling, and protection of cells against stress/apoptosis (Li & Srivastava, 2003).

HSPs are broadly classified by molecular weight into distinct families: Hsp100, Hsp90, Hsp70, Hsp60, Hsp40, small Hsp and Hsp10 (Richa et al., 2007). The functions of the Hsp are acting like chaperones to maintain and control protein and proteases native structure. All HSP families possess chaperone activity (Sikora & Grzesiuk, 2007). Many members are present constitutively in cells while some are only expressed after stress. Table 1 shows the HSP Family, localization and functions.

Table 1. HSP Family, localization and function in the tissue

Family	Localization	Function	Reference
Hsp100	Cytosol	Chaperones	Mogk et al., 2015
Hsp90	Cytosol	Chaperones	Pearl & Prodromou, 2006
Hsp70	Cytosol	Chaperones	Schuermann et al., 2008
Hsp60	Mitochondria, cytosol and nucleus	ATP-dependent chaperone	Okamoto et al., 2017.
Hsp40	Nervous system	Co-chaperones	Gorenberg & Chandra, 2017
sHsp	Ubiquitous, heart, muscle, placenta, cytoplasm.	Chaperones	Bakthisaran et al., 2015; Garrido et al., 2012.
Hsp10	Mitochondria, cytosol	Co-chaperones	Fan et al., 2017; Jia et al., 2011.

Hsp100

The Hsp100 family of molecular chaperones shows a unique capability to resolubilize and reactivate aggregated proteins. The Hsp100-mediated protein disaggregation is linked to the activity of other chaperones from the Hsp70 and Hsp40 families. The best studied members of the Hsp100 family are the bacterial ClpB and Hsp104 from yeast. Hsp100 chaperones are members of a large super-family known as AAA+ ATPases (Zolkiewski et al., 2012). The AAA (ATPases associated with various cellular activities) family is a large group of ATPases found in all biological kingdoms and characterized by the presence of one or two conserved ATP-binding domains of a type called the AAA motif. This 200–250 amino-acid domain is defined by sequence and structural properties (Bar-Nun & Glickman, 2012).

The Hsp100 forms a bi-chaperone system composed of ATP-dependent Hsp70 and hexameric Hsp100 (ClpB/Hsp104) chaperones, which rescues aggregated proteins and provides thermotolerance to cells (Mogk et al., 2015).

Hsp90

The 90 kDa heat shock proteins are a family of molecular chaperones found in bacteria and all eukaryotes, but are absent in archaea. Many eukaryotes possess multiple Hsp90 homologues, including endoplasmic reticulum and mitochondrial isoforms (Pearl & Prodromou, 2006). The Hsp90 chaperone family includes the eponymous Hsp90 (90 kDa heat shock protein) of the eukaryotic cytosol, termed variously Hsp90alpha and beta in humans (corresponding to a major and minor isoform), Hsp86 and Hsp84 in mice, Hsp83 in Drosophila, and Hsc82 and Hsp82 in yeasts. Other family members are HtpG in the bacterial cytosol, Grp94/gp96 in the endoplasmic reticulum of eukaryotes, and the Hsp75/TRAP1 in the mitochondrial matrix (Young et al., 2001).

The Hsp90 is a key regulator of proteostasis under both physiological and stress conditions in eukaryotic cells. As Hsp90 has several hundred protein substrates (or "clients"), it is involved in many cellular processes beyond protein folding, which include DNA repair, development of the immune response and neurodegenerative disease. Owing to the importance of Hsp90 in the regulation of many cellular proteins, it has become a promising drug target for the treatment of several diseases, which include cancer and diseases associated with protein misfolding (Schopf et al., 2017).

Hsp70

Hsp70s are a diverse family of chaperones involved in a variety of protein processing reactions, including import into the endoplasmic reticulum and mitochondrion, protein folding, and the remodeling of protein complexes (Schuermann et al., 2008). Hsp70 has two isoforms of 66 and 76 kDa, both called Hsp70, which is derived from a gene with the same name (Evans et al., 2010). HSPs prevent incorrect folding. Hsp70 plays a role in folding of proteins that have been destroyed and damaged by stress. The most important roles of Hsp70 are the management of folding of damaged and abnormal proteins (Valizadeh et al., 2017).

Hsp60

The eukaryotic heat shock protein Hsp60 (GroEL) belongs to the group I of chaperonins and is classically defined as an intramitochondrial protein that assists with the correct folding of other mitochondrial proteins, together with its co-chaperonin Hsp10 (GroES) (Bukau &Horwich, 1998).

Hsp60 plays an essential role in protein homeostasis through mediating protein folding and assembly. It was found that Hsp60 has different properties with respect to allostery, complex formation and protein folding activity, depending on the nucleoside triphosphate present (Okamoto *et al.,* 2017). Since the initial discovery of Hsp60 as the mitochondrial molecular chaperone many studies have shown that it is also localized outside mitochondria with perhaps both chaperoning and non-chaperoning activities. Therefore, it is not surprising that many different disease states, especially autoimmune diseases and cancers, have presented altered expression levels of Hsp60 (Meng et al., 2018). Recent investigations have suggested that Hsp60 functions and interactors may vary, depending on its cell and tissue localization, giving this protein the capacity to be a multifaceted molecule in the middle of the delicate balance between health and disease (Marino et al., 2018).

Hsp40

Ohtsuka et al., (1990) discovered a novel 40-kDa protein in addition to the classical heat shock proteins with molecular sizes of 110-, 90-, 70-, and 47-kDa. The p40 was induced not only by heat shock but also by arsenite and 2-azetidine carboxylic acid. It was induced in rat, mouse and chicken cells by these stresses. The p40 is a novel heat shock protein in mammalian and avian cells.

Heat shock proteins also called DnaJ/Hsp40 proteins have been preserved throughout evolution and are important for protein translation, folding, unfolding, translocation, and degradation, primarily by stimulating the ATPase activity of chaperone proteins, Hsp70s. DnaJ/Hsp40 proteins actually determine the activity of Hsp70s by stabilizing their interaction with substrate proteins (Qiu et al., 2006). Hsp40s are defined by their conserved J domain and typically have large variation in the rest of the protein. Classical Hsp40s have both a J domain as well as a zinc finger domain that regulates client protein binding. The J domain of Hsp40 proteins consists of a 70-amino acid sequence with four alpha helices (Gorenberg & Chandra, 2017).

sHsp

Small heat shock proteins (sHsp) were discovered in *Drosophila melanogaster* salivary glands as a set of sHsp, with small molecular weight (15–30 kDa) induced after a heat shock and accumulating in the cells after many different environmental, physiological or pathological stresses (Tissieres et al., 1974). The human sHsp family contains ten members and one related protein (HSP16.2/HSPB11) is ubiquitously expressed while others are only expressed in specific tissues (Garrido et al., 2012).

Small heat shock proteins (sHsps) are conserved across species and are important in stress tolerance. Many sHsps exhibit chaperone-like activity in preventing aggregation of target proteins (Bakthisaran et al., 2015). The sHsp participate in a large number of fundamental cellular processes such as controlling protein folding, F-actin-dependent processes, cytoprotection/anti-apoptosis, differentiation, cell proliferation, and gene expression, and therefore are involved in many pathologies, such as neurodegenerative diseases, cancer, and cardiovascular diseases (Guo & Wu, 2015).

Hsp10

Protein folding in mitochondria is mediated by the chaperonin Hsp60, the homologue of *E. coli* GroEL. Mitochondria also contain a homologue of the cochaperonin GroES, called Hsp10. Hsp10 is required for the folding and assembly of proteins imported into the matrix compartment, and is involved in the sorting of certain proteins, such as the Rieske Fe/S protein, passing through the matrix en route to the intermembrane space (Hohfeld & Ulrich, 1994).

Hsp10, an important member of Hsps family, classically works as a molecular chaperone folding or degradating of target proteins. Evolutionarily, Hsp10 is also reported to be involved in immunomodulation and tumor progression (Fan et al., 2017). Hsp10 appears to be related to pregnancy, cancer and autoimmune inhibition. It can be released to peripheral blood at a very early time point of pregnancy and received another name, early pregnancy factor (EPF), which seems to play a critical role in developing a pregnant niche (Jia et al., 2011).

HSP IN LACTIC ACID BACTERIA

LAB are important starter, commensal, or pathogenic microorganisms. The stress physiology of LAB has been studied in depth for over 2 decades, fueled mostly by the technological implications of LAB in the food industry (Papadimitriou et al., 2016).

Hartke et al., (1996) studied the lactic acid tolerance of *Lactococcus lactis* subsp. *lactis*. The results showed a dramatic increase in survival to severe acid stress during 30 min at pH 3.9. One protein extract analysis revealed that the major heat shock proteins DnaK and GroEL are produced during acid tolerance. Highly conserved chaperones, DnaK, DnaJ, GrpE, GroES, GroEL, and proteases Clp, HtrA, and FtsH, have been detected in LAB (Guchte et al., 2002). Lactobacilli respond to stress in specific ways dependent on the strain, species, and the type of stress.

Coucheney et al., (2005) studied the *Oenococcus (O.) oeni*, a lactic acid bacterium mostly responsible for malolactic fermentation in wine. This bacterium produces an 18 kDa sHSP, called Lo18 that is induced by multiple stresses, such as heat, ethanol, sulphites, and low pH, among others. *O. oeni* cells showed that purified Lo18 interacts with the liposomes and increases the molecular order of the lipid bilayer in these membranes when the temperature reaches 33.8°C. These data suggest that Lo18 could be involved in an adaptive response allowing the maintenance of membrane integrity during stress conditions in *O. oeni* cells.

Spano et al., (2005) using a molecular approach based on PCR, RT-PCR and Northern blot analysis, described a new member of the small heat shock family of wine, *Lactobacillus (Lb.) plantarum*. The protein had a calculated molecular of mass of 18.548 kDa and was named Hsp18.55. Expression of the newly identified small heat shock gene was induced by a wide range of abiotic stresses including heat, cold and ethanol, suggesting that the small family of heat shock genes is probably involved in the general stress response in *Lb. plantarum* from wine.

Capozzi et al., (2011) studied the *Lb. acidophilus* hps16 gene structure, genomic organization, and deduced amino acid sequence. The results indicated a strong evolutionary relationship within members of the *Lb. acidophilus* group. This species is used to drive dairy fermentations and in functional probiotic foods and is therefore commonly exposed to multiple physiological stresses. Wu et al., (2014) reported that a small heat shock protein (18 kDa) probably indicated the link between the osmotic tolerance of *Lb. paracasei* during fermentation due to salt stress response.

Pandi and Basher (2016) studied the stress resistance of *Lactobacillus* strains and concluded that the genus *Lactobacillus* contains many stress resistant organisms. They can produce stress resistant metabolites like trehalose and heat shock proteins during growth at high temperatures, which helps the growth of microorganisms at high temperatures. Serata et al., (2016) reported Hsp18 and GroEL as one gene involved in oxidative stress resistance in *Lb. casei* strain Shirota.

Wang et al., (2017) studied the characteristics of four isolated LAB and their effects on fermentation quality of Italian ryegrass (*Lolium multiflorum* Lam.), which is difficult to silage on the Tibetan Plateu due to the low temperature. The results showed that *Lb. plantarum* strains overproducing Hsp 18.5, Hsp 18.55, and Hsp 19.3 can grow better at low temperatures.

THERMOTOLERANT LACTIC ACID BACTERIA IN MEAT PRODUCTS

The three-dimensional structure of cellular proteins is sensitive to small increases when they are exposed to high temperature, and the misfolding and aggregation of unfolded proteins are the primary problems for cells subjected to heat stress (Varmanen and Savijoki, 2011).

LAB are employed as starter cultures for the fermentation of meat products. In some countries, cooked meat products undergo a heat treatment of 70°C (pasteurization temperature) in their processing; some LAB can survive this treatment due to the production of Hsp.

The role of LAB in both cooked and fresh meat products is controversial due to the spoilage characteristic of LAB. On one hand, several species are able to dominate the meat system in VP and MAP storage conditions and can release odor-impact molecules, which may alter the sensory profile of raw or cooked meat. On the other hand, LAB have a bioprotective function since they can provide favorable antagonistic activity against other undesired microorganisms (Pothakos et al., 2015).

Chenoll et al., (2006) analysed the presence of LAB associated with a variety of vacuum-packed fresh and spoiled meat products. From a total of 18 samples, *Lactobacillus Leuconostoc, Carnobacterium*, and *Enterococos* were the predominant species. Victoria-Leon et al., (2008) isolated eight strains of LAB from commercial sausages. The strains were heated in water baths to 50, 60 and 70°C, during 30-60 min. Four strains survived the treatments and were identified *as Lb. alimentarius, Lactococcus (L.). lactis, Lb. piscicola* and *Enterococcos sp.* These strains were considered thermotolerant. Ramírez-Chavarín et al., (2010) isolated a total of 68 presumptive LAB from commercial cooked sausages collected in Mexico City supermarkets. Based on biochemical tests, 22 strains were selected and their thermotolerant capacity determined. Only 10 strains were considered thermotolerant after surviving thermal treatments at 70°C during 30 min. The genera identified by phenotypic and genotyping analyses were *Pediococcos, Lactobacillus, Aerococcos* and *Enterococcos*.

Chávez-Martinez et al., (2016) quantified and identified LAB in thirty samples of sliced cooked ham, reporting that enterococci showed the lowest count and lactobacilli the highest counts; 23.8% of the strains were identified as thermophilic *Lactobacillus*. Saucedo-Briviesca et al., (2017) studied the effect of probiotic thermotolerant LAB on the physicochemical, microbiological and sensorial characteristics in a meat batter. Two thermotolerant probiotic LAB were used: *Pediococcus pentosaceus (Ped.)* and *Enterococcus (Ec.)* faecium, which were inoculated to 5% in a meat batter. The results showed that the stability to cooking, expressible moisture, hardness and cohesion increased during storage in the batters inoculated with the 2 strains of LAB. The count of LAB increased in the inoculated meat batters and the coliforms decreased overall. When the strain mixture was used, inhibition of coliforms was total at day 6. Sensory analysis showed that judges detected when *Ec. faecium* was inoculated. Thermotolerant BAL can be used as functional ingredients in meat batters.

When a LAB demonstrates its resistance to thermal stress it is very possible that this bacterium can resist another type of stress. Fragoso et al., (2016) studied this and demonstrated that the use of thermotolerant probiotic LAB in fat reduced ice cream with inulin as fat replacer could be a good alternative to formulate symbiotic foods.

PROBIOTICS. DEFINITION AND CHARACTERISTICS

Several specific LAB strains are marketed on the basis of their beneficial effects on the consumer's health, representing an explosively growing market for the products containing these so-called probiotics (Bron & Kleerebezem, 2011).

Etymologically, the term probiotic is derived from the Greek language meaning "for life." The first definition of probiotics was coined by Lilly and Stillwell (1965) to describe substances produced by one microorganism that stimulate the growth of another, thus meaning the opposite of antibiotics. Later, Fuller (1989) redefined the term probiotics:

"Probiotics are live microbial feed supplements which beneficially affect the host animal by improving its intestinal microbial balance."

The FAO/WHO (2002) describes them as:

"Live microorganisms which when administered in adequate amounts confer a health benefit on the host."

This is the more widely accepted definition.

The principal characteristics of probiotics include: tolerance to human gastric juice and bile, adherence to epithelial surfaces, persistence in the human gut, immune stimulation, antagonistic activity toward intestinal pathogens, production of antimicrobial substances, safety in food and clinical use, susceptibility to antibiotics and bile salt hydrolase activity (Parvez et al., 2006). Among the benefits of probiotics are improvements of intestinal health, enhancement of the immune response, reduction of serum cholesterol and cancer prevention (Kechagia et al., 2013).

The most common microorganisms used as probiotics are from the *Lactobacillus* and *Bifidobacterium* genera; other bacterial genera, including *Enterococcus*, *Streptococcus*, and *Escherichia*, are also used. Some strains of the fungus *Saccharomyces boulardii* are also considered probiotics (Mack, 2005).

THERMOTOLERANT LACTIC ACID BACTERIA AS PROBIOTICS

Ramírez-Chavarín et al., (2013) evaluated 10 thermotolerant LAB (*Ped. pentosaceus* (4 strains), *Lb. plantarum* (3 strains), *Ec. faecium* (2 strains) and *Aerococcus viridans* (1 strain) to identify their probiotic properties. The results showed that 50% of the strains were intolerant to low pH and simulated gastric juice. All strains grew in taurocholic acid and bile concentrations greater than 0.3%, indicating they are good probiotic candidates. Less than 20% of the strains coaggregated with an *E. coli* indicator while 30% coaggregated with a *Salmonella* ssp.. Eight of the strains exhibited good autoaggregation capacity at 24 h and all ten had a high adherence capacity for HEp-2 cells. The studied thermotolerant LAB are promising ingredients in the production of cooked meat products with probiotic potential. Alginate and other gelling hydrocolloids encapsulation of thermotolerant and probiotic LAB is a good alternative to protect them in their passage through the pH of the stomach (Totosaus et al., 2013).

Tatsinkou et al., (2017) studied the protective effect of minerals such as calcium and magnesium on probiotic strains. This protective effect is due to the fact that calcium and magnesium ions have the property of stabilizing certain structural proteins and even enzymatic ones. Their presence may prevent the rapid denaturation of membrane proteins. These results showed that mineral salts increase the thermotolerance and resistance to bile salts of probiotic strains. They concluded that calcium and magnesium could be used to monitor the viability of probiotic strains in probiotic products.

Hernández-Alcántara et al., (2018) isolated and evaluated the probiotic properties of six thermotolerant LAB. The bacteria were typed by determination of the DNA sequence of their 16S rRNA coding genes, as one *Ec. faecium* (UAM1 strain) and five *Ped. pentosaceus* (UAM2-UAM6 strains). *In vitro*, these strains showed desirable probiotic properties. These trials are promising for the application of *Ec. faecium* UAM1 as a novel probiotic strain in the food industry, since it can be employed as bioprotective culture due to its thermotolerant capacity in functional foods.

SYNBIOTICS

Synbiotics are used for the improved survival of beneficial microorganisms added to food or feed, and to stimulate the proliferation of specific native bacterial strains present in the gastrointestinal tract (Markowiak & Slizewska, 2017). Gómez-Chávez et al., (2011) studied the prebiotic effect of cacao flour on the growth of probiotic thermotolerant lactic acid bacteria in a meat batter. The in vitro growth of lactic acid bacteria using cacao flour as carbon source was acceptable compared with the growth obtained using glucose. No differences in physicochemical analysis were shown. Those authors concluded that the use of this kind of bacteria in cooked meat products leads to a foodstuff with high nutritional and symbiotic value that enhances microbial intestinal flora as a result of its intake.

Díaz-Vela et al., (2015) studied the effect of fruit peels as prebiotic and probiotic thermotolerant lactic acid bacteria in the structural characteristics of cooked meat products. Scanning electron microscopy showed the development of LAB in the sausages. In the control treatment (not inoculated with bacteria, no fruit peel flour added), some LAB colonies were visible between the components of the meat batter, above all after 20 days of storage. Analysis showed the production of exopolysaccharides by thermotolerant LAB. The incorporation of fruit peel flour into the formulation increased the number of bacteria in cooked meat emulsions.

Lactic acid bacteria viability was improved by the incorporation of different compounds during alginate ionotropic gel matrix coencapsulation, enhancing the resistance to acidic conditions as well. The strong prebiotic effect of inulin and apple marc flour was to enhance bacterial viability due to insoluble carbohydrate content, resulting in higher particle size with a higher number of viable bacteria. Symbiotic co-gelification of alginate gel matrix with prebiotic compounds to protect probiotics can be employed in order to ensure the delivery of probiotic strains in the colon throughout the gastrointestinal tract (Serrano-Casas et al., 2017).

CONCLUSION

Lactic acid bacteria have been used for many years in fermented meat products. It has been shown that LAB can produce heat shock proteins (Hsp) that make them capable of resisting the heat treatments suffered by meat products in their preparation. Additionally, these bacteria may have probiotic characteristics. The use of probiotic thermotolerant lactic acid bacteria in cooked meat products is a good alternative in the manufacture of functional meat foods.

REFERENCES

Axelsson L. 2004. Chapter 1. Lactic acid bacteria: classification and physiology. In: *Lactic Acid Bacteria*. Salminen, S., von Weight, A., Ouwehand, A. (Eds.) Third Edition. Marcel Dekker. New York, U.S.A. 1-66.

Bakthisaran R., Tangirala R., Mohan, R. Ch. (2015) Small heat proteins: Role in cellular functions and pathology. *Biochimica et Biophysica Acta*, 1854, 291-319.

Bar-Nun S., Glickman M. H. (2012). Proteasomal AAA-ATPases: Structure and function. *Biochimica et Biophysica Acta*, 1823, 67-82.

Bron P. A., Kleerebezem M. (2011). Engineering lactic acid bacteria for increased industrial functionality. *Bioengineered Bugs*, 282, 80-87.

Bukau B., Horwich A. L. (1998). The Hsp 70 and Hsp 60 Chaperone machines. *Cell*, 92, 351-356.

Burdon R. H. (1987). Thermotolerance and the heat shock proteins. *Symposia of the Society for Experimental Biology*, 41, 269-283.

Capozzi V., Aren, M. P., Crisett, E., Spano G., Fiocco D. (2011). The hsp16 gene of the probiotic *Lactobacillus acidophilus* is differently regulated by salt, high temperature and acidic stresses, as revealed by reverse transcription quantitative PCR (qRT-PCR) analysis. *International Journal of Molecular Sciences*, 12, 5390-5405.

Chavez-Martinez A., Estrada-Ganderilla M., rentería-Monterrubio A. L., Gallegos-Acevedo M. A. (2016). Prevalence of lactic acid bacteria in sliced cooked ham as an indicator of its shelf life. *Vitae Revista de la facultad de Ciencias Farmaceuticas y Alimentarias*, 23 (3), 167-172.

Chenoll E., Macián M. C., Elizaquivel P., Aznar R. (2006). Lactic acid bacteria associated with vacuum-packed cooked meat product spoilage: population analysis by rDNA-based methods. *Journal of Appled Microbiology*, 102, 498-508.

Coucheney F., Gal L., Beney L., Lherminie, J., Gervais P., Guzzo J. (2005). A small HSP, Lo 18, interacts with the cell membrane and modulates lipid physical state under heat shock conditions in a lactic acid bacterium. *Biochimica et Biophysica Acta*, 1720, 92-98.

Díaz-Vela J., Totosaus A., Pérez-Chabela, M. L. (2015). Integration of agroindustrial co-products as functional food ingredients: cactus pear (*Opuntia ficus indica*) flour and pineapple (*Ananas comosus*) peel flour as fiber source in cooked sausages inoculated with lactic acid bacteria. *Journal of Food Processing and preservation*, 39, 2630-2638.

Evans C. G., Chang L., Gestwicki J. E. (2010). Heat shock protein 70 (hsp70) as an emerging drug target. *Journal of Medicinal Chemistry*, 53, 4585-4602.

FAO/WHO Food and Agriculture Organization of the United Nations/ World Health Organization. (2002). *Guidelines for the evaluation of probiotics in foods*. Ontario, Canada 1-11.

Fan W., Fan S. S., Feng J., Xiao, D., Fan S., Luo J. 2017. Elevated expression of HSP 10 protein inhibits apoptosis and associates with poor prognosis of Astrocytoma. *PLOS ONE*, 12(10), 1-10.

Fragoso M., Pérez-Chabela M. L., Hernández-Alcántara A. M., Escalona-Buendía H. B., Pintor A., Totosaus A. 2016. Sensory, melting and textural properties of fat-reduced ice cream inoculated with thermotolerant lactic acid bacteria. *Carpathian Journal of Food Science and technology*, 8(2), 11-21.

Fuller R. (1989). Probiotics in man and animals. *Journal of Applied Bacteriology*, 66, 365-378.

Garrido C., Paul C., Seigneuric R., Kampinga H H. (2012). The small heat shock proteins family: the long forgotten chaperones. *The International Journal of Biochemistry & Cell Biology*, 44, 1588-1592.

Gómez-Chávez G., Pérez-Chabela M. L., Totosaus A. (2011). Properties of meat batters inoculated with probiotic thermotolerant lactic acid bacteria and cacao flour: symbiotic food. *Ingeneria Agricola y Sistemas*, 3(1), 5-10.

Gorenberg E. L., Chandra S. S. (2017). The role of co-chaperones in synaptic proteostasis and neurodegenerative disease. *Frontiers in Neurosciences*, 11(248), 1-16.

Guchte M., Serror P., Chervaux C., Smokvina T., Ehrlich S. D., Maguin, E. (2002). Stress response in lactic acid bacteria. *Antonie van Leewenhoek*, 82, 187-216.

Guo H., Wu T. (2015). Small HSP variants and human diseases. In: *The big book on small heat shock protein, heat shock proteins*. Tanguay, R., Hightower, L. (Eds.) Springer, 383-397.

Hartke A., Bouché S., Giard J. C., Benachour A., Boutibonnes P., Auffroy, F. (1996). The lactic acid stress response of *Lactococcus lactis* subsp. Lactis. *Current Microbiology*, 33(3), 194-199.

Hernández-Alcántara A. M., Wacher C., Goretti Llamas M., López P., Pérez-Chabela M. L. (2018). Probiotic properties and stress response of thermotolerant lactic acid bacteria isolated from cooked meat products. *LWT Food Science and Technology*, 91, 249-257.

Hohfeld J., Ulrich H. F. (1994). Role of the chaperonin cofactor Hsp 10 in protein folding and sorting in yeast mitochondria. *JCB Journal of Cell Biology*, 126 (2), 305-316.

Hugas M. (1998). Bacteriocionogenic lactic acid bacteria for the biopreservation of meat and meat products. *Meat Science*, 49, S139-S149.

Jia H., Halilou I., Hu L., Cai W., Liu J., Huang B. (2011). Heat shock protein 10 (Hsp 10) in immune-related diseases: one coin, two sides. *International Journal of Biochemistry and Molecular Biology*, 2(1), 47-57.

Kechagia M., Basoulis D., Konstantopoulou S., Dimitriadi D., Gyftopoulou K., Skarmoutsou N., Fakir, E. M. (2013). *Health benefits of probiotics. A review. ISRN Nutrition* 2013, 481651, 1-7.

Li Z., Srivastava P. (2003). Heat-Shock proteins. *Current Protocols in Inmunology*, 58 (1), 1-6.

Lilly D. M., Stillwell R. H. (1965). Probiotics: growth- promoting factors produced by microorganisms. *Science,* 147(3659), 747-748.

Mack D. R. (2005). Probiotics. *Canadian Family Physician*, 51(11), 1455-1457.

Marcowiak P., Slizewska K. (2017). Effects of probiotics, prebiotics and symbiotics on human health. *Nutrients,* 9 (1021), 2-30.

Marino G. A., Macaluso F., di Felice V., Capello F., Baron, R. (2018). Hsp60 in skeletal muscle fiber biogenesis and homeostasis: from physical exercise to skeletal muscle pathology. *Cells,* 224,: 1-15.

Meng Q., Li B. X., Xiao X. (2018). Towards developing chemical modulators of HSP60 as potential therapeutics. *Frontiers in Molecular Biosciences*, 5(35), 1-11.

Mogk A., Kummer E., Bukau B. (2015). Cooperation of Hsp70 and Hsp100 chaperone machines in protein disaggregation. *Frontiers in Molecular Biosciences*, 2(22), 1-10.

Okamoto T., Yamamoto H., Kudo, I., Matsumoto K., Odaka M., Grave E., Itoh H. (2017). HSP 60 possesses a GTPase activity and mediates protein folding with HSP 10. *Scientific Reports*, 7, 1-11.

Othsuka K., Masuda A., Nakai A., Nagata K. (1990). A novel 40-kDa protein induced by heat shock and other stresses in mammalian and avian cells. *Biochimica et Biophysica Research Communications*; 166(2), 642-647.

Pandi S., Basheer S. (2016). Adaptation of Lactobacillus sp. and Sacharomyces sp. to heat stress. *International Journal of Microbiology and Allied Sciences*, 283, 7-16.

Papadimitriou K., Alegría A., Bron P. A., de Angelis M., Gobbetti M., Kleerebezem M., Lemus J. A., Linares D. M., Ross P., Stanton C., Turroni F., van Sindren D., Varmanen P., Ventura M., Zúñiga M., Tsakalidou E., Kok J. (2016). Stress Physiology of lactic acid bacteria. *Microbiology and Molecular Revision*, 80, 837-890.

Parsell D. A., Lindquist, S. L. (1993). The function of heat-shock proteins in stress tolerance: degradation and reactivation of damaged proteins. *Annual Review of Genetics*, 27, 437-496.

Parvez S., Malik A., Kang A., Kim H-Y. (2006). Probiotics and their fermented food products are beneficial for health. *Journal of Applied Microbiology*, 100(6), 1171-1185.

Pearl, L. H., and Prodromou, C. (2006). Structure and mechanism of the Hsp90 molecular chaperone machinery. *The Annual Review of Biochemistry*, 75, 271-294.

Pothakos V., Devliegere F., Villan, F., Bjorkroth J., Ercolini D. (2015). Lactic acid bacteria and their controversial role in fresh meat spoilage. *Meat Science*, 109, 66-74.

Qiu X-B., Shao Y-M., Wang M. L. (2006). The diversity of DNAJ/Hsp40 family, the crucial partners for HSP70 chaperones. *Cellular and Molecular Life Sciences*, 63(22), 2560-2570.

Ramirez-Chavarín N. L., Wacher-Rodarte C., Pérez-Chabela M. L. (2010). Characterization and identification of thermotolerant lactic acid bacteria isolated from cooked sausages as bioprotective cultures. *Journal of Muscle Foods*, 21, 585-596.

Ramírez-Chavarín M. L., Wacher C., Eslava-Campos C. A., Pérez-Chabela M. L. (2013). Probiotic potential of thermotolerant lactic acid bacteria strains isolated from cooked meat products. *International Food Research Journal*, 20(2), 991-2000.

Richa A., Mallik M., Lakhotia S. C. (2007). Heat shock genes-integrating cell survival and death. *Journal of Biosciences*, 32(3), 595-610.

Ritossa F. (1962). A new puffing pattern induced by temperature shock and DNA in Drosophila. *Experientia*, 18(12), 571-573.

Saucedo-Briviesca N., Cuesta A. I., Pérez-Chabela M. L. (2017). Efecto de bacterias lácticas termotolerantes probióticas sobre las caracteristicas fisicoquímicas, microbiológicas y sensoriales en batidos cárnicos cocidos [Effect of probiotic thermotolerant lactic bacteria on physicochemical, microbiological and sensory characteristics in cooked meat shakes]. *Nacameh*, 11(1), 1-17.

Schopf F., Biebl M. M., Buchner J. (2017). The Hsp 90 chaperone machinery. Nature reviews *Molecular Cell Biology*, 18, 345-360.

Schuermann J,. P., Jiang J., Cuellar J., Llorca O., Wang L., Gimenez L. E., Jin S., Taylor A. B., Demejer B., Morano K. A., Hart P. J., Valpuesta J. M., Lafer E. M., Sousa R. (2008). Structure of the Hsp110: Hsc70 nucleotide exchange machine. *Molecular Cell*, 31(2), 232-243.

Serat, M., Kiwaki M., Lino T. (2016). Functional analysis of a novel hydrogen peroxide resistance gene in *Lactobacillus casei* strain Shirota. *Microbiology*, 162, 1885-1894.

Serrano-Casas V., Pérez-Chabela M. L., Cortés-Barberena E., Totosaus A. (2017). Improvement of lactic acid bacteria viability in acid conditions employing agroindustrial co-products as prebiotics on alginate ionotropic gel matrix co-encapsulation. *Journal of Functional Foods*, 38, 293-297.

Sikor, A.. Grzesiuk E. (2007). Heat shock response in gastrointestinal tract. *Journal of Phisiology and Pharmacology*, 58, 43-62.

Spano G., Beneduce L., Perrota C., Massa C. (2005). Cloning and characterization of the hsp 18.55 gene, a new member of the small heat shock gene family isolated from wine *Lactobacillus plantarum*. *Research in Microbiology*, 156, 219-224.

Tatsinkou F. B., Nchanji T. G., Ndjouenkeu R. (2017). Effect of inorganic salts on the thermotolerance and probiotic properties of Lactobacillus isolated from curdled milk traditionally produced in Mezan Division Cameroon. *Advances in Microbiology*, 7, 589-601.

Tissieres A., Mitchell H. K., Tracy U. M. (1974). Protein synthesis in salivary glands of *Drosophila melanogaster*: relation to chromosome puffs. *Journal of Molecular Biology*, 84(3), 384-392.

Totosaus A., Ariza-Ortega T. J., Pérez-Chabela M. L. (2013). Lactic acid bacteria microencapsulation in sodium alginate and other gelling hydrocolloids mixtures. *Journal of Food and Nutrition Research*, 52(2), 107-120.

Valizadeh A., Pakzad I. R., Khosravi A. (2017). Investigating the role of thermal shock proteins (DanK) HSP 70 in bacteria. *Journal of Bacteriology and Mycology*, 4(3), 1055-1060.

Varmanen P., Savijoki K. (2011). Chapter 3: Responses of lactic acid bacteria to heat stress. In: *Stress responses of lactic acid bacteria*. Tsakalidou, E. and Papadimitriou, K. (Eds.) Springer, Atenas, Grecia, 55-66.

Victoria-León T., Totosaus A., Guerrero I., Pérez-Chabela M. L. (2008). Efecto de bacterias ácido lácticas termoresistentes en salchichas cocidas. *Ciencia y Tecnología Alimentaria*, 5(2), 135-141.

Wang S., Yuan X., Dong Z., Li J., Guo, G., Bai Y., Zhang J., Shao T. (2017). Characteristics of isolated lactic acid bacteria and their effects on the silage quality. *Asian-Australasian Journal of Animal Sciences*, 30(6), 819-827.

Willey J. M., Sherwood L. M., Woolverton Ch. J. (2008). Chapter 23: Bacteria: grampositivas con bajo contenido de G + C. In: *Microbiología de Prescott, Harley y Klein*. Willey, Sherwood and Woolverton (Eds). Mc Graw Hill. 7ª. Edition. Madrid, España, 571-588.

Wu R., Xu X., Meng L., Zou T., Tang X., Wu J., Yue X. (2014). Identification of salt stress responsive protein in *Lactobacillus paracasei* LN-1 using SDS-PAGE. IERI Procedia, 8, 60-65.

Young J. C., Moarefi I., Hartl F. U. (2001). HSP90: a specialized but essential protein-folding tool. *The Journal of Cell Biology*, 154, 267-273.

Zolkiewski M., Zhang T., Nagy M. (2012). Aggregate reactivation mediated by the Hsp 100 chaperones. *Arch Biochem Biophys,* 520(1), 1-6.

In: The Many Benefits of Lactic Acid Bacteria
Editors: J. G. LeBlanc and A. de Moreno

ISBN: 978-1-53615-388-0
© 2019 Nova Science Publishers, Inc.

Chapter 3

MICROENCAPSULATED LACTIC ACID BACTERIA IN MEAT PRODUCTS

Daneysa L. Kalschne*, Rosana A. Silva-Buzanello, Marinês P. Corso, Eliane Colla and Cristiane Canan

Departamento de Alimentos, Universidade Tecnológica Federal do Paraná,
Medianeira, Paraná, Brazil

ABSTRACT

The lactic acid bacteria (LAB) that belongs to the *Lactobacillaceae* family is classified into fifteen different genera. Some LAB, including the *Lactobacillus casei, Lb. delbrueckii, Lb. acidophilus, Lb. fermentum, Lb. reuteri, Lb. sakei,* and *Lb. plantarum,* are recognized worldwide as probiotics. International agencies such as the United Nations' Food and Agricultural Organization and the World Health Organization have defined probiotics as "living microorganisms of which when administered in adequate amounts confer a health benefit on the host." LAB selection as probiotics must include criteria such as safety for human consumption, stability against acid and bile conditions, and the capacity to colonize the human gastrointestinal tract. Probiotic microorganisms have been associated with improvements in the host's health, which has stimulated the consumption of LAB-added functional foods. There are some concerns associated with probiotics survival during manufacturing, storage, and specific preparation processes for the final product. Different approaches have been studied in order to increase LAB feasibilty in meat products. Microencapsulation could be employed as an alternative since it is a LAB protective method consisting of isolating the microorganism to obtain spherical-shaped and micrometer-sized products, in which a membrane shields the LAB. The probiotics ought to survive and dominate other microorganisms present in the food product, evidencing the use of microencapsulation for cellular viability increase. The probiotics used in fermented meat products are primarily associated with salami, which

---

* Corresponding Author's Email: daneysa@hotmail.com.

differs in terms of raw material composition, ripening time, appearance, and flavor. Fermented meat products encounter some barriers such as pH, water activity, lower sugars content, temperature, and preparation form limitations. On the other hand, the instinctive search for better a quality of life demands practical and functional foods and the added value is imperative to make a difference in the meat food industry. In this context, microorganisms are added to food aiming at increasing its properties, creating new functional foods capable to promote healthy benefits to the host. This chapter focused on the microencapsulated probiotic LAB potentials and their application in fermented meat products.

INTRODUCTION

The concern about food and health is reflected in the lifestyle of the population. Food habits change according to what is offered and made available in the market, favoring and encouraging the consumption of foods with functional properties. Consumers' acceptance of functional foods is complex and influenced by product-related factors such as sensorial characteristics, price, healthy ingredients, related claims, and also by consumer-related factors (Shan et al., 2017).

Functional foods potentially promote health benefits, they can provide nutrients associated with probiotics that may contribute diminishing intestinal disorders (Kaprasob, Kerdchoechuen & Laohakunjit, 2018). Probiotics are defined as microorganisms that, under certain conditions, confer the host some health benefits. Many LAB are probiotic microorganisms and they are especially associated with milk-based products. The most important LAB probiotic genera are *Lactobacillus* and *Bifidobacterium* (Ferreira & Silva, 2016).

Probiotics are associated with intestinal microbiota control, pathogens population decrease, intestinal microbiota stabilization after the use of antibiotics, lactose digestion, constipation relief, host immune system stimulation, short chain fatty acids production (acetate and butyrate), inhibitory effects on mutagenicity, and colon cancer risk reduction (Ferreira& Silva, 2016).

Probiotics offer some limitations for their application in foods. There are some concerns associated with probiotics survival during manufacturing, storage, and specific preparation processes to obtain the final product. In this respect, microencapsulation ought to improve the probiotic stability, proving to be an interesting alternative for adverse conditions. Different microencapsulation methods and wall materials have been tested in order to ensure high stability.

Pork consumption has been on the rise and is expected to reach 43,619 tons worldwide by 2026 (Oecd/Fao, 2017). A great part of its consumption is related to industrialized products, and it should be diversified to achieve consumer's preference. Probiotic meat products are mainly associated with fermented meat products. Salami is an excellent alternative for probiotic addition because it could be consumed without preheating.

In this way, LAB microencapsulation appears as an alternative to producing probiotic meat products, by breaking technological barriers. Additionally, probiotics could improve food safety and preservation, sensory characteristics development, acceptability, nutritional value, and health benefits.

LACTIC ACID BACTERIA

LAB are Gram-positive ubiquitous microorganisms, which are found in plants, mammals` intestinal mucosa, as well as in fermented foods (Teneva-Angelova et al., 2018). LABs have a long history of use for food production and conservation, as they are comprised of a wide group of microorganisms. Originally, LAB group included four kinds of great importance genera for food industry, including *Lactobacillus, Leuconostoc, Pediococcus,* and *Streptococcus.* Currently, it consists of fifteen classified genera based on morphology, glucose fermentation mode, the configuration of lactic acid produced, growth at certain temperatures, high NaCl concentration and pH, and catalase activity (Table 1) (Kalschne et al., 2015).

LAB are mesophilic, non-sporogenic, devoid of cytochromes, facultative anaerobes, fastidious, acid tolerant, catalase, and oxidase negative (Khalid, 2011). However, their main characteristic is carbohydrates fermentation with lactic acid production. The differences between homofermentative and heterofermentative carbohydrate metabolism involves LAB genetics and their physiological basis. Homofermentative carbohydrate metabolism use aldolase and hexose isomerase enzymes by the Embden-Meyerhof-Parnas pathway to produce two lactic acid molecules from a one glucose one. In contrast, the heterofermentative metabolism uses phosphoketolase enzyme by hexose monophosphate pathway or pentose degradation of glucose to produce lactic acid, ethanol, and carbon dioxide from glucose (Kalschne et al., 2015).

The *Bifidobacterium* genus is classified as a LAB since it shares a few characteristics, despite being phylogenetically unrelated. *Bifidobacterium* differs from other LAB genera on glucose fermentation mode, which occurs via fructose-6-phosphate due to the presence of fructose-6-phosphate phosphoketolase enzyme (Mazo et al., 2009).

LAB are usually of great biotechnological interest due to their morphological, metabolic, and physiological characteristics. The addition of LAB to foods have some positive characteristics such as a probiotic appeal, desired sensory characteristics development in fermented meat products, and food safety due to bacteriocin production and/or also due to competition with the microbiota present in food (Domínguez, Agregán & Lorenzo, 2016; Juturu & Wu, 2018; Ahn, Kim & Kim, 2017; Ruggirello et al., 2018).

Table 1. LAB key phenotypic characteristics

Genera	Morphology	CO_2 from glucose	Lactic acid isomer	Growth at 10°C	Growth at 45°C	Growth in 18% NaCl	Growth at pH 4.4	Growth at pH 4.5	Catalase activity
Aerococcus	C[a]	-	L	+	-	-	-		-
Atopobium	C/R								-
Bifidobacterium	R	+[b]	L	-	+		-	-	-
Brochothrix	C/R[d]		L		-[f]				+
Carnobacterium	R	-[e]	L	+	-	-		-	-
Enterococcus	C	-	L	+	+	-	+		-
Lactobacillus	R	±	D, L, DL	±	±	-	±	+	-[f]
Lactococcus	C	-	L	+	-	-	±		-
Leuconostoc	C	+	D	+	±	-	±		-
Oenococcus	C	+	D	+	±	-	±		-
Pediococcus	C[a]	-	L, DL	±	±	-	+		-
Streptococcus	C	-	L	-	±	-	-		-
Tetragenococcus	C[a]	-	L	+	-	+	-[c]	-[c]	-
Vagococcus	C	-	L	+	-	-	±		-
Weissella	C/R	+	D, DL	+	-	-	±		-

C: cocci; R: rod-shape; L, D and LD: optical isomers; +: positive; -: negative; a: cocci may also be tetrad formation; b: differ from bacterial homo and heterofermentative on fermentation of glucose, which occured via fructose-6-phosphate; c: grow did not occur at pH 5.0 or lower; d: in older cultures the rod may give rise to coccoid forms, which will develop into rod forms when subcultured onto a suitable medium; e: small amounts of CO_2 from glucose might be produced; f: growth did not occur at 37 °C; f: can be present pseudocatalase activity.
Source: Adapted from Kalschne et al., (2015).

Probiotics

Probiotics have been getting various definitions since their discovery. According to Fuller (1989) probiotics are living microbial food supplements that beneficially affect the host by improving their intestinal microbiota balance. International agencies such as the United Nations' Food and Agricultural Organization and the World Health Organization defined probiotics as "living microorganisms that when administered in adequate amounts confer a health benefit on the host" (Hill et al., 2014; FAO/WHO, 2001). In this way, a food is considered functional in terms of probiotic activity when it generates a beneficial effect to the host that receives it, besides its nourishing ability.

Probiotics are microorganisms present in the intestine on a constant survival contest and they are associated with many beneficial effects at this level, as was explained above (Ferreira & Silva 2016). Additionally, a recent study about *Lactobacillus plantarum* KX881772, a new probiotic from camel milk, observed potential anti-diabetic activities in camel fermented sausages and beef by α-amylase (A) and α-glucosidase (B) inhibitions, and they also showed anti-hypertensive and cytotoxicity activities (Ayyash et al., 2019).

LAB selection as probiotics includes anti-pathogenic activity and other criteria for safety assessment such as taxonomy identification, virulence absence, enterotoxins production, hemolytic activity, and transferable antibiotic resistance genes. Additionally, probiotics must exhibit stress tolerance to salivary and gastric enzymes, body temperature, low pH, gastric juice and bile salts, and adhesion ability to colonize the human gastrointestinal tract demonstrating functional efficacy (Distrutti et al., 2016; Pereira et al., 2018).

Strains of probiotics for use in humans have been selected mainly among the genera *Lactobacillus* and *Bifidobacterium*. They are commonly found in the intestines and mucosal surfaces of healthy humans or in spontaneously fermented foods (Balakrishnan & Floch, 2012; Vinderola et al., 2017). However, Pereira et al., (2018) also emphasized that *Lactococcus, Streptococcus,* and *Enterococcus* have a long history of safe use as probiotics.

Lactobacillus is the main LAB genus present in the small intestine, and it is susceptible to changes due to the entry of exogenous bacteria; thus *Lactobacillus* consumption during several days might be necessary to achieve the probiotic effect. In contrast, *Bifidobacterium* is mainly present in the large intestine and it does not offer a practical use as probiotics due to difficulties of its isolating and manipulating (anaerobiosis) and also, it does not tolerate acidic environments, hampering its use when fermented products are used as carriers. The microbiota in the large intestine is more stable and suffers less influence from exogenous bacteria due to the effect of consumed prebiotics. However, the large intestine is the first one to be attacked by antibiotics (Ferreira & Silva, 2016).

In this sense, there are some concerns associated with probiotics survival during manufacturing, storage, and specific preparation processes for the final product. In general, the use of LAB in food products depends on the lactic acid amount reduced, the type of glucose fermentation (homo or heterofermentatives), growth at different temperatures, and the tolerability to acidic and alkaline substances.

The standard for health-claiming-probiotics-added foods follows FAO/WHO recommendations per plate count with a minimum 10^6 to 10^7 UFC/g (FAO/WHO, 2001). Similarly, Minelli et al., (2008) recommended 10^6 CFU/mL and 10^8 CFU/mL as minimum probiotic concentrations in the small intestine and in the colon, respectively.

There are some concerns associated with the survival of probiotic strains during the manufacturing, storage, and high quantification number in the final product (Leroy, Verluyten & De Vuyst, 2006). Different approaches have been studied in order to increase the resistance of these sensitive microorganisms to adverse conditions, being probiotics microencapsulation one of them. Many authors have reported the advantages of this method in prolonged preservation and controlled release of probiotics (Etchepare et al., 2015; Shori, 2017; Huq et al., 2017; Prisco & Mauriello, 2016; Arepally & Goswami, 2019; Mao et al., 2018).

MICROENCAPSULATION

Different approaches have been studied in order to increase LAB feasibility in meat products. Microencapsulation is a promising alternative that could be employed as a LAB protective method, consisting of the coating of microorganisms to obtain spherical-shaped and micrometer-sized products, in which a membrane shields the LAB (Arepally & Goswami, 2019). The probiotics can survive and dominate over other microorganisms present in the food product, emphasizing the use of microencapsulation to increase cellular viability.

Microencapsulation is a coating process of small particles or droplets; or the compounds incorporation in a homogeneous or heterogeneous matrix, obtaining small capsules. Microencapsulation promotes a physical barrier between the core and wall materials, enabling a controlled release (Prisco & Mauriello, 2016).

Probiotics can be kept viable by microencapsulation, depending on processing conditions both during food manufacturing process and in the final product. Microencapsulation allows the coating material, capable of withstanding extrinsic conditions adverse to microorganisms, to protect the culture. On the other hand, coating should be released upon reaching the intestines so that probiotics can be able to perform their functions there (Arslan et al., 2015; Martín et al., 2015).

Several factors affect the protection and survival of encapsulated probiotics during gastric transit. The most important factors are probiotic strains' acid resistance properties, wall materials and their concentrations, encapsulation methods, and polymers incorporation in the matrix (Shori, 2017).

Microencapsulation has different uses, namely the food industry has reported the implementation of increasingly complex formulations with microorganisms encapsulated in fermented meat; the addition of polyunsaturated fatty acids susceptible to auto-oxidation in milk, yogurts or ice creams; and the use of highly volatile flavor compounds in instant foods, which oftentimes can only be achieved by microencapsulation (Khan et al., 2011; Gharsallaoui et al., 2012).

Microencapsulation Techniques

The most common methods used for probiotic microencapsulation are emulsion, spray drying, and extrusion, due mainly to their efficiency and low costs (Santos et al., 2019; Holkem et al., 2017; Ramos et al., 2018; Shori, 2017; Prisco & Mauriello, 2016). Furthermore, complex coacervation is an emerging method also used for probiotic encapsulation that results in an increase of probiotics feasibility and great encapsulation efficiency (Bosnea, Moschakis & Biliaderis, 2014; Bosnea, Moschakis & Biliaderis, 2017). Gelation, including the alginate-based, has been investigated (Kia, Alizadeh &

Esmaiili, 2018; Huq et al., 2017). Often, probiotic microcapsules obtained by coacervation, emulsion or gelation methods are posteriorly freeze-dried to ensure preservation and microorganisms' viability (Silva et al., 2018; Halim et al., 2017; Huq et al., 2017).

Choosing microencapsulation for a specific process will depend on factors such as particle size, the microcapsules biocompatibility and biodegradability, core and coating physical and chemical properties, microcapsules application, proposed active core release mechanism, and process costs (Dubey, Shami & Bhasker Rao, 2009). For probiotic applications, it is important that microcapsules be water-insoluble so the structure could be maintained during food and beverages processing, and through the digestive system (Ramos et al., 2018).

The spray drying technique is viable and effectively applicable for microorganism's encapsulation. This microencapsulation technique has been largely used for microorganisms' preservation (Arslan-Tontul & Erbas, 2017; Arepally & Goswami, 2019; Santos et al., 2019; Loyeau et al., 2018; Mao et al., 2018; Prisco & Mauriello, 2016). It may offer advantages such as ease of operation and cost-effectiveness, and that it is already available for industrial use. The atomization process consists in numerous droplets formation by solution, emulsion or suspension spraying into the drying chamber where they make contact with the hot gas (air or nitrogen) for subsequent dehydration, transforming the liquid component in a powdered solid one, later recovered in the cyclone (Burgain et al., 2011).

The risk of cell death associated with high temperatures is one of the disadvantages of spray drying. However, some authors mention temperatures ranging from 70 to 130 °C in the drying chamber to promote microorganisms' microencapsulation (Anekella & Orsat, 2013), considering that the contact time between microorganisms and the heat source is quite short, and the chamber outlet temperature is around 30-40 °C lower than the inlet temperature (Mao et al., 2018).

Wall Materials

The correct selection of wall material is associated with microcapsules' stability, process efficiency, and active core protection level. The ideal wall material should be non-reactive with the core, capable to seal and keep it within the capsule, provide core maximum protection against adverse conditions, no unpleasant taste, and be economically feasible (Nazzaro et al., 2012).

For food applicable microencapsulation, the wall materials normally include proteins (whey protein, soy protein, caseinate, and gelatin), carbohydrates (starch, alginate, maltodextrin, pectin, inulin, gum arabic, chitosan, carrageenan, cellulose nanocrystal, β-cyclodextrin, dextrans, and fructo-oligosaccharides) or their mixtures (Etchepare et al.,

2015; Shori, 2017; Huq et al., 2017; Prisco & Mauriello, 2016; Arepally & Goswami, 2019; Mao et al., 2018; Loyeau et al., 2018; Arslan-Tontul & Erbas, 2017). Most wall materials do not have all the desired properties, thus the mixture of two or more is a common practice.

The release of microcapsules content at controlled rates could be triggered by shearing, heating, pH, and enzymatic action (Lam & Gambari, 2014) and are mainly related to the wall material employed. Carbohydrates are often mixed with proteins in order to improve emulsifying and film-forming properties during their encapsulation (Mao et al., 2018; Loyeau et al., 2018).

In addition to being biodegradable and biocompatible materials, the wall materials should release the microencapsulated probiotics in the intestines without earlier release due to the wall materials solubility in water. Thus, probiotic microcapsules pH dependence is an interesting alternative to release the microorganisms at the correct time in the small intestine. Most foods have acidic pH, and the conditions for probiotics in the digestive tract before reaching the small intestine are acidic; however, pH increases when the probiotics enter in the intestine, surpassing neutrality and reaching alkaline levels. In this way, pH-dependent wall materials, especially with probiotic release at alkaline pH are the most favorable ones.

PROBIOTIC LAB ADDED IN FERMENTED MEAT PRODUCTS

Several foods of various products categories have been added with probiotics such as bakery products (bread, biscuits, chocolate soufflé, and stuffed cakes), fruit and vegetable juices (acerola nectar, longan, cranberry, pomegranate, berries, carrot, orange, peach, and apple), dairy products (yoghurt, ice cream, and fermented soured milk beverage), cheese (feta, cheddar, white Iranian brined cheese, white brined cheese, kasar, mozzarella, fiordilatte, pecorino, and oxaca) and others (i.e., mayonnaise). For meat products, only fermented ones have been associated with probiotics (Prisco & Mauriello, 2016).

Fermented meat products such as salami have some technical limitations regarding probiotics addition. Barriers such as pH, water activity, lower sugars content, and preparation form limitations, mainly associated with high cooking, baking, and frying temperatures need to be surpassed. In order to overcome this problem, the addition of microencapsulated probiotics in fermented meat products has been discussed (Barbosa et al.,. 2015; Muthukumarasamy & Holley, 2006; Macedo et al., 2008). However, these food products added with microencapsulated probiotics should be consumed without pre heating at preparation in order to preserve the microorganisms.

Salami is prepared by mixing all of its ingredients (pork shoulders, pork fat, salt, curing salts, sugars, spices, antioxidant, and starter culture), stuffing in artificial casings, fermentation/drying for a few days, and ripening for some days up to reaching the desired water activity in the final product (Alamprese, Fongaro & Casiraghi, 2016).

The fermentation process is defined as the anaerobic metabolism of carbohydrates by microorganisms. It has been used to extend foods expiration date, with an important role in diets worldwide offering a variety of benefits. Starter cultures were initially used for food preservation, including sensory characteristics, acceptability, nutritional value, and food safety improvements while also providing diet diversification (Malo & Urquhart, 2016).

Salami starter cultures generally combine the *Micrococcaceae* family represented by *Staphylococcus (S.) xylosus* and *S. carnosus* species, and LAB *Pediococcus (Ped.) pentosaceus, Ped. acidilactici,* and *Lactobacillus (Lb.) sakei.* The *Debaryomyces hansenii* yeast and *Penicillium nalgiovense* mold were also included in some commercial starter cultures (Alves et al., 2018).

The cited *Staphylococcus* species are associated with bacteriocins production and lipolytic and proteolytic activities, while the LAB are associated with the lactic acid production. *Debaromyces hasenni* and *Penicillium nalgiovense* also have lipolytic and proteolytic activities. *Penicillium nalgiovense* provide a fungal mycelium on salami's external surface, offering a protective effect against some undesirable microorganisms during the ripening process, and it is a good quality indicator by manufacturers (Galvalisi et al., 2012; Mattei, 2012).

In terms of starter cultures and probiotics, two important issues need to be mentioned. First, not all LAB used in food products fermentation are considered probiotic, since the probiotic is related to the colonization of the intestines benefiting its host. Second, the fermentative LAB microorganisms are not always present in amounts required to obtain the beneficial properties of a functional food. Therefore, there is a close link between the possibilities that a fermenter LAB be a microorganism with probiotic characteristics.

Overall, during the fermentation and ripening processes, the sugar present in the salami is fermented by homofermentative *Lactobacillus* and converted to organic acids, mainly lactic acid by Embden-Meyerhof-Parnas, resulting in pH reduction (Krummenauer et al., 2015). The pH decreases and reaches the isoelectric point of proteins (neutral charge) promoting myofibrillar proteins insolubilization and the consequent water loss by dehydration.

Therefore, acid pH is one of the main physicochemical and sensorial characteristics of salami, which usually reach values around 5.0 and, increasing to 5.3 after proteolysis (Massaguer, 2006). Acid pH also works as a barrier to the development of undesirable microorganisms in the product, especially when associated with low water activity.

Probiotic microorganisms in salami must withstand the product's (pH from 4.5 to 6.0) and the human stomach's (pH from 1.0 to 3.5) acidic conditions. Microencapsulation allows greater probiotic stability against adverse pH conditions. Alternatively, the wall material could be pH dependent as it remains intact at acidic pHs but is disrupted by releasing the probiotics at the small intestine alkaline pH (pH around 7.4 in the jejunum and up to 7.7 in the ileum) (Ekmekcioglu, 2002). In this sense, more studies should be conducted to evaluate the probiotic effect of microencapsulated LAB applied in a fermented meat product.

CONCLUSION

The consumers' search for better quality of life and the increase in quality of life demand practical and functional foods and the added value is imperative to make a difference in the meat industry. Functional foods are an option in order to add value and make a difference in the meat industry for the consumers that search for better quality of life. Probiotic LAB are recognized to offer different possibilities in physicochemical and sensory innovations and microbiological safety. In this context, alternatives such as probiotic microencapsulation help to protect the microorganisms and promote a wide range of options to enrich complex food matrices with probiotics. Among the different microencapsulation techniques, spray drying is characterized as ease of operation and cost-effective for microorganism's microencapsulation. Different wall materials are available, and the combination of carbohydrates and proteins normally promote greater functional properties. Considering that the meat products industry has been growing year by year in line with an increasing demand for functional foods, efforts have been made to produce functional meat products. Fermented products such as salami are an excellent alternative for probiotics addition, as they are consumed without any preheating, a thermal process that could kill microorganisms.

REFERENCES

Ahn, H., Kim J. and Kim W. J. (2017). Isolation and characterization of bacteriocin-producing *Pediococcus acidilactici* HW01 from malt and its potential to control beer spoilage lactic acid bacteria. *Food Control*, 80: 59–66.

Alamprese, C., Fongaro L. and Casiraghi E. (2016). Effect of fresh pork meat conditioning on quality characteristics of salami. *Meat Science*, 119: 193–198.

Alves, L. L., Silva, M. S., Flores, D. R. M., Athayde, D. R., Ruviaro, A. R., Brum, D. S., Batista, V. S. F., Mello, R. O., Menezes, C. R., Campagnol, P. C. B., Wagner, R., Barin, J. S. and Cichoski A. J. (2018). Effect of ultrasound on the physicochemical and microbiological characteristics of Italian salami. *Food Research International*, 106: 363–373.

Anekella, K. and Orsat, V. (2013). Optimization of microencapsulation of probiotics in raspberry juice by spray drying. *LWT - Food Science and Technology,* 50 (1): 17–24.

Arepally, D. and Goswami, T. K. (2019). Effect of inlet air temperature and gum arabic concentration on encapsulation of probiotics by spray drying. *LWT - Food Science and Technology,* 99: 583–593.

Arslan-Tontul, S. and Erbas, M. (2017). Single and double layered microencapsulation of probiotics by spray drying and spray chilling. *LWT - Food Science and Technology*, 81: 160–169.

Arslan, S., Erbas, M., Tontul, I. and Topuz, A. (2015). Microencapsulation of probiotic *saccharomyces cerevisiae* var: boulardii with different wall materials by spray drying. *LWT - Food Science and Technology,* 63 (1): 685–690.

Ayyash, M., Liu, S. Q., Mheiri, A. A., Aldhaheri, M., Raeisi, B., Al-Nabulsi, A., Osaili, T. and Olaimat, A. (2019). In vitro investigation of health-promoting benefits of fermented camel sausage by novel probiotic *Lactobacillus plantarum*: a comparative study with beef sausages. *LWT - Food Science and Technology,* 99: 346–354.

Balakrishnan, M. and Floch, M. H. (2012). Prebiotics, probiotics and digestive health. *Current Opinion in Clinical Nutrition and Metabolic Care*, 15 (6): 580–585.

Barbosa, M. S, Todorov, S. D., Jurkiewicz, C. H. and Franco B. D. G. M. (2015). Bacteriocin production by *Lactobacillus curvatus* MBSA2 entrapped in calcium alginate during ripening of salami for control of *Listeria monocytogenes. Food Control,* 47: 147–153.

Bosnea, L. A., Moschakis, T. and Biliaderis C. G. (2017). Microencapsulated cells of *Lactobacillus paracasei* subsp. paracasei in biopolymer complex coacervates and their function in a yogurt matrix *Food and Function*, 8 (2): 554–562.

Bosnea, L., Moschakis, T. and Biliaderis, C. G. (2014). Complex coacervation as a novel microencapsulation technique to improve viability of probiotics under different stresses. *Food and Bioprocess Technology,* 7 (10): 2767–2781.

Burgain, J., Gaiani, G., Linder, M. and Scher, J. (2011). Encapsulation of probiotic living cells: from laboratory scale to industrial applications. *Journal of Food Engineering*, 104 (4): 467–83.

Distrutti, E., Monaldi, L., Ricci, P. and Fiorucci, S. (2016). Gut microbiota role in irritable bowel syndrome: New therapeutic strategies. *World Journal of Gastroenterology,* 22 (7): 2219–2241.

Domínguez, R., Agregán, R., Lorenzo, J. M. (2016). Role of commercial starter cultures on microbiological, physicochemical characteristics, volatile compounds and sensory properties of dry-cured foal sausage. *Asian Pacific Journal of Tropical Disease*, 6 (5): 396–403.

Dubey, R., Shami, T. C. and Rao, K. U. B. (2009). Microencapsulation technology and applications. *Defence Science Journal*, 59 (1): 82–95.

Ekmekcioglu C. (2002). A physiological approach for preparing and conducting intestinal bioavailability studies using experimental systems. *Food Chemistry,* 76 (2): 225–230.

Etchepare, M. A., Barin J. S., Cichoski, A. J., Jacob-Lopes E., Wagner, R., Fries, L. L. M. and Menezes, C. R. (2015). Microencapsulation of probiotics using sodium alginate. *Ciência Rural,* 45 (7): 1319–1326.

FAO/WHO. 2001. *Health and nutritional properties of probiotics in food including powder milk with live lactic acid bacteria.* Accessed December 10, 2018. http://www.fao.org/3/a-a0512e.pdf.

Ferreira, C. L. L. F. and Silva, A. C. (2016). Probiotics and prebiotics in child health." In: *Functional Foods Bioactive Components and Physiological Effects*, Costa, N. M. B. and Rosa, C. O. B. R. (Eds), Rio de Janeiro: Rubio, 79–89.

Fuller, R. (1989). Probiotics in man and animals. *Journal of Applied Bacteriology,* 66: 365–378.

Galvalisi, U., Lupo, S., Piccini, J. and Bettucci, L. (2012) *Penicillium* species present in Uruguayan salami. *Revista Argentina de Microbiología,* 44, 36–42.

Gharsallaoui, A., Roudaut, G., Beney, L., Chambin, O., Voilley, A. and Saurel, R. (2012). Properties of spray-dried food flavours microencapsulated with two-layered membranes: Roles of interfacial interactions and water. *Food Chemistry*, 132 (4): 1713–1720.

Halim, M., Mustafa, N. A. M., Othman, M., Wasoh, H., Kapri, M. R. and Ariff A. B. (2017). Effect of encapsulant and cryoprotectant on the viability of probiotic *Pediococcus acidilactici* ATCC 8042 during freeze-drying and exposure to high acidity, bile salts and heat. *LWT - Food Science and Technology,* 81: 210–216.

Hill, C., Guarner, F., Reid, G., Gibson, G. R., Merenstein, D. J., Pot, B., Morelli, L., Canani, R. B., Flint, H. J., Salminen, S., Calder, P. C. and Sanders, M. E. (2014). The International Scientific Association for Probiotics and Prebiotics consensus statement on the scope and appropriate use of the term probiotic. *Nature Reviews Gastroenterology & Hepatology,* 11 (8): 506-514.

Holkem, A. T., Raddatz, G. C., Barin, J. S., Flores, E. M. M., Muller, E. I., Codevilla, C. F., Jacob-Lopes E., Grosso, C. R. F., Menezes, C. R. (2017). Production of microcapsules containing *Bifidobacterium* BB-12 by emulsification/internal gelation. *LWT - Food Science and Technology*, 76: 216–221.

Huq, T., Fraschini, C., Khan, A., Riedl, B., Bouchard, J. and Lacroix, M. (2017). Alginate based nanocomposite for microencapsulation of probiotic: Effect of cellulose nanocrystal (cnc) and lecithin. *Carbohydrate Polymers,* 168: 61–69.

Juturu, V. and Wu, J. C. (2018). Microbial production of bacteriocins: Latest research development and applications. *Biotechnology Advances*, 36 (8): 2187–2200.

Kalschne, D. L., Womer, R., Mattana, A., Sarmento, C. M., Colla, L. M. and Colla, E. (2015). Characterization of the spoilage lactic acid bacteria in 'sliced vacuum-packed cooked ham'. *Brazilian Journal of Microbiology*, 46 (1): 173–181.

Kaprasob, R., Kerdchoechuen, O. and Laohakunjit, N. (2018). B vitamins and prebiotic fructooligosaccharides of cashew apple fermented with probiotic strains *Lactobacillus* spp., *Leuconostoc mesenteroides* and *Bifidobacterium longum*. *Process Biochemistry,* 70: 9–19.

Khalid, K. (2011). An overview of lactic acid bacteria." *International Journal of Biosciences*, 1 (3): 1-13.

Khan, M. I., Arshad, M. S., Anjum, F. M., Sameen, A., Rehman, A. and Gill, W. T (2011). Meat as a functional food with special reference to probiotic sausages. *Food Research International*, 44 (10): 3125–3133. Kia, E. M., Alizadeh, M., Esmaiili, M. (2018) "Development and characterization of probiotic of feta cheese containing *Lactobacillus paracasei* microencapsulated by enzyme based gelation method. *Journal of Food Science and Technology*, 55 (9), 3657–3664.

Krummenauer, E. P., Paranhos, G. O., Silva, J. F., Silva-Buzanello, R. A., Kalschne, D. L., Corso, M. P. and Canan, C. (2015). Partial replacement of backfat by mozzarella cheese in Milano type salami. *Revista Cultivando o Saber*, 8 (2): 143–161.

Lam, P. L. and Gambari, R. (2014). Advanced progress of microencapsulation technologies: *In vivo* and *in vitro* models for studying oral and transdermal drug deliveries. *Journal of Controlled Release,* 178 (28): 25-45.

Leroy, F., Verluyten, J. and De Vuyst L. (2006). Functional meat starter cultures for improved sausage fermentation. *International Journal of Food Microbiology,* 106 (3): 270–285.

Loyeau, P. A., Spotti, M. J., Braber, N. L. V., Rossi, Y. E., Montenegro, M. A., Vinderola, G. and Carrara, C. R. (2018). Microencapsulation of *Bifidobacterium animalis* subsp. *lactis* IN11 using whey proteins and dextrans conjugates as wall materials. *Food Hydrocolloids,* 85: 129–135.

Macedo, R. E. F., Pflanzer Junior, S. B., Terra, N. N and Freitas, R. J. S. (2008). Production of fermented sausage using probiotic *Lactobacillus* strains: quality characteristics. *Food Science and Technology,* 28 (3): 509–519.

Malo, P. M. and Urquhart, E. A. (2016). Fermented foods: Use of starter cultures. In: *Encyclopedia of Food and Health*, Caballero, B., Finglas, P. M. and Toldrá, F. (Eds), Cambridge: Academic Press, 681–685.

Mao, L., Pan, Q., Hou, Z., Yuan, F. and Gao, Y. (2018). "Development of soy protein isolate-carrageenan conjugates through maillard reaction for the microencapsulation of *Bifidobacterium longum*. *Food Hydrocolloids*, 84: 489–497.

Martín, M. J., Lara-Villoslada, F., Ruiz, M. A. and Morales, M. E. (2015). Microencapsulation of bacteria: a review of different technologies and their impact on the probiotic effects. *Innovative Food Science and Emerging Technologies*, 27: 15–25.

Massaguer, P. R. (2006). *Microbiologia dos processos alimentares*. São Paulo: Varela.

Mattei, F. J. (2012). *Starter cultures (Lactobacillus plantarum AJ2 and Staphylococcus xylosus U5) on technological properties of sausage produced with a combination of pork and chicken meat*. Ph.D. Universidade Federal de Pelotas. Accessed December 10, 2018. http://dctaufpel.com.br/ppgcta/manager/uploads/documentos/dissertacoes/dissertacao_-_fabio.pdf.

Mazo, J. Z., Ilha, E. C., Arisi, A. C. M. and Sant'Anna, E. S. (2009). *Bifidobacteria:* Isolation, identification, and application in probiotic foods. *Boletim Cepa*, 27 (1): 119–134.

Minelli, E. B., Benini, A. (2008). Relationship between number of bacteria and their probiotic effects. *Microbial Ecology in Health and Disease*, 20: 180–183.

Muthukumarasamy, P. and Holley, R. A. (2006). Microbiological and sensory quality of dry fermented sausages containing alginate-microencapsulated *Lactobacillus reuteri*. *International Journal of Food Microbiology*, 111 (1): 164–169.

Nazzaro, F., Orlando, P., Fratianni, F. and Coppola, R. (2012). Microencapsulation in food science and biotechnology. *Current Opinion in Biotechnology*, 23 (2): 182–186.

OECD/FAO. (2017). Meats - OECD-FAO Agricultural Outlook 2017-2026." Accessed December 10, 2018. https://stats.oecd.org/Index.aspx?DataSetCode=HIGH_AGLINK_2017.

Pereira, G. V. M., Coelho, B. O., Magalhães Júnior A. I., Thomaz-Soccol, V. and Soccol C. R. (2018). How to select a probiotic? A review and update of methods and criteria. *Biotechnology Advances,* 36 (8): 2060–2076.

Prisco, A. and Mauriello, G. (2016). Probiotication of foods: A focus on microencapsulation tool. *Trends in Food Science and Technology*, 48: 27–39.

Ramos, P. E., Cerqueira, M. A., Teixeira, J. A. and Vicente, A. A. (2018). Physiological protection of probiotic microcapsules by coatings. *Critical Reviews in Food Science and Nutrition,* 58 (11): 1864–1877.

Ruggirello, M., Nucera, D., Cannoni, M., Peraino, A., Rosso, F., Fontana, M., Cocolin, L. and Dolci, P. (2019). Antifungal activity of yeasts and lactic acid bacteria isolated from cocoa bean fermentations. *Food Research International*, 115: 519-525.

Santos, D. X., Casazza, A. A., Aliakbarian, B., Bedani, R., Saad, S. M. I. and Peregoa, P. (2019). Improved probiotic survival to in vitro gastrointestinal stress in a mousse containing *Lactobacillus acidophilus* La-5 microencapsulated with inulin by spray drying. *LWT - Food Science and Technology*, 99: 404-410.

Shan, L. C., Henchion, M., Brún, A., Murrin, C., Wall, P. G., Monahan, F. J. (2017). Factors that predict consumer acceptance of enriched processed meats. *Meat Science* 133: 185–193.

Shori, A. B. (2017). Microencapsulation improved probiotics survival during gastric transit. *HAYATI Journal of Biosciences,* 24 (1): 1–5.

Silva, T. M., Lopes, E. J., Codevilla, C. F., Cichoski, A. J., Flores, E. M. M., Motta, M. H., Silva, C. B., Grosso, C. R. F. and Menezes, C. R. (2018). Development and characterization of microcapsules containing *Bifidobacterium* Bb-12 produced by complex coacervation followed by freeze drying. *LWT - Food Science and Technology*, 90: 412–417.

Teneva-Angelova, T., Hristova, I., Pavlov, A. and Beshkova, D. (2018). Lactic acid bacteria - from nature through food to health." In: *Advances in Biotechnology for Food Industry Handbook of Food Bioengineering*. Holban, A. M. and Grumezescu, A. M. (Eds), 4th ed., San Diego: Elsevier, 91–133.

Vinderola, G., Gueimonde, M., Gomez-Gallego, C., Defederico, L. and Salminen, S. (2017). Correlation between *in vitro* and *in vivo* assays in selection of probiotics from traditional species of bacteria. *Trends in Food Science & Technology*, 68: 83-90.

BIOGRAPHICAL SKETCHES

Daneysa Lahis Kalschne, PhD

Affiliation: Programa de Pós-Graduação em Tecnologia de Alimentos, Departamento de Alimentos, Universidade Tecnológica Federal do Paraná, Medianeira, Paraná, Brazil.

Research and Professional Experience: The researcher has experience with lactic acid bacteria, meat products and microencapsulation.

Professional Appointments: Researcher and Post-doctorate student

Publications:

Corso, M. P., Canan, C., Silva-Buzanello, R. A., Kalschne, D. L., Biasuz, T., Scremin, F. R., Bittencourt, P. R. S., Dias, T. and Vasconcelos, L. I. M. (2018). *Processo de produção de microrganismos probióticos microencapsulados em poli(metacrilato de metila-co-ácido metacrílico) e produto obtido* [*Process for the production of microencapsulated probiotic microorganisms in poly (methacrylic acid methacrylate) and the product obtained*]. 2018, Brasil. Patente: Privilégio de Inovação. Número do registro: BR10201800053. Instituição de registro: INPI - Instituto Nacional da Propriedade Industrial. Depósito: 10/01/2018.

Silva-Buzanello, R. A., Kalschne, D. L., Heinen, S. M., Pertum, C., Schuch, A. F., Corso, M. P. and Canan, C. (2017). Pork: profile of the West of Paraná consumers and physical evaluation of chop. Semina. *Ciências Agrárias*, 38 (6): 3563-3578.

Rosana Aparecida da Silva-Buzanello, PhD

Affiliation: Departamento de Alimentos, Universidade Tecnológica Federal do Paraná, Medianeira, Paraná, Brazil

Research and Professional Experience: The researcher has experience with meat products and microencapsulation.

Professional Appointments: Lecturer at the Food Science Department

Publications:

Corso, M. P., Canan, C., Silva-Buzanello, R. A., Kalschne, D. L., Biasuz, T., Scremin, F. R., Bittencourt, P. R. S., Dias, T. and Vasconcelos, L. I. M. (2018). *Processo de produção de microrganismos probióticos microencapsulados em poli(metacrilato de metila-co-ácido metacrílico) e produto obtido* [*Process for the production of microencapsulated probiotic microorganisms in poly (methacrylic acid methacrylate) and the product obtained*]. 2018, Brasil. Patente: Privilégio de Inovação. Número do registro: BR10201800053. Instituição de registro: INPI - Instituto Nacional da Propriedade Industrial. Depósito: 10/01/2018.

Silva-Buzanello, R. A., Schuch, A. F., Nogues, D. R. N., Melo, P. F., Gasparin, A. W., Torquato, A. S., Canan, C. and Soares, A. L. (2018). Physicochemical and biochemical parameters of chicken breast meat influenced by stunning methods. *Poultry Science*, 97 (11): 3786-3792.

Canan, C., Silva-Buzanello, R. A., Veiga, R. S., Colla, E., Corso, M. P., Scremin, F. R., Bittencourt, P. R. S. and Flores, E. L. M. (2017). *Processo de produção de microcápsulas a partir de proteína do farelo de arroz, albumina sérica bovina e carragena, e produto obtido* [*A process for the production of microcapsules from rice bran protein, bovine serum albumin and carrageenan, and the product obtained*]. 2017, Brasil. Patente: Privilégio de Inovação. Número do registro: BR10201700078. Instituição de registro: INPI - Instituto Nacional da Propriedade Industrial. Depósito: 13/01/2017.

Scremin, F. R., Veiga, R. S., Silva-Buzanello, R. A., Becker-Algeri, T. A., Corso, M. P., Torquato, A. S., Bittencourt, P. R. S., Flores, E. L. M. and Canan, C. (2017). Synthesis and characterization of protein microcapsules for eugenol storage. *Journal of Thermal Analysis and Calorimetry*, 131 (1): 653-660.

Silva-Buzanello, R. A., Kalschne, D. L., Heinen, S. M., Pertum, C., Schuch, A. F., Corso, M. P. and Canan, C. (2017). Pork: profile of the West of Paraná consumers and physical evaluation of chop. Semina. *Ciências Agrárias*, 38 (6): 3563-3578.

Marinês Paula Corso, PhD

Affiliation: Programa de Pós-Graduação em Tecnologia de Alimentos, Departamento de Alimentos, Universidade Tecnológica Federal do Paraná, Medianeira, Paraná, Brazil

Research and Professional Experience: The researcher has experience with meat products and microencapsulation.

Professional Appointments: Lecturer at the Food Science Department

Publications:
Corso, M. P., Canan, C., Silva-Buzanello, R. A., Kalschne, D. L., Biasuz, T., Scremin, F. R., Bittencourt, P. R. S., Dias, T. and Vasconcelos, L. I. M. (2018). *Processo de produção de microrganismos probióticos microencapsulados em poli(metacrilato de metila-co-ácido metacrílico) e produto obtido [Process for the production of microencapsulated probiotic microorganisms in poly (methyl methacrylate-methacrylic acid methacrylate) and the product obtained]*. 2018, Brasil. Patente: Privilégio de Inovação. Número do registro: BR10201800053. Instituição de registro: INPI - Instituto Nacional da Propriedade Industrial. Depósito: 10/01/2018.

Canan, C., Silva-Buzanello, R. A., Veiga, R. S., Colla, E., Corso, M. P., Scremin, F. R., Bittencourt, P. R. S. and Flores, E. L. M. (2017). *Processo de produção de microcápsulas a partir de proteína do farelo de arroz, albumina sérica bovina e carragena, e produto obtido [A process for the production of microcapsules from rice bran protein, bovine serum albumin and carrageenan, and the product obtained]*. 2017, Brasil. Patente: Privilégio de Inovação. Número do registro: BR10201700078. Instituição de registro: INPI - Instituto Nacional da Propriedade Industrial. Depósito: 13/01/2017.

Scremin, F. R., Veiga, R. S., Silva-Buzanello, R. A., Becker-Algeri, T. A., Corso, M. P., Torquato, A. S., Bittencourt, P. R. S., Flores, E. L. M. and Canan, C. (2017). Synthesis and characterization of protein microcapsules for eugenol storage. *Journal of Thermal Analysis and Calorimetry*, 131 (1): 653-660.

Silva-Buzanello, R. A., Kalschne, D. L., Heinen, S. M., Pertum, C., Schuch, A. F., Corso, M. P. and Canan, C. (2017). Pork: profile of the West of Paraná consumers and physical evaluation of chop. Semina. *Ciências Agrárias*, 38 (6): 3563-3578.

Eliane Colla, PhD

Affiliation: Programa de Pós-Graduação em Tecnologia de Alimentos, Departamento de Alimentos, Universidade Tecnológica Federal do Paraná, Medianeira, Paraná, Brazil.

Research and Professional Experience: The researcher has experience with lactic acid bacteria and biotechnology.

Professional Appointments: Lecturer at the Food Science Department

Publications:

Siepmann, F. B., Canan, C., Jesus, M. M., Pazuch, C. M. and Colla, E. (2018). Release optimization of fermentable sugars from defatted rice bran for bioethanol production. *ACTA Scientiarum. Technology*, 40: 1-9.

Colla, L. M., Reinehr, C. O., Colla, E., Margarites, A. C. L. and Costa, J. A.V. (2017). Production of Lipases by a Newly Isolate of Aspergillus niger Using Agroindustrial Wastes by Solid State Fermentation. *Current Biotechnology*, 6 (4): 301-307.

Oliveira, T., Ramalhosa, E., Nunes, L., Pereira, J. A., Colla, E. and Pereira, E. L. (2017). Probiotic potential of indigenous yeasts isolated during the fermentation of table olives from Northeast of Portugal. *Innovative Food Science & Emerging Technologies*, 44: 167-172.

Wochner, K. F., Becker-Algeri, T. A., Colla, E., Badiale-Furlong, E. and Drunkler, D. A. (2017). The action of probiotic microorganisms on chemical contaminants in milk. *Critical Reviews in Microbiology*, 44 (1):112-123.

Canan, C., Silva-Buzanello, R. A., Veiga, R. S., Colla, E., Corso, M. P., Scremin, F. R., Bittencourt, P. R. S. and Flores, E. L. M. (2017). Processo de produção de microcápsulas a partir de proteína do farelo de arroz, albumina sérica bovina e carragena, e produto obtido. 2017, Brasil. Patente: Privilégio de Inovação. Número do registro: BR10201700078. Instituição de registro: INPI - Instituto Nacional da Propriedade Industrial. Depósito: 13/01/2017.

Cristiane Canan, PhD

Affiliation: Programa de Pós-Graduação em Tecnologia de Alimentos, Departamento de Alimentos, Universidade Tecnológica Federal do Paraná, Medianeira, Paraná, Brazil

Research and Professional Experience: The researcher has experience with meat products, microencapsulation and microbiology.

Professional Appointments: Lecturer at the Food Science Department

Publications:

Corso, M. P., Canan, C., Silva-Buzanello, R. A., Kalschne, D. L., Biasuz, T., Scremin, F. R., Bittencourt, P. R. S., Dias, T. and Vasconcelos, L. I. M. (2018). *Processo de produção de microrganismos probióticos microencapsulados em poli(metacrilato de metila-co-ácido metacrílico) e produto obtido* [*Process for the production of microencapsulated probiotic microorganisms in poly (methyl methacrylate-methacrylic acid methacrylate) and the product obtained*]. 2018, Brasil. Patente: Privilégio de Inovação. Número do registro: BR10201800053. Instituição de registro: INPI - Instituto Nacional da Propriedade Industrial. Depósito: 10/01/2018.

Silva-Buzanello, R. A., Schuch, A. F., Nogues, D. R. N., Melo, P. F., Gasparin, A. W., Torquato, A. S., Canan, C. and Soares, A. L. (2018). Physicochemical and biochemical parameters of chicken breast meat influenced by stunning methods. *Poultry Science*, 97 (11): 3786-3792.

Canan, C., Silva-Buzanello, R. A., Veiga, R. S., Colla, E., Corso, M. P., Scremin, F. R., Bittencourt, P. R. S. and Flores, E. L. M. (2017). *Processo de produção de microcápsulas a partir de proteína do farelo de arroz, albumina sérica bovina e carragena, e produto obtido* [*A process for the production of microcapsules from rice bran protein, bovine serum albumin and carrageenan, and the product obtained*]. 2017, Brasil. Patente: Privilégio de Inovação. Número do registro: BR10201700078. Instituição de registro: INPI - Instituto Nacional da Propriedade Industrial. Depósito: 13/01/2017.

Scremin, F. R., Veiga, R. S., Silva-Buzanello, R. A., Becker-Algeri, T. A., Corso, M. P., Torquato, A. S., Bittencourt, P. R. S., Flores, E. L. M. and Canan, C. (2017). Synthesis and characterization of protein microcapsules for eugenol storage. *Journal of Thermal Analysis and Calorimetry*, 131 (1): 653-660.

Silva-Buzanello, R. A., Kalschne, D. L., Heinen, S. M., Pertum, C., Schuch, A. F., Corso, M. P. and Canan, C. (2017). Pork: profile of the West of Paraná consumers and physical evaluation of chop. Semina. *Ciências Agrárias*, 38 (6): 3563-3578.

In: The Many Benefits of Lactic Acid Bacteria
Editors: J. G. LeBlanc and A. de Moreno

ISBN: 978-1-53615-388-0
© 2019 Nova Science Publishers, Inc.

Chapter 4

LACTIC ACID BACTERIA RESPONSES TO OSMOTIC CHALLENGES

Maria K. Syrokou, Spiros Paramithiotis[*]
and Eleftherios H. Drosinos
Laboratory of Food Quality Control and Hygiene,
Department of Food Science and Human Nutrition,
Agricultural University of Athens, Athens, Greece

ABSTRACT

Lactic acid bacteria (LAB) stress responses and adaptation mechanisms have been the epicenter of intensive study over the last decades due to their importance for food industry. Among the stresses encountered, cells are subjected to osmotic pressure during production of starter cultures as well as during food fermentations. Cells react to fluctuations in external osmotic pressure by accumulating or releasing compatible solutes, which aid the microorganism to support a constant positive turgor, by balancing the osmotic difference between cell surroundings and cytosol. The function of cells is determined by the semipermeable attributes of cell membrane, functioning as a barrier for most solutes. The use of specific osmoprotectants is facilitated by certain transporters, whose intercommunication with membrane lipids defines their response to osmotic up- and down- shifts. Glycine betaine (GB) is considered the most efficient osmoprotectant in restricting the repressive effects of NaCl, while carnitine (CA) offers protective action to LAB in a more moderate way. In case the osmotic pressure drops, compatible solutes are released via mechanosensitive channels (MSC), acting as emergency release valves. In the present chapter current knowledge on the LAB osmoprotection mechanisms along with aspects of their genomic and transciptomic organization is assembled and critically reviewed.

---
[*] Corresponding Author's E-mail: sdp@aua.gr.

INTRODUCTION

Lactic acid fermentation has been used for centuries as a method to preserve foodstuff. It has been exercised on practically any type of substrate, either of plant or animal origin. Through this biotransformation, the final product gains enhanced shelf-life, sensorial properties and nutritional value. Lactic acid bacteria that drive this process have been in the epicenter of intensive study over the last decades. Their responses to environmental stimuli as well as their metabolic properties are constantly assessed and our understanding of their physiology accordingly improved.

During fermentation, lactic acid bacteria have to address a series of stresses, such as the osmotic stress due to NaCl that is used as a selective agent, acidic stress due to the acidity that develops as a consequence of their metabolic activity, oxidative stress due to the production of reactive oxygen species (ROS), starvation stress due to the depletion of the available nutrients, etc. It can be claimed that growing in stressful environments is the routine for LAB. In order to address effectively these challenges, a series of tuning activities, collectively referred to as stress response, is required.

Osmotic stress is among the most important ones from a technological perspective. It is the result of dissolved solute abundance and quite different from the desiccation stress, which is the result of lack of water. It may be encountered during production of starter cultures; their preservation may include dehydration and/or freezing. In addition, as already mentioned, the presence of NaCl creates an osmotic pressure to the cells. Finally, the increased osmolarity that characterizes the upper small intestine needs to be effectively addressed in order to exert all the health benefits through the probiotic action.

The aim of the present chapter is to collect all available information regarding the osmoprotection mechanisms of LAB as well as aspects of their genomic and transciptomic organization and present them in a comprehensive way.

BACTERIAL RESPONSES TO OSMOTIC STRESS

Bacteria protect themselves against detrimental fluctuations of external osmotic pressure by accumulating compatible solutes, i.e., water soluble compounds that may accumulate intracellularly without affecting negatively any cellular functions. In that way the osmotic difference between the environment and cytosol is counter-balanced, and water loss and plasmolysis are prevented. These osmolytes may be either *de novo* synthesized or transported from the environment through specific permeases; depending on the biosynthetic capacity of the bacterium and the growth medium. The latter constitutes the most rapid and effective response, since the former may require additional time and is therefore slower, especially if protein synthesis is involved. Compatible solutes can be classified into three categories: a) sugars and polyols, such as trehalose and glycerol; b) amino acids, such as glutamate and proline; c) amino acid derivatives, such as betaines, ectoines and N- acetylated amino acids. Adjustment of their intracellular concentration is achieved through the coordinated action of specific transporters and mechanosensitive channels (Figure 1). Apart from their osmoprotective action, additional roles have been claimed for glycine betaine, ectoine and proline. The first has been reported to assist *in vivo* protein folding (Bourot et al., 2000) as well as protection against cold and heat stress (Holtmann & Bremer, 2004; Hoffman & Bremer, 2011); ectoine against cold and heat stress (Bursy et al., 2008; Kuhlmann et al., 2011) and proline is involved in protein stability, ROS scavenging, redox homeostasis and signaling (Liang et al., 2013).

Cell wall of Gram positive bacteria is comprised of a peptidoglycan layer (PG), also known as murein, enclosing the cytoplasmic membrane. This polymer consists of N-acetylglucosamine (GlcNAc) and N-acetylmuramic acid (MurNAc), cross linked by peptide side chains (Chapot-Chartier & Kulakauskas, 2014). The penicillin binding proteins (PBPs) play a crucial role in PG synthesis, which are involved in glycosyltransferase and transpeptidase reactions. Except for the presence of PG sacculus, cell wall of Gram positive bacteria includes teichoic acid polymers as well, being separated into wall teichoic acid (WTA) and lipoteichoic acid (LTA). The former is covalently linked to the PG molecule, while the latter is attached to the membrane by a lipid ligand. The backbone chains of both WTA and LTA can be replaced by D- alanine residues, which reduce the negatively charged phosphate groups of TAs by offering protonated amino groups, thus regulating ionic attributes of TAs. Finally, the existence of O- acetyl groups in PG molecule is highly related to resistance to lysis by hydrolases. LAB lysis/autolysis is a defining parameter for controlling and accelerating ripening events and contributes to the sensory profile of the end product.

In hypertonic conditions, where the concentration of solutes in the environment is higher than that of the cytosol, osmotic pressure stimulates water efflux, resulting in loss of turgor or plasmolysis. Conversely, in hypotonic circumstances, where the culture media is characterized by lower concentration of solutes, compared to cytosol, water diffuses into the cell, increasing its volume, thus leading to cell lysis. The cell membrane plays a crucial role in osmotic regulation; it acts as selective barrier, hindering the passage of solutes, while allowing water permeation in both directions. An alternative response to osmotic gradient includes water transportation through channel like structures, called aquaporins of the MIP (major intrinsic protein) family. A subset of aquaporins, aquaglyceroporins are responsible for glycerol transit. There is evidence that except for water and glycerol, some aquaporins transport CO_2, NH_3, NO, NO_2, H_2O_2 and even ions, K^+, Cl^- (Verkman, 2014).

COMPATIBLE SOLUTES ACCUMULATION IN LAB

The main response of LAB to osmotic challenges, involves the transport of compatible solutes from the surrounding media, since they are characterized by rather narrow biosynthetic potentials. The selection of specific osmoprotectants is facilitated by certain transporters, whose intercommunication with membrane lipids defines their response to osmotic up- and down- shocks.

Table 1. Compatible solutes assessed for their osmoprotective effect on LAB

	Lb. plantarum[a, d, e, f]	*Lc. lactis*[g]	*Ped. pentosaceus*[a]	*Ec. faecalis*[c]	*T. halophila*[a, b]	*O. oeni*[h]
carnitine	+	+	+	+	+	-
choline	+	-	+	nd	+	-
DMSA	nd	+	+	+	+	-
DMSP	nd	+	-	+	+	nd
ectoine	-	nd	-	-	+	-
glycine betaine	+	+	+	+	+	-
pipecolate	nd	-	-	-	-	nd
proline	+	-	+	-	-	-
sarcosine	-	nd	-	nd	-	nd
taurine	nd	nd	-	nd	-	nd
arsenobetaine	nd	+	nd	+	nd	nd
γ-butyrobetaine	nd	nd	nd	+	nd	nd
DMG	nd	nd	nd	+	nd	nd
glutamate	+	nd	nd	nd	nd	-
alanine	+	nd	nd	nd	nd	-
glycine	+	nd	nd	nd	nd	nd
betaine	+	nd	nd	nd	nd	nd
acetylcholine	+	nd	nd	nd	nd	nd
aspartate	nd	nd	nd	nd	nd	+

Lb.: *Lactobacillus*; *Lc.*: *Lactococcus*; *Ped.*: *Pediococcus*; *Ec.*: *Enterococcus*; *T.*: *Tetragenococcus*; *O.*: *Oenococcus*
DMSA, dimethylsulfonioacetate; DMSP, dimethylsulfoniopropionate; nd: not determined, DMG: dimethyl glycine.
+: compound exerted osmoprotective action; -: compound did not exert osmoprotective action; nd: not determined
[a]Baliarda et al. (2003); [b]Robert et al. (2000); [c]Pichereau et al. (1999); [d]Glaasker et al. (1996b); [e]Kets et al. (1997); [f] Kets et al. (1997); [g]Uguen et al. (1999); [h]Le Marrec et al. (2007).

The beneficial effect of a number of potential osmoprotectant solutes has been experimentally evaluated for a rather restricted number of LAB species, namely *Enterococcus (Ec). faecalis, Lactobacillus (Lb.) plantarum, Lactococcus (Lc.) lactis, Oenococcus (O.) oeni, Pediococcus (Ped.) pentosaceus* and *Tetragenococcus (T.) halophila*, and the results are summarized in Table 1. In all cases, except for *O. oeni*, glycine betaine and carnitine have been reported to effectively restore growth under osmotic pressure. Further examination revealed that the *O. oeni* genome lacked genes homologous to the ones encoding the transporters of glycine betaine and proline that have been reported for *Lb. plantarum* and *Lc. lactis* (Le Marrec et al., 2007). Except from the accumulation of quaternary ammonium compounds, adjustment of specific amino acids concentration inside the cytosol represents a different strategy to alleviate the adverse effects of osmotic challenges. Robert et al. (2000) reported the accumulation of glutamate, proline and aspartate in *T. halophila* under osmotic constraint. However, uptake of the specific amino acids was restricted in the presence of glycine betaine in a high- osmolarity medium. Similar effect, where concentrations of alanine, glutamate and proline increased in the absence of glycine betaine, was also observed in *Lb. plantarum* (Glaasker et al., 1998). In contrast to the previously mentioned studies, glycine betaine had no protective effects on *O. oeni*, while exogenously supplemented aspartate was able to alleviate the restrictive action of osmotic up- shifts on the cell turgor (Le Marrec et al., 2007). Finally, ectoine, a compatible solute found in halophilic microorganisms, exerted protective effects on *T. halophila* stressed cells (Baliarda et al., 2003b). One last approach reported, includes exogenously supplied osmolytes, such as proline and glutamate-containing peptides conferring protective effects on *O. oeni* sugar induced stressed cells (Le Marrec et al., 2007). Its role as osmoprotectant has been also reported for protease, PrtP, being activated during growth of *Lb. casei* in hyperosmotic conditions (Piuri et al., 2005).

LAB CYTOPLASMIC MEMBRANE AND OSMOTIC STRESS

As previously mentioned the cell cytoplasmic membrane functions as a barrier for most solutes, while is highly permeable to water molecules, thus its coordinated response to osmotic challenges regulates growth restoration. Degree of saturation, fatty acid (FA) composition and size of cells are attributes directly connected to membrane stability, and their behavior in hypertonic media determines the osmotic adaptation of cytoplasmic membrane. It has been shown that proline transport on the hyperosmotic medium of *T. halophila* led to an increase in the amounts of unsaturated FA, as well as the levels of cyclopropanoic FA (He et al., 2017), determining its crucial part in the regulation of fatty acid composition. However, it has also been reported that salt stress conditions favor the saturation degree of FA, often accompanied by FA cyclization, with the latter one being perceived as a process of controlling membrane fluidity (Machado et al., 2004). The presence of saturated FA was also validated by Tymczyszyn et al. (2005), as a basic component of membrane rigid structure of *Lb. bulgaricus* grown in MRS, supplemented with polyethylene glycol (PEG). In addition a smaller shrinkage of cells, grown in MRS-PEG, was observed due to a higher anisotropy rate (a capability to exhibit variations in physical properties along different molecular axes). Similarly the existence of larger cells of *Lb. casei* in osmotically stressed conditions compared to those under unstressed, was also documented by Piuri et al. (2005).

Growth under high salt circumstances has been shown to facilitate the access of antibiotics to their target, validated by the increased sensitivity of cells to nisin (Piuri et al., 2005; Palomino et al., 2013). This observation was attributed to a decrease in peptidoglycan (PEPG) cross- linkage, caused either by a lower expression of PBP genes or by discrepancies in their appropriate folding in the cell envelope. Another alteration during growth of *Lb. casei* under hypertonic media was the decreased production of a cell wall component, namely lipoteichoic acid (LTA). The specific change was followed by a decline in LTA D-alanylation, generating a negatively charged cell wall, thus hindering the passage of Na^+ to the cytoplasm. Decreased LTA substitution by D-alanine residues in combination with the participation of LTA in attachment of autolysines develops a PEPG more vulnerable to lysis. Lysis of LAB is crucial for the development of texture and flavor in food products and the harmony between autolysed and intact cells determines the fate of various ripening events.

Osmoprotectants are accumulated in the cytosol via active transport (Figure 1), upon osmotic up- shifts; however, upon decrease of the osmotic pressure, they are released from the cell via mechanosensitive channels (MSC). In fact, MSC function as biological emergency release valves, by facilitating a compatible solute efflux, having been accumulated in the cytosol, upon osmotic stimulation of the OpuA transport system (Booth & Blount, 2012). As the OpuA uptake system is activated by sensing the interactions between protein module and lipid membrane, so do MSC, which sense possible distortion of the lipid bilayer. The osmoprotectant role of MSC was determined with the presence of two MSC in *E. coli*, namely MscS (mechanosensitive channel small conductance) and MscL (mechanosensitive channel large conductance) (Levina et al., 1999). The MSC study has been restricted in LAB, from which only the *Lc. lactis* mechanosensitive solute transfer system has been reported (Folgering et al., 2005). Their results revealed the presence of two different genes, *yncB* and *mscL*, regulating MscS and MscL, respectively, with MscL being preferentially activated by *Lc. lactis* under osmotic down- shock. Despite the fact that *yncB* gene was transcribed, MscS stimulation was not confirmed. MscS and MscL detected in *Lc. lactis* shared many similarities with those identified in *E. coli*; nevertheless the medium ionic strength had no effect on the transcription process of the corresponding genes in the case of *Lc. lactis*, but was determinant for the gene expression in *E. coli*.

Mechanosensitive ion channels play a crucial role in sensing membrane tension within a cell's surroundings and transducing mechanical constrain into an electrochemical response. The *E. coli* MscS homoheptameric structure (MscS-Ec) is strongly dependent on the TM1-TM2 region, which senses membrane distortions and its interaction with TM3 helix, forms specific channels, comprised of Gly and Ala residues (Booth & Blount, 2012). The block complex, identified in the open structure of A106V mutant of the MscS-Ec channel, is based on the presence of 2 rings of Leu residues and is highly hydrophobic, hindering the passage to ions and water. The transition of close to open state of MscS starts with the TM1-TM2 helix being triggered by ionic changes in lipid membrane. Then, a further stimulation of the TM3 domain follows, which is accompanied by a slight rotation, resulting in the discharge of Leu residues and generating the open configuration. As for MscL, they respond to osmotic challenges, sensed by alterations in membrane tension, by opening a large water- filled pore, thus aiding as osmoprotective emergency valves. MscL was first identified in *E. coli* constituting a non- selective ion pathway, but a three dimensional (3D) conformation of MscL from *Mycobacterium tuberculosis* (MtMscL) revealed the basic elements defining its structure. In brief, crystal structure of MtMscL indicates a highly α- helical homopentamer, with N- and C- terminal regions facing the cytoplasm, TM1, TM2 transmembrane helices and a periplasmic loop connecting them (Chang et al., 1998; Pivetti et al., 2003; Bavi et al., 2017). Questionable was the role of C- terminal domain in the regulation of the MscL opening, with many studies supporting its complete dissociation, and others arguing that the helical bundle remained intact, all along its

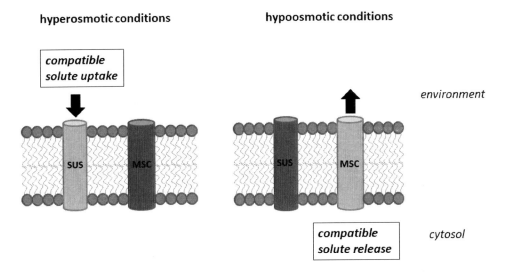

Figure 1. Schematic representation of a generic LAB osmotic adaptation mechanism. Upon hyperosmotic stress, osmolytes are accumulated in the cytoplasm, via a compatible solute uptake system (SUS), while they are released to the culture media via mechanosensitive channels (MSC), upon osmotic downshifts. Green color designates open configuration to the respective system; red color designates close configuration.

transition pathway from close to open state. Bavi et al. (2017) reported the partial dissociation of the MscL-Ec channel under different conditions, favoring the open channel conformation. More analytically, the results showed that the gating of the MscL channel was accompanied by a restricted dissociation of the top part of the C- terminal region, due to weak hydrogen bonds between subunits, while the residual domain of the central and the end bundle remained associated. A study employing various experimental circumstances for studying 3 different states of MscL, close, intermediate and open was performed by Perozo et al. (2002). The conversion pathway of close- intermediate conformation initiates imperceptible shifts in the transmembrane helix, TM1; while TM2 does not rotate about its central axis. Further transition to the open state, is followed by crucial structural changes, with TM1 actively participating in the channel gating. The transition to the open MscL conformation, escorted by the presence of an open pore, 30-40 A° in diameter, was triggered by adding lysophosphatidylcholine (LPC) into phosphatidylcholine (PC) vesicles.

Lactococcus lactis

Lc. lactis posseses an uptake system for the osmoprotectant GB, known as BusA/OpuA, linked with the ATP- binding cassette (ABC) superfamily of transporters. The system has high compatibility with GB and a lower one with proline. It is comprised of 2 nucleotide- binding domains (NBD) integrated into cystathionine-β-synthase (CBS) domains, forming the OpuAA subunit and 2 substrate binding domains embedded in transmembrane domains, constituting the OpuABC subunit (Mahmood et al., 2009). The osmotic stimulation of OpuA is based on regulating the on/off mode of the transporter, through (inter) mediate action of a protein module (CBS domains) with charged membrane surface. Its activation involves intensity levels of ionic strength, sensed by tandem CBS domains, found at the C- terminal end of OpuAA complex (Biemans-Oldehinkel et al., 2006). The negatively charged C-terminal tail coordinates the sensing of internal ionic strength. Briefly, below the intensity levels of ionic strength, the transporter is kept in an inactivate mode, while at higher medium osmolality, the osmosensing mechanism shifts the "electrostatically sealed mode" of the transporter to an "active" one.

BusR protein has been in the epicenter of intensive study due to its repressor character on the osmotically activated system, BusA, of *Lc. lactis*. Upon osmotic down-shock, BusR attaches the BusA promoter (BusAp) in a manner determined by the ionic concentration. In more detail, BusR reacts to hypoosmotic conditions, by forming a ternary structure with RNA polymerase, found at the promoter site. When cytoplasmic ionic strength increases, BusR detaches from BusAp, facilitating the initation of busA operon transcription.

An osmosensitive phenotype in *Lc. lactis* subsp. *cremoris* has been described by Obis et al. (2001), having been associated with a limited capability of betaine uptake. This contrasting behavior with the rest *Lc. lactis* strains could be attributed to a deficiency of the *busA* operon or moderate expression or even restricted/poor activity of the BusA transporter.

Lactobacillus plantarum

As previously mentioned, LAB are characterized by restricted biosynthetic capabilities, making the transport of osmoprotectants from the surrounding media, critical for their survival under osmotic constraints. *Lb. plantarum* possesses a transport system, termed QacT (quaternary ammonium compound transporter), driven by ATP, with high affinity for GB and CA and a lower one for proline (Glaasker et al., 1998). As the GB uptake system is stimulated at elevated osmolality conditions until osmotic balance is achieved, so does the GB efflux from the cytosol, upon hypoosmotic circumstances. GB release from the cells include an accelerated initial phase, facilitated by a stress regulated mechanosensitive channel and a second slower one, enhanced by a carrier- like protein (Glaasker et al., 1996a; b). Therefore the up and down- regulation of GB transport system in *Lb. plantarum* is crucial for osmostasis control.

The addition of chlorpromazine, a cationic amphipath, reinforces the open state of MSCs and imitates osmotic down- shifts, by facilitating GB efflux from the cytosol (Glaasker et al., 1998). The release of osmoprotectants is accompanied with a lag time of 15 sec, when chlorpromazine is added, whereas it is extremely rapid upon osmotic down-shifts, highlighting the characteristics differentiating the QacT system from the common transport mechanisms.

Tetragenococcus halophila

T. halophila (formerly *Pediococcus halophilus*) is a moderately halophilic bacterium, characterized by the capacity to tolerate high salt concentrations. Sharing analogous biosynthetic attributes with other LAB, *T. halophila* cannot synthesize *de novo* effective compatible solutes, therefore it is strongly dependent on their presence in the defined media. Except from the compatible solute uptake system, activated under hyperosmotic conditions, *T. halophila* is the first lactic acid bacterium, in which the existence of a choline- GB pathway, under aerobic conditions, has been reported (Robert et al., 2000). Baliarda et al. (2003a) documented the presence of two GB accumulation systems, the former carrying GB only and the latter facilitating the transport of betaines, choline and carnitine. The identification of ButA protein from the osmotolerant *T. halophila* has been linked with the betaine choline carnitine transporter (BCCT) family, involving divergent osmoprotectant uptake systems, for instance BetL from *Listeria* (*L.*) *monocytogenes* and OpuD from *Bacillus* (*B.*) *subtilis*. Except from the osmoprotective role of GB on *T. halophila*, the exogenously supplied DMSA and DMSP (GB structural analogues) also relieve inhibitory effects of NaCl stress. In addition, presence of tetrahydropyrimidine ectoine in the surrounding media of the halophilic bacterium has been found to enhance its osmoprotection (Baliarda et al., 2003b).

Additional Lactic Acid Bacteria

Although the aid of compatible solutes like GB, CA and choline was previously described and highlighted for a variety of LAB, *Lb. bulgaricus* was not able to accumulate those osmolytes under stressed conditions (Chun et al., 2012). In other words, GB and proline uptake system was not activated, which was associated with a lack of genes regulating the transport mechanism in the genome of *Lb. bulgaricus*. More analytically, *B. subtilis* uptake system, also known as OpuA, preferentially accumulates GB; however, the genes encoding the specific transport system were not identified in *Lb. bulgaricus*. The amino acids aspartate and alanine could counterbalance the osmotic suppression caused under NaCl stress.

Similar osmotic response was observed in *O. oeni*, the predominant bacterium of malolactic fermentation (MLF). GB accumulation system was not up- regulated, probably due to a lack of uptake mechanisms related to the BusA/OpuA- BusA/OpuC GB transporters (Le Marrec et al., 2007). Inability to concentrate choline was also reported, validating the absence of choline- GB pathway in *O. oeni*. Nevertheless, the exogenously added aspartate exhibited its osmoprotective role, restoring the ionic imbalance between cell's surroundings and cytosol.

As for *Streptococcus (Str.) mutans* UA159, resided in oral cavity, an *opc/opu* gene array was identified (Abranches et al., 2006). The specific gene complex, belonging to the Opu family of ABC transport systems, seems homologous with corresponding genes in *Str. agalactiae* as well as the OpuC transport mechanism of *L. monocytogenes*.

CONCLUSION

The response of some LAB with technological significance to osmotic changes has been studied to some extent. Current available data, although revealing, are based on a rather limited number of strains. Therefore it is practically impossible to assess strain diversity and include it to the already discovered differences between species. In addition, identification of the transcription factors that are activated by this environmental stimulus as well as their regulons is still to be explored. Taking into consideration the importance of an effective osmotic stress response, it is reasonable to expect our knowledge to expand within the next few years.

REFERENCES

Abranches, J., Lemos, J. A. & Burne, R. A. (2006). Osmotic responses of *Streptococcus mutans* UA159. *FEMS Microbiol. Lett.*, *255*, 240–246.

Baliarda, A., Robert, H., Jebbar, M., Blanco, C., Deschamps, A. & Le Marrec, C. (2003b). Potential osmoprotectants for the lactic acid bacteria *Pediococcus pentosaceus* and *Tetragenococcus halophila*. *International Journal of Food Microbiology*, *84*, 13–20.

Baliarda, A., Robert, H., Jebbar, M., Blanco, C. & Le Marrec, C. (2003a). Isolation and characterization of ButA, a secondary glycine betaine transport system operating in *Tetragenococcus halophila*. *Current Microbiology*, *47*, 347–351.

Bavi, N., Martinac, A. D., Cortes, D. M., Bavi, O., Ridone, P., Nomura, T., Hill, A. P., Martinac, B. & Perozo, E. (2017). Structural dynamics of the MscL C-terminal domain. *Science Report*, *7*, 17229.

Biemans-Oldehinkel, E., Mahmood, N. A. & Poolman, B. (2006). A sensor for intracellular ionic strength. *Proceedings of the National Academy of Sciences of the United States of America*, *103*, 10624–10629.

Booth, I. R. & Blount, P. (2012). The MscS and MscL families of mechanosensitive channels act as microbial emergency release valves. *Journal of Bacteriology*, *194*, 4802–4809.

Bourot, S., Sire, O., Trautwetter, A., Touzé, T., Wu, L. F., Blanco, C. & Bernard, T. (2000). Glycine betaine-assisted protein folding in a *lysA* mutant of *Escherichia coli*. *Journal of Biological Chemistry*, *14*, 275, 1050-1056.

Bursy, J., Kuhlmann, A. U., Pittelkow, M., Hartmann, H., Jebbar, M., Pierik, A. J. & Bremer, E. (2008). Synthesis and uptake of the compatible solutes ectoine and 5-hydroxyectoine by *Streptomyces coelicolor* A3(2) in response to salt and heat stresses. *Applied and Environmental Microbiology*, *74*, 7286-7296.

Chang, G., Spencer, R. H., Lee, A. T., Barclay, M. T. & Rees, D. C. (1998). Structure of the MscL homolog from *Mycobacterium tuberculosis*: a gated mechanosensitive ion channel. *Science*, *282*, 2220–2226.

Chapot-Chartier, M. P. & Kulakauskas, S. (2014). Cell wall structure and function in lactic acid bacteria. *Microbial Cell Factor*, 2014, *13*(Suppl 1), S9.

Chun, L., Li-bo, L., Di, S., Jing, C. & Ning, L. (2012). Response of osmotic adjustment of *Lactobacillus bulgaricus* to NaCl stress. *The Journal of Northeast Agricultural University*, *19*, 66-74.

Folgering, J. H., Moe, P. C., Schuurman-Wolters, G. K., Blount, P. & Poolman, B. (2205). *Lactococcus lactis* uses MscL as its principal mechanosensitive channel. *Journal of Biological Chemistry*, *280*, 8784–8792.

Glaasker, E., Heuberger, E. H., Konings, W. N. & Poolman, B. (1998). Mechanism of osmotic activation of the quaternary ammonium compound transporter (QacT) of *Lactobacillus plantarum. Journal of Bacteriology, 180,* 5540–5546.

Glaasker, E., Konings, W. N. & Poolman, B. (1996a). Glycine betaine fluxes in *Lactobacillus plantarum* during osmostasis and hyper- and hypo-osmotic shock. *Journal of Biological Chemistry, 271,* 10060–10065.

Glaasker, E., Konings, W. N. & Poolman, B. (1996b). Osmotic regulation of intracellular solute pools in *Lactobacillus plantarum. Journal of Bacteriology, 178,* 575–582.

He, G., Wu, C., Huang, J. & Zhou, R. (2017). Effect of exogenous proline on metabolic response of *Tetragenococcus halophilus* under salt stress. *Journal of Microbiology and Biotechnology, 27,* 1681–1691.

Hoffmann, T. & Bremer, E. (2011). Protection of *Bacillus subtilis* against cold stress via compatible-solute acquisition. *Journal of Bacteriology, 193,* 1552-1562.

Holtmann, G. & Bremer, E. (2004). Thermoprotection of *Bacillus subtilis* by exogenously provided glycine betaine and structurally related compatible solutes: involvement of Opu transporters. *Journal of Bacteriology, 186,* 1683-1693.

Kets, E. P. W. & de Bont, J. A. M. (1997). Effects of carnitine on *Lactobacillus plantarum* subjected to osmotic stress. *FEMS Microbiology Letters, 146,* 205–209.

Kets, E. P. W., Nierop Groot, M., Galinski, E. A. & de Bont, J. A. M. (1997). Choline and acetylcholine: novel cationic osmolytes in *Lactobacillus plantarum. Applied Microbiology and Biotechnology, 48,* 94–98.

Kuhlmann, A. U., Hoffmann, T., Bursy, J., Jebbar, M. & Bremer, E. (2011). Ectoine and hydroxyectoine as protectants against osmotic and cold stress: uptake through the SigB-controlled betaine-choline- carnitine transporter-type carrier EctT from *Virgibacillus pantothenticus. Journal of Bacteriology, 193,* 4699-4708.

Le Marrec, C., Bon, E. & Lonvaud-Funel, A. (2007). Tolerance to high osmolality of the lactic acid bacterium *Oenococcus oeni* and identification of potential osmoprotectants. *International Journal of Food Microbiology, 115,* 335–342.

Levina, N., Tötemeyer, S., Stokes, N. R., Louis, P., Jones, M. A. & Booth, I. R. (1999). Protection of *Escherichia coli* cells against extreme turgor by activation of MscS and MscL mechanosensitive channels: identification of genes required for MscS activity. *EMBO Journal, 18,* 1730–1737.

Liang, X., Zhang, L., Natarajan, S. K. & Becker, D. F. (2013). Proline mechanisms of stress survival. *Antioxidant and Redox Signaling, 19,* 998–1011.

Machado, M. C., Lopez, C. S., Heras, H. & Rivas, E. A. (2004). Osmotic response in *Lactobacillus casei* ATCC 393: biochemical and biophysical characteristics of membrane. *Archives of Biochemistru and Biophysics, 422,* 61–70.

Mahmood, N. A., Biemans-Oldehinkel, E. & Poolman, B. (2009). Engineering of ion sensing by the cystathionine beta-synthase module of the ABC transporter OpuA. *Journal of Biological Chemistry, 284,* 14368–14376.

Obis, D., Guillot, A. & Mistou, M. Y. (2001). Tolerance to high osmolality of *Lactococcus lactis* subsp. *lactis* and *cremoris* is related to the activity of a betaine transport system. *FEMS Microbiology Leters*, *202*, 39–44.

Palomino, M. M., Allievi, M. C., Grundling, A., Sanchez-Rivas, C. & Ruzal, S. M. (2013). Osmotic stress adaptation in *Lactobacillus casei* BL23 leads to structural changes in the cell wall polymer lipoteichoic acid. *Microbiology*, *159*, 2416–2426.

Perozo, E., Cortes, D. M., Sompornpisut, P., Kloda, A. & Martinac, B. (20012). Open channel structure of MscL and the gating mechanism of mechanosensitive channels. *Nature*, *418*, 942–948.

Pichereau, V., Bourot, S., Flahaut, S., Blanco, C., Auffray, Y. & Bernard, T. (1999). The osmoprotectant glycine betaine inhibits salt induced cross-tolerance towards lethal treatment in *Enterococcus faecalis*. *Microbiology*, *145*, 427–435.

Piuri, M., Sanchez-Rivas, C. & Ruzal, S. M. (2005). Cell wall modifications during osmotic stress in *Lactobacillus casei*. *Journal of Applied Microbiology*, *98*, 84–95.

Pivetti, C. D., Yen, M. R., Miller, S., Busch, W., Tseng, Y. H., Booth, I. R. & Saier, Jr. M. H. (2003). Two families of mechanosensitive channel proteins. *Microbiology and Molecular Biology Reviews*, *67*, 66–85.

Robert, H., Le Marrec, C., Blanco, C. & Jebbar, M. (2000). Glycine betaine, carnitine, and choline enhance salinity tolerance and prevent the accumulation of sodium to a level inhibiting growth of *Tetragenococcus halophila*. *Applied and Environmental Microbiology*, *66*, 509–517.

Tymczyszyn, E. E., Gomez- Zavaglia, A. & Disalvo, E. A. (2005) Influence of the growth at high osmolality on the lipid composition, water permeability and osmotic response of *Lactobacillus bulgaricus*. *Archives of Biochemistry and Biophysics*, *443*, 66–73.

Uguen, P., Hamelin, J., Le Pennec, J. P. & Blanco, C. (1999). Influence of osmolarity and the presence of osmoprotectant on *Lactococcus lactis* growth and bacteriocin production. *Applied and Environmental Microbiology*, *65*, 291–293.

Verkman, A. S., Anderson, M. O. & Papadopoulos, M. C. (2014). Aquaporins: important but elusive drug targets. Nature Reviews. *Drug Discovery*, *13*(4), 259-277.

BENEFICIAL COMPOUND PRODUCTION BY LACTIC ACID BACTERIA

Chapter 5

LACTIC ACID BACTERIA AS SOURCE OF BIOSURFACTANTS

Carmen A. Campos[1,2,], Virginia M. Lara[1] and María F. Gliemmo[1,2]*

[1]Departamento de Industrias, Facultad de Ciencias Exactas y Naturales,
Universidad de Buenos Aires, Buenos Aires, Argentina
[2]Instituto de Tecnología de Alimentos y Procesos Químicos (ITAPROQ).
CONICET - Universidad de Buenos Aires, Buenos Aires, Argentina

ABSTRACT

Nowadays consumers demand for healthier foods and for chemical free additives products. Lactic acid bacteria (LAB) have a generally recognized as safe (GRAS) status according to Food and Drug Administration Agency. Furthermore, they are an important source of bioactive compounds with wide applications in the food industry. Some of these compounds are *in situ* produced while others are *ex situ*. LAB have been used to produce fermented products since ancient times and the safety and sensory characteristics of these products rely on a diversity of metabolites produced *in situ* by LAB. In the last decades, the possible application of LAB metabolites as preservatives has increased; mainly bacteriocins, which were even approved worldwide. However, the use of LAB metabolites as additives is still emerging, particularly in the case of biosurfactants. These can exhibit emulsifying, antiadhesive and antimicrobial activity. The main objective of this chapter is to review the current status of the knowledge and possible applications of biosurfactants as additives for the formulation of foods.

*Corresponding Author's Email: carmen@di.fcen.uba.ar.

INTRODUCTION

Biosurfactants are amphiphilic compounds produced by microorganisms, mainly bacteria and yeasts. They are multipurpose additives since they can reduce surface and interfacial tension, act as emulsifying, antiadhesive and antimicrobial agents. Furthermore, they are environmental friendly, biodegradable, and show low toxicity. Consequently, their search has attracted the attention in the last decades. Most studies were focused on bioremediation and few were done about the food industry, although the increasing demands for replacing synthetic additives by naturals (Nitschke & Costa, 2007).

Biosurfactants can be classified according to their chemical composition and microbial source. Regarding their chemical composition, some of them are low-molecular mass compounds while others are high molecular mass ones. The first group includes glycolipids, lipopeptides and phospholipids that are efficient for lowering surface and interfacial tension, the second ones are of polymeric nature and act as emulsions stabilizers (Nitschke & Costa, 2007).

Most biosurfactants are produced by bacteria belonging to *Pseudomonas, Bacillus, Acinetobacter* and by lactic acid bacteria (LAB). The latter, include different genera, particularly, *Lactobacillus*, which constitutes part of the indigenous flora of the gastrointestinal track of humans, and is generally recognized as safe. For this reason, its application is considered safe.

The main drawbacks from LAB biosurfactants are that they are less effective to reduce the surface tension, they have higher critical micelle concentrations (CMC), they are produced in lower amounts and finally, they are cell-bound being necessary an extraction process for their recovery (Nitschke & Sousa e Silva, 2018).

Regarding the chemical composition of biosurfactants produced by lactobacilli, they are glycolipids, glycoproteins, lipopeptides, glycopeptides or phosphoglycoproteins, independently of the strain and culture conditions used (Hajfarajollah et al., 2018).

The role of biosurfactants *in vivo* is still a matter of debate. One proposal is that they are implicated in the emulsification of water-insoluble substrates improving their absorption. Also, they are considered as components of cellular metabolism, motion and defense since they take part in cell adhesion to interfaces (Sharma et al., 2016).

Food products have pH values within the range of 3.0 to 7.0, they content salts that modify ionic strength and during processing and storage they are exposed to different temperatures. Therefore, the biosurfactant must be selected taking into account that it will have to be stable in the conditions of use. For example, Rodrigues et al., (2006 a) evaluated surface tension within the range of pH 4 to 10 for the crude and purified biosurfactant produced by *Streptococcus (Strep.) thermophilus* A. They found that the ability to reduce surface tension remained stable in the range studied but a purified fraction precipitated at pH 4, limiting its use in this condition.

Optimal conditions for production depend on the producing microorganism and also on the nature of biosurfactant. In most cases, the amount of biosurfactant released by *Lactobacilli* is maximum during the stationary growth phase (Velraeds et al., 1996; Golek et al., 2009). In the case of *Acinetobacter* (*Ac.*) *calcoaceticus*, biosurfactant production was directly associated with the growth of the microorganism (Desai & Desai, 1993). Furthermore, the C:N ratio seems to be a critical factor in the production of biosurfactants and a 1:1 ratio would be optimal (Bakhshi et al., 2018).

It is well known that one of the drawbacks for biosurfactant uses is the high cost of production and particularly, for LAB biosurfactants, their low productivity. For this reason, many investigations have been done looking for the conditions to optimize production. Moreover, alternatives culture media using renewable and low-cost raw materials such as agricultural wastes were evaluated. For example, hemicellulosic sugars from vineyard pruning waste were successfully applied to obtain a glycolipopeptide with promissory emulsifying capacity from *Lactobacillus* (*Lb.*) *pentosus* fermentation (Vecino et al., 2015). Moreover, rice, wheat bran and straw were evaluated for biosurfactant production by *Lb. plantarum* subsp. *plantarum* PTCC 1896 and it was reported that only the rice bran hydrolysate was adequate as substrate for biosurfactant production (Bakhshi et al., 2018). Finally, biosurfactant production by *Lb. pentosus* using whey as culture media was assayed and compared to production using MRS and it was found that biosurfactant production was similar in both media. However, the strain did not growth well in the whey media, probably due to the lack of nutrients (Rodrigues et al., 2006 b). The satisfactory results of the use of wastes for biosurfactant production encourage the continuous search for the optimal conditions of production and the study of the structure–function relationship in food systems. According to Mouafo et al., (2018) a reduction of 30-50% of production cost can be obtained by using alternative substrates.

Regarding the use of different culture media, it must be stressed that most of the biosurfactants are cell bound; consequently, their extraction from culture medium is necessary. After the incubation of LAB at selected conditions for production and centrifuging to remove the supernatant, bound to cells biosurfactant can be extracted by re-suspension in PBS and gentle stirring or by re-suspension in distilled water and extraction using an ultrasonic cell disintegrator. Use of phosphate buffer saline (PBS) together with gentle stirring is the most common procedure (Hajfarajollah et al., 2018).

Main biosurfactants can be applied as emulsifying, antiadhesive and antimicrobial agents as it will be discussed in the next sections. In addition to this, they can be used to control flat globules agglomeration, stabilize foams and improve bakery products texture (Nitschke & Costa, 2007). Finally, some recent studies proposed their use as antioxidant agents, for example, a liopopeptide from *Bacillus subtilis* exhibited good activity to scavenge free radicals (Nitschke & Sousa e Silva, 2018).

APPLICATIONS OF BIOSURFACTANTS AS EMULSIFIER AGENTS

Emulsions are colloidal dispersions of lipids and water that need the addition of a surfactant to lower the interfacial tension in order to decrease the superficial energy between the two phases. Different types of surfactants are used in the food industry, some of them are synthetic compounds and, as previously mentioned, there is a trend to replace them by natural products. There are a large number of excreted biosurfactants derived from *Pseudomonas* and *Bacillus* strains that have potential capacities to be used as emulsifiers, such as rhamnolipids, emulsan and liposan. However, their origin limits their use in the food industry. Therefore, the biosurfactants derived from LAB are an alternative option, although they are less effective in reducing surface tension of water than the former (Mouafo et al., 2018).

The potential application of biosurfactants as emulsifying agents is measured by their effectiveness in reducing superficial tension, in having low CMC and high emulsification capacity. Table 1 shows parameters of several biosurfactants obtained from LAB. Reductions of surface tension of water higher than 8 mM/m are considered as the minimum for a biosurfactant to be effective. Furthermore, CMC expresses the concentration which corresponds to the minimum value of the surface tension that can be obtained with each compound. Important surface tension reductions (41.90 and 46.20 mN/m), low CMC (8 and 15 mg/mL) and high emulsification indexes (77.25 and 89.00% after 24 h) were observed with *Lb. delbrueckii* N2 using molasses or glycerol as substrates (Mouafo et al., 2018). Morais et al., (2017) extracted surface active compounds from *Lb. jensenii* P6A and *Lb. gasseri* P65 with potential application as bioemulsifiers in food (Table 1).

Biosurfactant production is not a simple matter. The extraction yields and structures are dependent on the right balance of culture media nutrients, microbial growth phase, and conditions of growth such us fermentation time, pH and temperature. These factors should be controlled to assure biosurfactant production with well-defined and consistent functionality and quality. Sugar cane molasses or glycerol were evaluated for biosurfactants production by *Lb.cellobiosus* TM1, *Lb.delbrueckii* N2, and *Lb. plantarum* G88. Mainly glycoproteins were obtained with molasses and glycolipids with glycerol indicating that biosurfactant composition is also substrate dependent. Yields obtained with both substrates were higher than those obtained with MRS broth whereas high surface tension reductions and emulsification indexes were observed with molasses as substrate (Mouafo et al., 2018).

Table 1. Compilation of some uses of biosurfactants as emulsifier agents

LAB Strain	Substrate used for production	Emulsion formation	Assay performed and trends observed	Reference
Lb. pentosus	Hemicellulosic sugars from vineyard pruning waste	Rosemary oil/water emulsions	- CMC*: 2 mg/L (measured as surfactin equivalents, 25°C) - Relative emulsion volume of 55.5% after 24 h of emulsion formation Emulsions of water and rosemary oil stabilized with the biosurfactant were more stable than those stabilized with polysorbate 20.	Vecino et al., (2015)
Lb. plantarum subsp. *plantarum* PTCC 1896	Rice and wheat bran and straw	Canola oil emulsion	- Maximum emulsification index was 64.70% - Maximum emulsion stability was 97.73%	Bakhshi et al., (2018)
Lb. delbrueckii N2, *Lb. cellobiosus* TM1, *Lb. Plantarum* G88	Sugar cane molasses or glycerol	Refined palm oil emulsion	- STR* ranged from 49.00 to 41.90 mN/m; - Emulsification index ranging from 41.81 to 81.00%. - CMC: 8 to 20 mg/mL	Mouafo et al., (2018)
Lb. jensenii P6A *Lb. gasseri* P65	MRS broth	Emulsification of several hydrocarbons and cotton, olive and sunflower oils.	For *Lb. jensenii* P6A: - STR to 42.5 mN/m - CMC: 8.58 mg/mL - Maximum emulsifying activity: at 5 mg/mL - Emulsification activity values ranged from 71 to 75% for cotton, olive, and sunflower oils For *Lb. gasseri* P65: - STR to 42.5 mN/m - CMC: 8.58 mg/mL - Maximum emulsifying activity: at 5 mg/mL - Emulsification activity values ranged from 71 to 75% for cotton, olive, and sunflower oils.	Morais et al., (2017)
Lb. Plantarum CFR 2194	MRS broth	Coconut and sunflower oils emulsions	- STR to 44.3 mN/m - CMC: 6 g/L - Emulsification index: 37.9% for coconut oil and 19.43% for sunflower oil. Coconut and sunflower oils emulsions were more stable than emulsions formed with hydrocarbons.	Madhu and Prapulla (2013)
Lb. helveticus MRTL	MRS-Lac (glucose replaced by lactose)	Emulsification with kerosene	- STR to 39.5 mN/m - CMC: 2.5 mg/mL - Emulsification index of 65% Biosurfactant was stable at pH range (4.0-12.0), it was more effective at pH 7 and surface properties were kept after heat treatment at 125°C for 15 min.	Sharma and Saharan (2016)

Table 1. (Continued)

LAB Strain	Substrate used for production	Emulsion formation	Assay performed and trends observed	Reference
Lb. pentosus	Residual stream after tartaric acid extraction from vinasses	Emulsification with kerosene	-Emulsion stability from 70.4 to 93.3% depending on growth media	Salgado et al., (2013)
Lb. coryniformis ssp. *torquens* CECT 25600 *Lb. plantarum* A14 *Leuconostoc mesenteroides* *Lb. paracasei* ssp. *paracasei* A20	MRS-Lac Different nitrogen sources in MRSLac broth (peptone, meat extract, and yeast extract) on *L. paracasei*ssp. *paracasei*A20	——	- STR from 69.6 to 50.0 mM/m after 8 h of fermentation for all strains. *Lb. paracasei* ssp. *paracasei* A20 showed the higher reduction. Higher biosurfactant production in presence of peptone and meat extract for *Lb. paracasei* ssp. *paracasei* A20, with a STR of 24.5 mN/m.	Gudiña et al., (2011)
Eight strains of *Lactobacillus* spp.	*Pendidam* (fermentative acid milk)	Emulsification with refined palm oil	- STR from 45.09 to 53.68 mM/m - Emulsification index from 10 to 56.80%	Mbawala and Mouafo (2012)
Lactobacillus acidophilus CECT-4179 (ATCC 832) Debaryomyceshansenii NRRL Y-7426	Synthetic hemicellulosic sugars Hemicellulosic vine-shoot trimming hydrolysates	——	- STR of 24.5 mN/m *Lb. acidophilus* STR in synthethic sugars and 18 mM/m in trimming vine shoots. - Dilutions (biosurfactant extract volume/total volume) to achieve CMC were 0.138 for synthetic sugars and 0.152 for trimming vine shoots.	Portilla et al., (2008)
Lactobacillus pentosus	Hydrolysates of vine pruning waste	gasoline/water emulsions	Studied the effect of pH, temperature and salinity on surface-active properties. - STR from 53.8 to 69.3 mM/m - Emulsion volume up to 47.48% - Emulsion stability up to 100% Maximum emulsifying capacity and emulsions stability were achieved at temperatures below 30°C, salinity below 3%, and pH higher than 5.	Vecino et al., (2012)

*STR: Surface tension reduction, CMC: Critical Micellar Concentration.

In relation to microbial growth phase and growth conditions, Madhu and Prapulla (2013) observed that biosurfactant released by *Lb. plantarum* CFR 2194 was higher under stationary than shaking conditions and it was maximum after 72 h of fermentation showing a decrease of surface tension of 44.3 mN/m, a CMC of 6 g/L and high emulsification indexes against coconut and sunflower oils (Table 1). These emulsions were more stable than emulsions formed with hydrocarbons and they could be of interest for application in food formulations.

Regarding the effects of other additives and conditions used for biosurfactant production, Vecino et al., (2015) studied the effect of NaCl concentrations, time and temperature of extraction on emulsifying ability of the biosurfactant produced by *Lb. pentosus* through a Box-Behnken response surface methodology. The extraction was carried out at the beginning of the stationary phase, after 24 h of fermentation, when biomass concentration was at its maximum. All variables affected the ability of the biosurfactant to act as emulsifier being the most influential variables the time of extraction, followed by temperature, and salt concentration. The optimal conditions were 9 g/kg salt, 45°C and 120 min. From the chemical point of view, the biosurfactant extracted was a mixture of carbohydrate, protein and lipid in a combination of 1:3:6, respectively. Then, this biosurfactant was successfully used to stabilized a rosemary oil/water emulsion with most droplets diameter < 100 μm and which exhibited a creamy consistency. On the contrary, the same emulsion stabilized with polysorbate 20 was less consistent and more unstable.

Sharma et al., (2014) studied the effect of pH (4, 6, 7, 8 y 10), temperature (4, 22 y 37°C), NaCl concentration (4, 6, 8 y 10%), carbon and nitrogen sources on emulsification index of different biosurfactants produced by *Lactobacillus* strains isolated from milk samples. The decrease in salt level and temperature promoted an increase in emulsification index. Strains preferred yeast extract as nitrogen source instead of urea, and there was no marked preference between glucose and sucrose as a carbon source.

Since most of chemical structure of biosurfactants contains proteins, the thermal stability is an important property to be studied. Producing a biosurfactant with high thermostability extends its range of application. Vecino et al., (2015) observed that as the production temperature increases, the emulsifying capacity of the biosurfactant obtained from *Lb. pentosus* increases, therefore, being thermostable makes it have better functionality. Bakhshi et al., (2018) extracted a water soluble extracellular high molecular weight α-glucan bound with protein, with high melting point (182.0°C) and high degradation temperature (211.74°C). Morais et al., (2017) evaluated the effect of pH, temperature (100°C) and salt concentrations (NaCl, KCl and $NaHCO_3$) on the activity of biosurfactants extracted from *Lb. jensenii* P6A and *Lb. gasseri* P65. They obtained stable emulsions after 60-min of incubation at 100°C, at pH 2–10, and after the addition of KCl and $NaHCO_3$.

APPLICATIONS OF BIOSURFACTANTS AS ANTIADHESIVE AGENTS

A large number of foodborne bacteria are capable of adhering to and forming biofilm on different surfaces, such as stainless steel, plastic, glass (Wang et al., 2014) and even food surfaces, like fruits and vegetables (Bilek et al., 2013). The adhesion of bacteria on surfaces compromises food safety, and produces food spoilage, generating huge economic losses (Van Houd & Michiels, 2010). Consequently, there is a focus of interest around the development of innovative techniques to overcome this issue.

One possible application of biosurfactants produced by LAB is as an antiadhesive agent (Table 2). It could be used as a surface conditioner, preventing the adhesion of microorganisms and the subsequent formation of biofilm, avoiding the use of synthetic disinfectants with considerable toxicity levels. For this reason, LAB surfactants present a great potential in food industry. Nevertheless, it is notorious that in most cases the applications studied are closely related with the biomedical area and used as conditioners of silicone rubber normally used in catheters and prosthesis (Rodrigues et al., 2004; Ceresa et al., 2015; Sharma et al., 2016). Furthermore, it was proved the anti-adhesion effect against strains isolated from voice prosthesis (Rodrigues et al., 2006, a), from the mouth and blood (Fracchia et al., 2010), uropathogenic bacteria (Velraedes et al., 1998) and skin pathogens (Vecino et al., 2018). To our knowledge only one study has reported the use of LAB biosurfactant as stainless steel conditioner, material commonly used in food industry (Meylheuc et. al, 2006), and none on the effects against film forming microorganisms isolated from food environments, nevertheless there is an expansion in the knowledge in these promising fields.

The genuses of LAB biosurfactant producers showing antiadherent capability are *Lactobacillus*, *Lactococcus*, and *Streptococcus*. Many researchers have shown that biosurfactants produced by *Lactobacillus* strains are capable to reduce the adherence of bacteria to different surfaces. An early investigation on this subject has reported the ability for biosurfactant production of fifteen *Lactobacillus* strains and investigated further to determine their capacity to inhibit the initial adhesion of *Enterococcus* (*Ec.*) *faecalis* 1131 to glass. This research showed that *Lb. acidophilus* RC14 and *Lb. fermentum* B54 were able to produce a biosurfactant, which reduced the deposition of *Ec. faecalis* by an average of 70% and the adhesion by an average of 77% (Velraeds et al., 1998). Those strains of *Lactobacillus* biosurfactant producers were two of many others found in the last decades capable to produce biosurfactants which reduced the adhesion of bacteria to surfaces. *Lb. paracasei* isolated in a Portuguese daily industry produced a biosurfactant capable to reduce the adhesion of *Staphylococcus* (*S.*) *aureus* to polystyrene by a percentage of 72.0%, 62.1%, for *S. epidermidis* and *Strep. agalactiae*, respectively in a biosurfactant concentration of 25 mg/mL. However, a lower activity was observed for decreasing the adhesion of *Pseudomonas* and *Escherichia* strains, at the same biosurfactant concentration (Gudiña et al., 2010 a). On the other hand, biosurfactants produced by *Lb. helveticus* inhibit the adhesion of *Escherichia* (*E.*) *coli* ATCC 25922 in a 50.1% at a concentration of 25 mg/mL (Sharma & Saharan, 2016). Hence, the antiadhesive capability of a biosurfactant against a pathogenic strain is different according to the bacterium that is producing the biosurfactant and also the target of the bacterium which is desirable to control.

Table 2. Compilation of some uses of biosurfactants as antiadhesive agents

LAB strain/source	Target m.o. for antiadhesive activity	Assay performed	Effect	Reference
Lc. lactis 53 / Nutricia (The Netherland)	*S. epidermidis* GB 9/6, *Strep. salivarius* GB 24/9, *S. aureus* GB 2/1, *Rothia (R.) dentocariosa* GBJ 52/2B, *Candida (C.). albicans* GBJ 13/4A, *C. tropicalis* GB 9/9	Anti-adhesion assay in polystyrene 96 wells plate	Adhesion reduction of 70% for *S. aureus* GB 2/1 and *S. epidermidis* GB 9/6 at concentration 25g/L	Rodrigues et al., (2006 a)
Strep. thermophilus A / NIZO	*S. epidermidis* GB 9/6, *Strep. salivarius* GB 24/9, *S. aureus* GB 2/1, *R. dentocariosa* GBJ 52/2B, *C. albicans* GBJ 13/4A, *C. tropicalis* GB 9/9	Parallel-plate flow chamber on silicone rubber and image analysis	Adhesion reduction of 89-97% for bacteria and 67-70% for yeasts at concentration 3g/L	Rodrigues et al., (2006 b)
Lb. helveticus 1181 / Cheese industry	*L. monocytogenes* CIP103574, *L. monocytogenes* CIP104794, *L. monocytogenes* CIP103573 *L. monocytogenes* CIP78.39	Co-incubation assays on metallic supports	Adhesion reduction of 93.62–99.77% at concentration 5g/L.	Meylheuc, et al., (2006)
Strep. mitis / Human oral cavity	*Strep. sobrinus* HG1025 *Strep. mutans* ATCC 25175 *Strep. mitis* BMS	Co-incubation of enamel particle and atomic force microscopy	The biosurfactant generate an increase in the repulsive forces between the surface and the microorganism.	Van Hoogmoed, et al., (2006)
Lb. acidophilus - H-1, *Lb. acidophilus* - 336, *Lb. acidophilus* -Ch-2 / Rhodia Food Biolacta Company, Olsztyn, Poland	*S. aureus* - ATCC 29213, *S. aureus* -1474/01, *S. aureus* - A3, *S. epidermidis* -6756/99, *S. epidermidis* -A4c, *S. epidermidis* -RP12	Initial adherence evaluation by MTT*-reduction assay/ Confocal microscopy	Biofilm covered less surface area, decrease in their total biovolume and have a lower mean biofilm thickness. LAA strain H-1 produced the best anti-adhesion active agent.	Walencka et al., (2008)
Lb. paracasei / Portuguese dairy industry	*E. coli, P. aeruginosa, S. aureus S. epidermidis, Strep. agalactiae Strep. pyogenes*	Anti-adhesion assay in polystyrene 96 wells plate	Adhesion reduction of 72.0% for *S. aureus* at a concentration 25 g/l.	Gudiña et al., (2010, a)
Lb. paracasei ssp. paracasei A20	*Lb. casei* 36, *Lb. Casei* 72, *Lb. reuteri* 104R, *Lb. reuteri* ML1, *S. mutans* NS, *Strep. mutans* HG985, *Strep. oralis* J22, *Strep. sanguis* 12, *E. coli, P. aeruginosa, S. aureus, S. epidermidis, Strep. agalactiae, Strep. pyogenes, C. albicans, Malassezia sp., Trichophyton (T.) mentagrophytes, T.* rubrum	Anti-adhesion assay in polystyrene 96 wells plate	Adhesion reduction of 87,7% for *L. reuteri* ML1 at a concentration 50 g/L	Gudiña et al., (2010, b)
Lactobacillus sp. CV8LAC / Fresh Cabbage	*C. albicans* DSMZ, *C. albicans* CA-2894	Pre-coating and co-incubation assay in polystyrene 96 wells plate	Adhesion reduction of 86% for *C. albicans* CA-2894 at a concentration 25g/L in the pre-coated surface and 86% for *C. albicans* DSMZ 1122 in the co-incubate plate at concentration 160.5 µg/well.	Fracchia et al., (2010)

Table 2. (Continued)

LAB strain/source	Target m.o. for antiadhesive activity	Assayperformed	Effect	Reference
Lb.plantarum CFR 2194 Arenahalli / kanjika	*E. coli* ATCC 31075, *E. coli* MTCC 108, *Salmonella typhi, Yersinia (Y.) enterocolitica* MTCC 859, *S. aureus* F 722	Anti-adhesion assay in polystyrene 96 wells plates.	Adhesion reduction of 67.18% for *S. aureus* at concentration 25 g/L	Madhu and Prapulla (2013)
Lb.brevis CV8LAC / Fresh cabbage	*C. albicans*	Co-incubation assays on silicone surface and precoating assays in silicone surface SEM analysis	Adhesion reduction of 90% at the concentration 4g/L, using co-incubation assay Adhesion reduction of 50% at the concentration 3g/L, using precoating assay	Ceresaet al., (2015)
Lb. helveticus / Yak milk cheese	*E. coli* ATCC 25922, *P. aeruginosa* ATCC 15442, *S. typhi* MTCC 733, *Shigella flexneri* ATCC9199, *S. aureus* ATCC 6538P, *S. epidermidis* ATCC12228, *L. monocytogenes* MTCC 657, *L. innocua* ATCC 33090, *Bacillus (B.). cereus* ATCC 11770	Anti-adhesion assay in polystyrene 96 wells plate and co- incubation assays on silicon tubes.	Adhesion reduction of 87% *B. cereus* at concentration 25 g/L. Decrease of adhesion on silicone tube at concentration 25 mg/mL for all strains	Sharma & Saharan (2016)
Lb. pentosus CECT-4023T (ATCC-8041) / Spanish Type Culture Collection (Valencia, Spain) *Lb. paracasei* / Portuguese dairy industry	*E. coli, P. aeruginosa, S. aureus, S.epidermidis, Strep. agalactiae, Strep. pyogenes, C. albicans*	Anti-adhesion assay in polystyrene 96 wells plate	Adhesion reduction of 77% for using biosurfactant produced by *L. pentosus* and extracted with PB, at concentration 25g/L Adhesion reduction of 81% for *S. agalactiae* using biosurfactant produced by *L. paracasei* extracted with PBS, at concentration 25g/L	Vecino et al., (2018)

*MTT: 4,5-dimethylth iazol-2-yl)-2,5-diphenyltetrazolium bromide.

As previously commented, there are several studies reporting *Lactobacillus* strains as producers of biosurfactants, but there are few studies about the ability to produce biosurfactants by strains of *Lactococcus*. Three *Lactococcus* strains were reported as biosurfactant producers, *Lc. lactis* 53 (Rodrigues et al., 2004; Rodrigues et al., 2006 b), *Lc. lactis* CECT-4434 (Rodriguez et al., 2010; Souza et al., 2017), *Lc. lactis* (Saravanakumari & Mani, 2010), but only for *Lc. Lactis 53* the anti-adhesive potential had been studied. According to Rodrigues et al., (2006 b), the biosurfactant produced by *Lc. lactis* 53 shown three fractions, isolated by hydrophobic interaction chromatography. All fractions shown antiadherent activity, but the best result was provided by the fraction A, inhibiting the adhesion to polystyrene of all microorganisms target even in a concentration of 5 mg/mL. The crude biosurfactant produced by *Lc. lactis 53* has been also proved as conditioner of silicone rubber (Rodrigues et al., 2004). This study has been performed in a parallel-plate flow chamber equipped with a microscope allowing *in situ* observation and images analysis. In these experimental conditions, the surfactant adsorbed on the silicone rubber reduced the deposition rate above 90% for *S. epidermidis* GB 9/6, *Strep. Salivarius* GB 24/9, and *S. aureus* GB 2/1, the same percentage of reduction was reached for the cells adhering after 4h.

Some *Streptococcus* strains have been identified as producer of biosurfactants exhibiting anti-adherent activity. Rodrigues et al., (2006 a) studied the biosurfactant produced by *Strep. thermophilus* A, reporting a reduction of adhesion after 4 hours of 89-97% for bacteria and 67-70% for yeasts on silicon rubber, using the biosurfactant at a concentration on 3 mg/mL. Moreover, Van Hoogmoed et al., (2006), Busscher and Van der Mei (1997) have reported two strains of *Streptococcus* biosurfactant producers with antiadhesive activity. But all these studies were focused on biomedical applications.

It was demonstrated that not only the strain of the bacterium is responsible for the biosurfactant characteristics; also the method of extraction can change the anti-adherent properties of a biosurfactant. For example, when extraction of biosurfactants produced by *Lb. pentosus* and *Lb. paracasei* were done using phosphate buffer (PB), the adherence of *S. epidermis* and *Strep. pyogenes* were decreased in 57 and 69% respectively. But, when extraction was done using PBS, the decrease in adherence was 38% and 52%, respectively (Vecino et al., 2018).

One emerging trend is the incorporation of anti-adhesive agents into the packaging material. According to Abdul et al., (2014), the addition of a biosurfactant produced by *Lb. rhamnosus* to a polyvinyl alcohol film prevented the biofilm formation of *S. aureus* and *Pseudomona* (*P.*) *aeruginosa*.

The main pathogenic bacteria related to the food industry that is of interest to prevent biofilm formation are *Listeria* (*L.*) *monocytogenes, E. coli* (*enteropathogenic strains*) and *S. aureus.*

Listeria monocytogenes is linked with many outbreaks and cause a food illness that presents a mortality rate of 25% worldwide (de Noordhout et al., 2014). It can be found in a high concentration in foods, especially ready-to-eat foods (Buchanan et al., 2017). It is capable to form biofilm even at refrigeration temperatures (Moltz & Martin, 2005) and can persist in food process environments for years (Hoelzer et al., 2015). For these reasons it is necessary to develop strategies to combat the adhesion of cells to surfaces and the subsequent formation of biofilm. So far, the effect of two biosurfactants produced each one by a *Lb. helveticus* strains were reported showing very promising results (Meylheuc et al., 2006; Sharma & Saharan, 2016). The biosurfactant produced by *Lb. helveticus* (Sharma & Saharan, 2016) at a concentration of 25 mg/mL, inhibited the adhesion of *L. monocytogenes* MTCC 657 and *L. innocua* ATCC 33090 to polystyrene in a percentage of 84% and 82%, respectively.

Escherichia coli is identified as responsible for outbreaks of foodborne illness (Greig & Ravel, 2009; Bélanger et al., 2015). It is a strong biofilm former that persists in the equipment of food industries (Yang et al., 2018), produced in presence of food residues (Dutra et al., 2018). Therefore, it is of interest to find approaches that prevent *E. coli* biofilm formation. Some researchers studied the use of LAB as inhibitors of adherence to surfaces for different *E. coli* strains (Gudiña et al., 2010 a, b; Madhu & Prapulla, 2013; Sharma et al., 2016; Vecino et al., 2018). The results by the time, present LAB as substances capable to produce a low degree in the adhesion of *E. coli* to polystyrene, been the highest reduction percentage scored of 56.78% using the biosurfactant produced by *Lb. plantarum* CFR 2194 in a concentration of 25 mg/mL, against the strain *E. coli* MTCC 108 (Madhu & Prapulla, 2013).

Staphylococcus aureus is well known as a foodborne pathogen, which produces enterotoxins that can poison food (Le Loir et al., 2003), form biofilm on surfaces used in food industries (Dutra et al., 2018) and under food-related stress conditions (Rode et al., 2007). Many researchers reported positive results for the anti-adhesive LAB properties against different *S. aureus* strains in two type of surfaces, polystyrene (Rodrigues et al., 2006 a, b; Madhu & Prapulla, 2013; Gudiña et al., 2010 a, b; Sharma et al., 2016; Vecino et al., 2018) and silicone rubber (Rodrigues et al., 2004; Sharma et al., 2016). In polystyrene, the highest inhibition of adhesion reported was 83%, conditioning the surface with a biosurfactant produced by *Lb. helveticus* in a concentration of 25 mg/mL, against the strain *S. aureus* ATCC 6538P (Sharma and Saharan, 2016). In silicone rubber, the initial deposition of *S. aureus* GB 2/1 was reduced in 90%, conditioning the surface with the biosurfactant produced by *Lc. lactis* 53 (Rodrigues et al., 2004). Eventhough LAB look to be great surface conditioners in polystyrene and silicon rubber, no results are reported for the effect of the adherence of *S. aureus* in common surfaces used in food industry such as stainless steel.

Regarding the spoilage bacteria, *P. aeruginosa* has the ability to form biofilm and exhibits several virulence factors coordinately by quorum sensing when it is contaminating foods (Viola et al., 2018). It is closely related with waterborne diseases, wet environments are ideal for the formation and proliferation of biofilm, as different components of water systems (Walker & Moore, 2015). Some researchers studied the anti-adherence capability of LAB against the adhesion of *P. aeruginosa* to polystyrene surfaces, and all of them obtained positives results but with different degrees of inhibition (Gudiña et al., 2010 a; Sharma et al., 2016; Vecino et al., 2018). The best result was obtained using the biosurfactant produced by *Lb. paracasei* and extracted with phosphate buffer saline in a concentration of 25 mg/mL, obtaining an inhibition of 72% (Vecino et al., 2018). This degree of inhibition was higher than the previously reported values of 49% (Sharma et al., 2016) and 16.5% (Gudiña et al., 2010 a) for the same biosurfactant concentration.

According to results previously commented, there is a huge potential in the use of biosurfactants produced by LAB as anti-adherence agents. However, there are not many studies about the effective reduction of adhesion of many foodborne bacteria. The lack of these studies opens a door for novel investigations in the field of food microbiology.

APPLICATIONS OF BIOSURFACTANTS AS ANTIMICROBIAL AGENTS

The search for natural antimicrobials is of increasing concern to mitigate the development of antibiotics resistance against clinical pathogens and also to satisfy consumer demands for chemical additives free and healthier foods. Antimicrobials obtained from microbial sources present a huge potential due to their high biological activity and the possibility of production at low cost in an industrial scale.

The main advances in the knowledge of antimicrobial action of biosurfactantes were done in the field of biomedical sciences (Nitschke & Costa, 2007; Sharma et al., 2016). Conversely, they are very few research focused on the food industry.

Biosurfactants present antimicrobial activity; they exhibit action against virus, algae, bacteria, yeast and molds. Furthermore, several hypotheses are proposed to explain the mechanism of action: i) adhesion to cell surfaces damaging cell membrane integrity; ii) insertion of fatty acids from biosurfactant into cell membrane promoting changes in its composition and functionally; iii) ability to form pores in the membrane altering its functions (Sharma et al., 2016).

The main genera within LAB producing biosurfactants with antimicrobial activity was *Lactobacilli*. Different studies reported that they were able to inhibit the growth of bacteria and yeasts. Gram positive bacteria were more sensitive than Gram negative and yeasts were the more resistant. These trends were related to the fact that Gram negative present a polysaccharide layer in the external membrane which acts as a barrier to biosurfactant. In addition, yeasts present higher phospholipid content into the membrane which facilitates the interaction with biosurfactant (Mouafo et al., 2018).

Table 3 shows a compilation of biosurfactant producing LAB exhibiting antimicrobial activity. It can be observed that most of the strains were isolated from food. In addition, the target microorganisms evaluated were pathogens, in some cases isolated from medical uses. Finally, in all cases the activity was only tested *in vitro*.

Very few studies were done proposing the use of biosurfactants as antimicrobials compounds *in vivo*. For example, Abruzzo et al., (2018) evaluated successfully phosphatidylcholine based vesicles containing a biosurfactant produced by *Lb. gasseri* and econazole for the treatment of vaginal Candida infections. Mbawala et al., (2017) applied a crude surfactant extracted from a fermented milk to improve the microbial quality of yellow achu soap, a typical African meal. They used the crude surfactant as a dual agent, emulsifier and antimicrobial and they could decrease total aerobic and coliform populations. Also, a complete inhibition of *Salmonella* spp. and faecal streptococci was reached.

Table 3. Compilation of some uses of biosurfactants as antimicrobial agents

LAB strain/source	Target m.o. for antiamicrobial activity	Assay performed	Effect	Reference
Strep. thermophilus A/ Nizo (The Netherlands)	*S. epidermidis* GB 9/6, *Strep. salivarius* GB 24/9, *S. aureus* GB 2/1, *R. dentocariosa* GBJ 52/2B, *C. albicans* GBJ 13/4A, *C. tropicalis* GB 9/9 Isolated from explanted voice prostheses	Agar diffusion assay	Antimicrobial action depended on concentration. At 40 g/L all microorganisms were inhibited. Most sensitive was *C. tropicalis*	Rodrigues et al., (2006 a)
Lc. lactis 53 / Nutricia (The Netherland)	*S. epidermidis* GB 9/6, *Strep. salivarius* GB 24/9, *S. aureus* GB 2/1, *R. dentocariosa* GBJ 52/2B, *C. albicans* GBJ 13/4A, *C. tropicalis* GB 9/9 Isolated from explanted voice prostheses	Agar diffusion assay	Antimicrobial action depended on biosurfactant fraction concentration.	Rodrigues et al., (2006 b)
Lb. casei 8/4/ Culture collection University of Warmia, Poland	*S. aureus* 11, *B. subtilis*, *Micrococcus (M.) roseus*	Agar diffusion assay	Antimicrobial action depended on incubation time.. Maximum activity was found after 18 h where all bacteria growth was inhibited	Golek et al., (2009)
Lb. paracasei / Portuguese dairy industry	*E. coli*, *P. aeruginosa*, *S. aureus*, *S. epidermidis*, *Strep. agalactiae*, *Strep. pyogenes*	Broth microdilution method and absorbance reading at 600nm	Crude biosurfactant showed antimicrobial activity against all tested bacteria	Gudiña et al., (2010 a)
Lb.plantarum CFR 2194 / kanjika	*E. coli* ATCC 31075, *E. coli* MTCC 108, *Salmonella typhi*, *Y. enterocolitica* MTCC 859, *S. aureus* F 722	Agar diffusion assay	Antimicrobial action depended on concentration. All bacteria assayed were inhibited except *S. thyphi*	Madhu & Prapulla (2013)
Lb. casei MRTL3/ Raw milk	*S. aureus* ATCC 6538P, *S. epidermidis* ATCC 12228, *Shigella flexneri* ATCC 9199, *Salmonella typhi* MTCC 733, *P. aeruginosa* ATCC 15442, *B. cereus* ATCC 11770, *L. monocytogenes* MTCC 657, and *L. innocua* ATCC 33090	Agar diffusion assay	Crude biosurfactant inhibited all pathogens assayed	Sharma & Saharan (2014)
Lb. agilis CCUG31450/ Culture Collection of University of Gothenburg (Sweden)	*E. coli*, *P. aeruginosa*, *S. aureus*, *Strep. agalactiae* and *C. albicans*	Broth microdilution method and absorbance reading at 600nm	No antimicrobial activity was observed against *E. coli* and *C. albicans*	Gudiña et al., (2015)
Lb. helveticus / Yak milk cheese	*E. coli* ATCC 25922, *P. aeruginosa* ATCC 15442, *Salmonella typhi* MTCC 733 *Shigella flexneri* ATCC9199, *S. aureus* ATCC 6538P, *S. epidermidis* ATCC12228, *L. monocytogenes* MTCC 657, *L. innocua* ATCC 33090, *B. cereus* ATCC 11770	Broth microdilution method and absorbance reading at 600nm	Biosurfactant concentrations between 1.56 and 25 mg/mL promoted different degree of inhibition of all bacteria evaluated	Sharma & Saharan (2016)

Table 3. (Continued)

LAB strain/source	Target m.o. for antiamicrobial activity	Assay performed	Effect	Reference
Pediococcus acidililacticil Artisanal milk cheese	*S. aureus* CMCC 26003	Broth microdilution method and visual evaluation of turbidity and time–kill assay	MIC was higher than 100 mg/mL and a 50 mg/mL affected the growth.	Yan et al., (2019)
Lactobacillus acidophilus, L. pentosus, L. fermentum, L plantarum, L. lactis, L. casei / Fermented foods	*P. fluorescens, P aeruginosa, E coli, Salmonella typhimurium*	Agar diffusion assay	Biosurfactants obtained from *L. plantarum* inhibited all bacteria tested	Abdalsadiq Nagea & Zaiton (2018)
Lactobacillus jensenii P6A and *Lactobacillus gasseri* P65/ Women vaginal fluids	*E. coli, Klebsiella pneumoniae, Enterobacter aerogenes, S. saprophyticus, C. albicans* ATCC 18804, *C. krusei* ATCC 20298, and *C. tropicalis* ATCC 750	Broth microdilution method and visual evaluation of turbidity	No antimicrobial activity was observed against *C. krusei* and *C. tropicalis*. The rest of microorganisms tested were inhibited.	Morais et al., (2017)

CONCLUSION

Regarding the attempts made to apply biosurfactants produced by LAB, the following facts can be remarked:

- The number of studies on this field has increased in the last years, especially in relation to medical uses. However, the cost is still a problem to solve since chemical surfactants are cheaper, even using agricultural wastes as substrates.
- In the food applications, up to now only one product, named Pre-liminate® is produced at an industrial scale, this formulation can be applied to surfaces in contact with food and it is designed to prevent biofilm formation.
- To increase food uses, *in vivo* studies must be performed in order to understand how the biosurfactants interact with other additives and/or food components. Also it is essential to know the sensorial impact of the biosurfactant and its stability during the food shelf life.
- In relation to the antimicrobial activity of biosurfactants, no information is available about their action on food spoilage microorganisms. It can be stressed

that growth of spoilage microorganisms is responsible for high economic losses in the food industry.

- Studies about the interaction with other biosurfactants or with other additives in order to find synergic combinations are useful approachs to facilitate biosurfactant uses.

Outcomes shown herein highlight that biosurfactants from LAB present a huge potential and there are many unexplored areas that need to be studied in order to develop novel applications for the food industry.

REFERENCES

Abdalsadiq Nagea K. A., Zaiton H. (2018). Biosurfactant and antimicrobial activity of lactic acid bacteria isolated from different sources of fermented foods. *Asian Journal of Pharmaceutical and Development*, 6 (2), 60-73.

Abdul J., Salman S., Kadhemy M. F. H. (2014). Effect of PVA, PVA/biosurfactant on some pathogenic bacteria in glass and plastic plates. *Internationa Journal of Current Microbiology. App. Sci.* 3, 301–309.

Abruzzo A., Giordani B., Parolin C., Vitali B., Protti M., Mercolini L., Cappelletti M., Fedi S., Bigucci F., Cerchiara T., Luppi B. (2018). Novel mixed vesicles containing lactobacilli biosurfactant for vaginal delivery of an anti-Candida agent. *European Journal of Pharmaceutical Sciences*, 112, 95-101.

Bakhshi N., Sheikh-Zeinoddin M., Soleimanian-Zad S. (2018). Production and Partial Characterization of a Glycoprotein Bioemulsifier Produced by *Lactobacillus plantarum* subsp. *Plantarum* PTCC 1896. *Journal of Agricultural Science and Technology*, 20, 37-49.

Bélanger P., Tanguay F., Hamel M., Phype M. (2015). An overview of foodborne outbreaks in Canada reported through Outbreak Summaries: 2008-2014. *Canada Communicable Disease Report (CCDR)*, 41–11, 254–262.

Bilek S. E., Turantaş F. (2013). Decontamination efficiency of high power ultrasound in the fruit and vegetable industry, a review. *International Journal of Food Microbiology*, 166 (1), 155–162.

Buchanan R. L., Gorris L. G., Hayman M. M., Jackson T. C., Whiting R. C. (2017). A review of *Listeria monocytogenes*: an update on outbreaks, virulence, dose-response, ecology, and risk assessments. *Food Control*, 75, 1-13.

Busscher H. J. Van der Mei H.C. (1997). Physico-chemical interactions in initial microbial adhesion and relevance for biofilm formation. *Advances in Dental Research*, 11(1), 24–32.

Ceresa C., Tessarolo F., Caola I., Nollo G., Cavallo M., Rinaldi M., Fracchia L. (2015). Inhibition of *Candida albicans* adhesion on medical-grade silicone by a *Lactobacillus*-derived biosurfactant. *Journal of Applied Microbiology*, 118 (5), 1116–1125.

de Noordhout C. M., Devleesschauwer B., Angulo F. J., Verbeke G., Haagsma J., Kirk M., ... Speybroeck N. (2014). The global burden of listeriosis: a systematic review and meta-analysis. *The Lancet Infectious Diseases*, 14 (11), 1073-1082.

Dutra T. V., Fernandes M. da S., Perdoncini M. R. F. G., Anjos M. M. dos, Abreu Filho B. A. de. (2018). Capacity of *Escherichia coli* and *Staphylococcus aureus* to produce biofilm on stainless steel surfaces in the presence of food residues. *Journal of Food Processing and Preservation*, 42(4), 1–6.

Fracchia L., Cavallo M., Allegrone G., Martinotti M. G. (2010). A Lactobacillus -derived biosurfactant inhibits biofilm formation of human pathogenic *Candida albicans* biofilm producers. *Technology and Education Topics in Applied Microbiology and Microbial Biotechnology*, 3, 827–837.

Gołek P., Bednarski W., Brzozowski B., Dziuba B. (2009). The obtaining and properties of biosurfactants synthesized by bacteria of the genus *Lactobacillus*. *Annals of Microbiology*, 59 (1) 119-126.

Greig J. D., Ravel A. (2009). Analysis of foodborne outbreak data reported internationally for source attribution. *International Journal of Food Microbiology*, 130(2), 77–87.

Gudiña E. J., Teixeira J. A., Rodrigues L. R. (2010 a). Isolation and functional characterization of a biosurfactant produced by *Lactobacillus paracasei*. *Colloids and Surfaces B: Biointerfaces*, 76 (1), 298-304.

Gudiña E. J., Teixeira, J. A., Rodrigues L. R. (2010 b). Antimicrobial and antiadhesive properties of a biosurfactant isolated from *Lactobacillus paracasei* ssp. *Paracasei* A20. *Letter of Applied Microbiology*, 50 (4), 419-424.

Gudiña E. J., Teixeira J.A. Rodrigues L. R. (2011). Biosurfactant-producing Lactobacilli: screening, production profiles, and effect of medium composition. *Applied and EnvironmentalSoil Science*, Article ID 201254, 9 pages.

Gudiña E. J., Fernandes E. C., Teixeira J.A. Rodrigues L. R. (2015). Antimicrobial and anti-adhesive activities of cellboundbiosurfactant from *Lactobacillus agilis*. RSC *Advances*, 5, 90960–90968 | 90961.

Hajfarajollah H., Eslami, P., Mokhtarani B., Akbari Noghabi K. (2018). Biosurfactants from probiotic bacteria: A review. *Biotechnol Appl Biochem*. Aug 18. doi: 10.1002/bab.1686.

Hoelzer K., Pouillot R., Dennis S., Gallagher D., Kause J. (2015). Update on *Listeria monocytogenes:* reducing cross-contamination in food retail operations. *Advances in Microbial Food Safety* (pp. 149-194).

Le Loir Y., Baron F., Gautier M. (2003). *Staphylococcus aureus* and food poisoning. *Genetics and Molecular Research*, 2(1), 63-76.

Madhu A. N., Prapulla S. G. (2013). Evaluation and functional characterization of a biosurfactant produced by *Lactobacillus plantarum* CFR 2194. *Applied Biochemistry and Biotechnology*, 172(4), 1777-1789.

Mbawala A. Mouafo T. H. (2012). Screening of biosurfactants properties of cell-free supernatants of cultures of *Lactobacillus* spp. isolated from a local fermented milk (Pendidam) of Ngaoundere (Cameroon). *International Journal of Engineering Research and Applications*, 2 (5): 974-985.

Mbawala A., Elysé R. Mouafo T. H., Hervé. T. M. (2017). Antimicrobial activity of crude biosurfactants extracted from a locally fermented milk (pendidam) on yellow Achusoup produced in Ngaoundere Cameroon. *International Journal of Applied Microbiology and Biotechnology Research*, 559-567.

Meylheuc T., Renault M., Bellon-Fontaine M. N. (2006). Adsorption of a biosurfactant on surfaces to enhance the disinfection of surfaces contaminated with *Listeria monocytogenes*. *International Journal of Food Microbiology*, 109 (1–2), 71–78.

Moltz A. G., Martin S. E. (2005). Formation of biofilms by *Listeria monocytogenes* under various growth conditions. *Journal of Food Protection*, 68(1), 92-97.

Morais I. M. C., Cordeiro, A. L., Teixeira G. S., Domingues V. S., Nardi R. M. D., Monteiro A. S., Alves R. J., Siqueira E. P., Santos V. L. (2017). Biological and physicochemical properties of biosurfactants produced by *Lactobacillus jensenii* P6A and *Lactobacillus gasseri* P65. *Microbial Cell Factory*, 16:155.

Mouafo T.M., Mbawala A., Ndjouenkeu R. (2018). Effect of different carbon sources on biosurfactants'production by three strains of *Lactobacillus* spp. BioMed Research International, Volume 2018, Article ID 5034783, 15 pages.

Nitschke M., Costa S. (2007). Biosurfactants in food industry. *Trends Food Sciences and Technology*, 18 (5), 252–259.

Nitschke M., Sousa e Silva S. (2018). Recent food applications of microbial surfactants, *Critical Reviews in Food Science and Nutrition*, 58:4, 631-638.

Portilla O. M., Rivas B., Torrado A., Moldes A. B., Dominguez, J. M. (2008). Revalorisation of vine trimming wastes using *Lactobacillus acidophilus* and *Debaryomyces hansenii*. *Journal of Science Food and Agriculture*, 88, 2298–2308.

Rodrigues L. R., Teixeira J. A., van der Mei H. C., Oliveira R. (2006 a). Isolation and partial characterization of a biosurfactant produced by *Streptococcus thermophilus* A. *Colloids and Surfaces B: Biointerfaces*, 53 (1), 105–112.

Rodrigues L. R., Teixeira J. A., van der Mei H. C., Oliveira R. (2006b). Physicochemical and functional characterization of a biosurfactant produced by *Lactococcuslactis* 53. *Colloids and Surfaces B: Biointerfaces*, 49 (1), 79-86.

Rodrigues L., Van Der Mei H., Teixeira J. A., Oliveira R. (2004). Biosurfactant from *Lactococcuslactis* 53 inhibits microbial adhesion on silicone rubber. *Applied Microbiology and Biotechnology*, 66 (3), 306–311.

Rodríguez N., Salgado J. M., Cortés S., Domínguez J. M. (2010). Alternatives for biosurfactants and bacteriocins extraction from *Lactococcuslactis* cultures produced under different pH conditions. *Letters in Applied Microbiology*, 51(2), 226-233.

Salgado J. M., Vázquez-Araújo L., Cortés S., Moldes A., Domínguez J. M. (2013). Valorisation of vinasses by recovery tartaric acid and bioproduction of lactic acid and emulsifiers by *Lactobacillus pentosus*. *Winery 2013-6th IWA International Specialized Conference "Viticulture and Winery Wastes: Environmental Impact and Management."*

Saravanakumari P., Mani K. (2010). Structural characterization of a novel xylolipid biosurfactant from *Lactococcuslactis* and analysis of antibacterial activity against multi-drug resistant pathogens. *Bioresource Technology*, 101 (22), 8851–8854.

Sharma A.; Soni, J.; Kaur G., Kaur J. 2014. A Study on biosurfactant production in *Lactobacillus* and *Bacillus* sp. *International Journal of Current Microbiology Applied Sciences*, 3 (11):723-733.

Sharma D., Saharan B. S. (2014). Simultaneous Production of Biosurfactants and Bacteriocins by Probiotic *Lactobacillus casei* MRTL3. *International Journal of Microbiology*, ID, 698713, 7 pages.

Sharma D., Saharan B. S. (2016). Functional characterization of biomedical potential of biosurfactant produced by *Lactobacillus helveticus*. *Biotechnology Reports*, 11, 27–35.

Sharma D., Saharan B. S., Kapil S. (2016). *Biosurfactants of Probiotic Lactic Acid Bacteria*, Springer.

Souza E. C., Azevedo P. O. de S. de, Domínguez J. M., Converti A., Oliveira R. P. de S. (2017). Influence of temperature and pH on the production of biosurfactant, bacteriocin and lactic acid by *Lactococcuslactis* CECT-4434. *CyTA - Journal of Food*, 15(4), 525–530.

Van Hoogmoed C. G., Dijkstra R. J. B., Van der Me, H. C., Busscher H. J. (2006). Influence of biosurfactant on interactive forces between mutans streptococci and enamel measured by atomic force microscopy. *Journal of Dental Research*, 85 (1), 54-58.

Van Houdt R., Michiels C. W. (2010). Biofilm formation and the food industry, a focus on the bacterial outer surface. *Journal of Applied Microbiology*, 109 (4), 1117–1131.

Vecino X., Rodríguez-López L., Ferreira D., Cruz J. M., Moldes A. B., Rodrigues L. R. (2018). Bioactivity of glycolipopeptide cell-bound biosurfactants against skin pathogens. *International Journal of Biological Macromolecules*, 109, 971-979.

Vecino X., Barbosa-Pereira L., Devesa-Rey R., Cruz J. M., Moldes A. B. (2015). Optimization of extraction conditions and fatty acid characterization of *L. pentosus* cell bound biosurfactant/bioemulsifier. *Journal of the Science of Food and Agriculture*, 95 (2), 313-320.

Vecino X., Devesa-Rey R., Cruz J. M., Moldes A. B. (2012). Study of the synergistic effects of salinity, pH, and temperature on the surface-active properties of biosurfactants produced by *Lactobacillus pentosus*. *Journal of Agricultural and Food Chemistry*, 60 (5), 1258-1265.

Velraeds M. M., Van de Belt-Gritter B., Van der Mei H. C., Reid G., Busscher H. J. (1998). Interference in initial adhesion of uropathogenic bacteria and yeasts to silicone rubber by a *Lactobacillus acidophilus* biosurfactant. *Journal of Medical Microbiology*, 47(12), 1081-1085.

Velraeds M., van der Mei H., Reid G., Busscher H. (1996). Physico-chemical and biochemical characterization of biosurfactants released by *Lactobacillus* strains, *Colloid Surface* B 8, 51–61.

Viola C. M., Torres-Carro R., Cartagena E., Isla M. I., Alberto M. R., Arena M. E. (2018). Effect of Wine Wastes Extracts on the Viability and Biofilm Formation of *Pseudomonas aeruginosa* and *Staphylococcus aureus* strains. *Evidence-Based Complementary and Alternative Medicine*, 2018.

Walencka E., Różalska S., Sadowska B., Różalska B. (2008). The influence of *Lactobacillus acidophilus*-derived surfactants on staphylococcal adhesion and biofilm formation. *Folia Microbiologica*, 53(1), 61.

Walker J., Moore G. (2015). *Pseudomonas aeruginosa* in hospital water systems: Biofilms, guidelines, and practicalities. *Journal of Hospital Infection*, 89(4), 324–327.

Wang Y., Lee S. M., Dykes G. (2014). The physicochemical process of bacterial attachment to abiotic surfaces: Challenges for mechanistic studies, predictability and the development of control strategies. *Critical Reviews in Microbiology*, 41(4), 452–464.

Yang X., Wang H., He A., Tran F. (2018). Biofilm formation and susceptibility to biocides of recurring and transient *Escherichia coli* isolated from meat fabrication equipment. *Food Control*, 90, 205–211.

In: The Many Benefits of Lactic Acid Bacteria
Editors: J. G. LeBlanc and A. de Moreno

ISBN: 978-1-53615-388-0
© 2019 Nova Science Publishers, Inc.

Chapter 6

BIOACTIVE PEPTIDES GENERATED BY LACTIC ACID BACTERIA

Marcela P. Castro[1,2]*, Noelia Z. Palavecino Prpich[1,2]
and María E. Cayré[1]

[1]Laboratorio de Microbiología de Alimentos. Departamento de Ciencias Básicas y
Aplicadas, Universidad Nacional del Chaco Austral. Sáenz Peña, Chaco. Argentina.
[2]Consejo Nacional de Investigaciones Científicas y Técnicas (CONICET)
de la República Argentina.

ABSTRACT

Numerous chronic diseases, including heart disease, stroke, cancer, and diabetes, have a significant influence on the health of people worldwide. Along with numerous preventive and therapeutic drug treatments, important advances have been achieved in the identification of bioactive peptides (BP) that may contribute to long-term health. Biopeptides with various biological activities have been found in several food matrices and by-products. Since microbial fermentation is one of the mechanisms for BP generation, lactic acid bacteria (LAB) arise as an ideal cell factory for their production. This chapter reviews the generation and biofunctional properties of various BP derived from meat and meat products, milk and dairy products, together with some other food substrates as beans and pseudo-cereals, which had been generated by several strains of LAB. Many apprehensions regarding their functionality and health-claims have been also considered.

* Corresponding Author's Email: mcastro@uncaus.edu.ar.

INTRODUCTION

Bioactive peptides (BP) have been defined as specific protein fragments that have a positive impact on body functions and may influence health. They are encrypted within the sequence of the parent protein and can be released through hydrolysis by proteolytic enzymes. Hydrolysis can be performed either by endogenous enzymes in ripened foods, or by the combined action of endogenous and microbial enzymes when fermented, and/or by exogenous proteolytic enzymes (e.g., enzymes derived from microorganisms and plants) on isolated food proteins (Toldrá et al., 2016; Brown et al., 2017). Health benefits of BP are continuously being explored and discovered, hence, there are still controversies regarding their magnitude and significance. Lafarga & Hayes (2014) indicated that BP have antimicrobial, antioxidative, antithrombotic, antihypertensive, anticancerogenic, satiety regulating and immunomodulatory activities and may affect the cardiovascular, immune, nervous and digestive systems. Peptides may also be effective in the treatment of mental health diseases, cancer, diabetes and obesity. Furthermore, many known BP are multifunctional and can present two or more health promoting activities which may or may not be related (Di Bernardini et al., 2011).

Among hydrolysis processes used to generate BP, microbial fermentation is the one addressed herein by lactic acid bacteria (LAB). The proteolytic system of LAB consists of a three steps breakdown process which comprises: I) a cell envelope-associated proteinase (CEP) - a key factor since it is responsible for the first breakdown of high molecular weight proteins, such as casein in milk-, II) specialized transport systems enabling the uptake of the resulting peptides, and III) numerous intracellular peptidases, which degrade peptides to amino acids. The activity of these biofunctional peptides relies on their inherent amino acid composition and sequence. The size of active sequences may vary from 2 to 30 amino acid residues, which can have one function or multifunctional health-beneficial properties. Once released, the BP may provide different functions that can be reproduced *in vitro* with biochemical assays or *in vivo* in cell or animal models and humans. Bioactive peptides that are being discovered are reported in several databases (BIOPEP, PEPBANK, ERP-Moscow and Brainpeps, among others), together with data about their chemical and structural characteristics, IC50 (concentration necessary to inhibit 50% of enzymatic activity), protein of origin, etc. (Van Dorpe et al., 2012; Toldrá et al., 2018). The degree of proteolysis and the amount of generated BP depends on numerous variables including raw materials, type of enzyme activity, microbial population, and processing conditions. Up to the present time, the most studied biological functions have been antioxidant activity, and ACE-inhibitory activity, while antimicrobial activity, opioid activity, immunomodulation and antithrombotic activities have been lightly examined.

The production and commercialization of hydrolysates containing one or more BP have proved to be successful; however, many industrial obstacles limit the sales of nutraceutical or functional products. One of the main drawbacks is the regulation aspect and the application for a health claim authorization (Arroume et al., 2016). In this sense, foods fermented by LAB are of special interest to food manufacturers and consumers because of their additional probiotic health-promoting and therapeutic benefits.

Along with the advent of increasing analytical technology comes the discovery of 'functional foods' which -unsurprisingly- have been nurturing human beings since ancient times. Such is the case of a broad spectrum of food sources, i.e., animal, plant, microalgae and fungi, which have been delivering BP through microbial fermentation. Lactic acid bacteria comprise a widespread microbiota responsible for many of these fermentations. In the following sections, an overview of this topic is presented focusing primarily on the bioactivity displayed by the BP generated in each food substrate.

ANTIOXIDANT ACTIVITY

The mechanism of antioxidant activity of peptides is yet not sufficiently elucidated in the literature. Based on current knowledge, it is supposed that they may exert their antioxidant effect through the scavenging of free radicals and reactive oxygen species (ROS), inhibiting lipid peroxidation and chelating transition metal ions. Their action may be determined by factors such as position of the peptide in the protein structure, hydrophobicity, protein isolation method, degree of hydrolysis, type of enzymes used, concentration of peptide and amino acid configuration (Standik & Keska, 2015). The antioxidant activity of the peptides depends on the type of their component amino acids (Toldrá et al., 2018). As regards to amino acid types, aromatic amino acids such as *Tyr*, *His*, *Trp*, and *Phe* are prone to donate protons contributing to the radical-scavenging properties (Rajapakse et al., 2005). Besides, the hydrophobic amino acids have been described to be able to increase the presence of peptides at the water-lipid interface and then access to scavenge free radicals from the lipid phase (Ranathunga et al., 2006). Lastly, acidic amino acids utilize carbonyl and amino groups in the side chain which function as chelators of metal ions (Suetsuna et al., 2000).

In the last decade, several studies have evaluated the potential effect of LAB strains in the production of antioxidant peptides in different food matrices aiming at endorsing the health benefits associated with functional foods. Keska & Standik (2018) evaluated the stability of antiradical activity of protein extracts from LAB-inoculated dry-cured pork loins during long-term aging before and after simulated gastrointestinal digestion with pepsin and pancreatin. The microorganisms tested were *Lactobacillus* (*Lb.*) *rhamnosus* LOCK900, *Lb. acidophilus* BAUER L0938, and *Bifidobacterium* (*B.*) *animalis* ssp. *lactis* BB12. The highest antiradical effects were achieved for loins with *Lb. acidophilus* BAUER L0938 after 90 and 180 days of aging and *B. animalis* ssp. *lactis* BB12 in 270 days in the water soluble protein fraction. The highest biological activity was reached after 90 days of aging due to the hydrolytic degradation of proteins under the action of pepsin. During this period, *B. animalis* ssp. *lactis* BB12 showed higher values of biological activity. Mejri et al., (2017) evaluated the generation of antioxidant peptides in dry fermented camel sausages inoculated with different starters cultures: *Staphylococcus (S.) xylosus* and *Lb. plantarum; S. xylosus* and *Lb. pentosus*; *S. carnosus* and *Lb. sakei.* They observed a significant increase in the production of antioxidant peptides during camel sausage ripening for 28 days, particularly in the release of peptides of molecular weight above 3 kDa. The accumulation of antioxidant peptides achieved the highest antioxidant values at the end of ripening, measured by ABTS, DPPH, inhibition of hydroxyl radicals, and FRAP assays. This bacterial growth resulted in an increasing amount of peptides with molecular weights below 3 kDa. In general, antioxidant capacities in non-inoculated sausages were lower than the observed for inoculated sausages, especially for the sausage inoculated with *S. xylosus* and *Lb. plantarum.* In keeping with these results, Ayyash et al., (2019a) found that the smallest peptides (<3kDa) were the ones linked with the most potent antiradical activity; being these peptides mostly generated by *Lb. plantarum* KX881772 in camel meat dry sausages.

Regarding milk as a substrate, several authors reported antioxidant activity peptides generated by LAB. Aguilar-Toalá et al., (2017) described that the growth of *Lb. plantarum* strains in whole bovine milk led to the production of crude extracts rich in antioxidants. After the isolation of two fractions (<3 kDa and 3-10 kDa), these authors observed a strong relationship among small peptides (<3 kDa) with antioxidant activity measured by DPPH and ABTS assays. Nevertheless, they found that peptides of molecular weight in the range 3-10 kDa also contributed to antioxidant potential of fermented whole bovine milk. Moreover, other studies support the use of different LAB, such as *Lb. helveticus* (Elfahriv et al., 2016); *Lb. rhamnosus, Lb. paracasei,* and *Lb. casei* (Solieri et al., 2015), to induce the release of antioxidant peptides in milk. Mohanty et al., (2016) reported *Lb. delbrueckii* subsp. *bulgaricus* IFO13953 as a producer of an antioxidant peptide whose amino acid sequence was identified. Besides, antioxidant activities were reported from cow and camel milks fermented by specific strains of *Lb. plantarum, Lb. paraplantarum, Lb. kefiri, Lb. gasseri,* and *Lb. paracasei* (Soleymanzadeh et al., 2016). Ayyash et al., (2017) also reported antioxidant activity in camel and cow milks fermented with *Lb. plantarum* and *Lb. reuteri.* El-Fattah et al., (2016) reported ACE inhibition and antioxidant activities after milk fermentation in 14 commercial dairy starters, which exhibited different degrees of proteolysis.

On wheat and other cereals, antioxidant peptides were described as the result of LAB activity (Coda et al., 2012). Generally, the high proline content of gluten alpha gliadins prevents hydrolysis by enzymes of the gastrointestinal tract, whereas LAB possess proline-specific peptidase systems. Alternatively, peptide degradation was achieved by the combined action of peptidase from different LAB strains simultaneously present in fermented food (Gerez et al., 2008). Ayyash et al., (2019b) investigated antioxidant activities on solid-state fermented (SSF) whole grain lupin, quinoa and wheat using *Lb. reuteri* KX881777 and *Lb. plantarum* KX881779. All quinoa fractions had an increase in DPPH after 24 h of fermentation. *Lb. reuteri* had the highest DPPH % in quinoa fraction compared with lupin and wheat fraction. Likewise, quinoa fermented by *Lb. reuteri* and *Lb. plantarum* had the greatest ABTS % compared with lupin and wheat fraction. The higher antioxidant activities in fermented quinoa fraction may be attributed to the nature of proteolytic products released and to other non-phenolic compounds with antioxidant activities. The antioxidant results in lupin fraction were in agreement with those reported by Fernández-Orozco et al., (2008) who run a similar fermentation with a strain of *Lb. sakei*. Besides, the fermentation of quinoa flour by *Lb. plantarum* (22 strains), *Lb. rossiae* (1 strain) and *Pediococcus* (*Ped.*) *pentosaceus* (3 strains) produced fermented quinoa flour extracts of high antioxidant activity (Rizzello et al., 2017). The authors observed that antioxidant activity was strain-dependent wherein *Lb. plantarum* strains displayed the highest capacity to breakdown quinoa proteins to antioxidant. The antioxidant activities in quinoa fermented for 24 h by *Lb. plantarum* was attributed to short chain peptides (<9.0 kDa).

ANTIHYPERTENSIVE ACTIVITY

The BP with antihypertensive activity - which mainly takes place through inhibition of the angiotensin I-converting enzyme (ACE) - has become of particular interest to the scientific community over the last few years. Within the regulation of blood pressure, ACE is a key enzyme that catalyzes the conversion of angiotensin I into angiotensin II, a potent vasoconstrictor, and inactivates bradykinin, a vasodilator (Ondetti & Cushman 1977). The higher the ACE activity, the higher the angiotensin II level, and the more likely that malignant hypertension (high blood pressure) develops. Hypertension is one of the risk factors for coronary heart disease and stroke, being closely linked to the eating habits of industrialized countries (Lin et al., 2012).

Despite the fact that herein BP antioxidant and ACE inhibitory activities are presented separately, they usually appear together in the literature since the same sample fraction is implicated in both determinations. Hence, many of the studies aforementioned will sound familiar for the reader who had read the previous section.

The study of biopeptides generated in meat and meat products implies the acceptance of the presence of existent BP in the matrix before fermentation, hampering the determination of those BP generated exclusively by inoculated LAB. As an illustration, the increased ACE inhibition observed in fermented camel sausages suggested that the strain *Lb. plantarum* KX881772 improved the bioactivity of this meat product (Ayyash et al., 2019). The higher ACE inhibition in fermented camel sausages, compared with beef sausages, was attributed by the authors to a high degree of hydrolysis and/or the nature of camel meat peptides released during fermentation and storage. Mejri et al., (2017) also studied camel meat sausages whereas inoculated with different starter bacteria. They reported that the percentage of ACE inhibition was increased significantly in the whole extract and in all fractions during the 28 days of ripening. It could be noted that the highest ACE inhibition capacity was always observed for the fraction with peptides below 3 kDa. This assumption was in keeping with Escudero et al., (2013) and Mora et al., (2015) results, which stated that low molecular weight peptides (<3 kDa), contributing to the development of flavor in dry fermented meat products, could also exert antihypertensive activity. Moreover, ACE inhibition through ripening was significantly affected by the addition of a starter *(S. xylosus* and *Lb. plantarum).* These results were in accordance with those of Flores & Toldrá (2011) and Mora et al., (2015) who found that the action of microorganisms such as LAB could influence the last period of fermentation, contributing to the generation of small peptides and free amino acids.

Considering milk, the number of BP increases during the production of fermented milk products compared to raw milk and the composition of the peptide fraction changes due to the proteolytic action of the employed microorganisms (Hayes et al., 2007). Some studies investigated BP in traditional fermented milk products and identified, for instance, that the major ACE inhibitory peptide of Manchego cheese was attributed to the di-peptide FP, originated from hydrolysis of various casein fractions. A single peptide was responsible for the ACE inhibitory activity, which had an IC50 value of 592µm (Gómez-Ruiz et al., 2002). Casein, as well as whey milk proteins, in particular α-lactalbumin and β-lactoglobulin, are a rich source of ACE-inhibitory peptides. These BP are naturally present in dairy products, generated during processing and ripening due to endogenous enzymes and microbial fermentation; besides, they are also released from milk protein sequences during gastrointestinal digestion by gastric and pancreatic enzymes (Seppo et al., 2003). The most significant examples of BP produced by *Lactobacillus* species in cheese are the antihypertensive tripeptides, Ile-Pro-Pro (IPP) and Val-Pro-Pro (VPP) -also called lactotripeptides-, generated from casein hydrolysis by different *Lb. helveticus* strains (Pihlanto, 2013). Moreover, sour milk fermented by *Lb. helveticus* contained ACE inhibitory tri-peptides (IPP and VPP) that lowered blood pressure in mild hypertensive subjects (Fekete et al., 2015). Jin et al., (2016) and Rutella et al., (2016) reported the presence of antihypertensive BP in yogurts and fermented milk products which are already being marketed (Hafeez et al., 2014; Raveschot et al., 2018). The magnitude of proteolysis would depend on the strain used. A good example of this situation is the work reported by Ahn et al., (2009) where whey fermented by *Lb. brevis* had a stronger ACE inhibitory ability than those fermented with *Lb. acidophilus, Lb. bifermentans, Lb. casei, Lb. helveticus, Lb. paracasei, Lb. plantarum* and *Lb. reuteri.* Consequently, the functionality of protein hydrolysates may differ between cultures since microorganisms have different proteolytic systems. Even within the same species, different extent of hydrolysis, titratable acidity, free amino nitrogen and ACE inhibitory activities can be expected, as reported in skim milk fermented with 37 different *Lb. helveticus* strains (Chen et al., 2014).

In addition to the milk products already mentioned, kefir comprises another source of BP. Quirós et al., (2005) revealed two ACE inhibitory peptides in caprine kefir. Compared to some other fermented dairy products, kefir showed a higher rate of proteolysis and could, therefore, be a good source of physiologically active peptides (Ebner et al., 2015). Results from the latter study demonstrated that fermentation caused the additional release of peptides compared to unfermented milk. It could be concluded that the fermentation process improved the health promoting effects of milk that arose from the presence of bioactive peptides, namely 236 of the identified peptides were unique for kefir. Consequently, not only the concentration of bioactive peptides may increase during the fermentation process, but the bioactivity profile may also be completely different in kefir compared to milk. While the milk peptides are generated mainly by the action of endogenous proteinases like plasmin, cathepsins B, D and G, microbiological degradation plays a key role in fermented products. Two general specificity classes of proteinases are reported for lactococci, which are merely one part of the complex kefir microflora. As reported by Ebner et al., (2015), two of the identified bioactive peptides were of particular importance, i.e., casecidin 17 -a multifunctional peptide exerting antimicrobial, ACE-inhibitory, immunomodulating, antioxidant, and antithrombotic activities (Birkemo et al., 2009; Rojas-Ronquillo et al., 2012), and the peptide V-β-casomorphin-9 another multifunctional peptide showing ACE-inhibitory, antioxidant, and opioid activity (Eisele et al., 2013).

Soybean proteins have also been used as a source of ACE inhibitory proteins. According to Daliri et al., (2018) soy protein isolates consists of 90% protein; their components being 34% glycininand and 27% β-conglycinin. The many benefits of soybean proteins to reduce cholesterol levels (Hoie et al., 2007), reduce the risk of cardiovascular disease (Xiao, 2008), reduce the risk of breast cancer (Nagata et al., 2007), and lower blood pressure (Aluko, 2015) are well established and correlated with intensive studies. Vallabha & Tiku (2014) used a *Lb. casei* spp. *pseudoplantarum* strain to produce fermented soy protein concentrates. Their results showed a strong ACE inhibitory potency (IC50 = 3.7 mg/mL) of the fermented concentrate, while its isolated ACE inhibitory peptide LIVTQ displayed a higher potency (IC50 = 0.087 µM). ACE inhibitory peptides (IFL and WL) isolated from tofu also displayed strong ACE inhibitory abilities (IC50 values of 44.8 and 29.9 µM, respectively) indicating their potential to reduce high blood pressure (Kuba et al., 2003). Moreover, fermentation *Ped. pentosaceus* SDL1409 on soybean protein isolates for 48 h at 37°C generated a strong angiotensin I-converting enzyme inhibitory ability of 65.1 ± 0.78% (IC50 = 0.123 ± 0.02 mg protein/mL). In this sense, Daliri et al., (2018) demonstrated that *Ped. pentosaceus* SDL 1409 could be useful for developing antihypertensive functional foods from soybean proteins.

EMERGING BIOACTIVITIES

In addition to both main health-promoting activities previously mentioned, several other bioactivities have been related to peptides generated by LAB fermentation. Hypocholesterolemic and antithrombotic peptides have been described to modulate the cardiovascular system whereas mineral binding and immunomodulatory peptides act in gastrointestinal and immune systems, respectively. Some groups of peptides are able to participate in multiple system reactions. Thus, opioid agonist and antagonists can act on nervous, gastrointestinal, and immune systems, whereas antimicrobial peptides can modulate gastrointestinal and immune systems (Lafarga & Hayes, 2014). The main problem of naturally generated peptides is the difficulty in controlling the hydrolysis conditions because many endogenous enzymes are acting at the same time and a wide profile of peptides showing different sizes and characteristics is generated (Mora et al., 2015). Despite the well documented evidence supporting these peptides bioactivities, trials were mainly conducted *in vitro,* giving scarce information about the availability and functionality of BP *in vivo.* For this reason, only few studies herein have been selected to illustrate the emerging bioactivities that had been investigated until the present time, taking into consideration that this scenario will evolve quickly, and that deeper comprehensive investigations would have to enrich this description in the near future.

The antithrombotic peptide YQEPVLGPVRGPFPIIV was generated by *Lb. casei* from bovine casein. This peptide not only presented antithrombotic activity but also inhibited ACE when tested *in vitro.* Besides, the peptide was able to maintain its bioactivity after enzymatic treatment with pepsin and trypsin (Rojas-Ronquillo et al., 2012). Very few meat-derived antithrombotic peptides have been described in the literature, being the few ones derived from the hydrolysis of papain, namely not from bacterial activity (Morimatsu et al., 1996; Shimizu et al., 2009).

Beta-caseins (β-CNs) harbor β-casomorphins (β-CMs) which are opioid peptides encrypted in an inactive form and can be released by enzymatic hydrolysis during fermentation and *in vivo* or *in vitro* gastrointestinal digestion (Asledottir et al., 2017; De Noni et al., 2015). Beta-CN fragments, β-CM5 (Tyr-Pro-Phe-Pro-Gly) and β-CM7 (Tyr-Pro-Phe-Pro-Gly-Pro-Ile) have strong opioid activity and are currently being extensively studied for their biological activity and their occurrence in a wide range of dairy products (Boutrou et al., 2013; De Noni et al., 2015; Nguyen et al., 2015). Nguyen et al., (2018) found that there were significant differences between the peptides generated by either *Streptococcus (Strep.). thermophillus* or *Lb. delbrueckii* ssp. b*ulgaricus* fermenting UHT milk alone. They also determined that even the peptide profile generated by both strains as a mixed culture had a different peptide composition. These findings suggested that a new peptide was released by *Lb. delbrueckii* ssp. b*ulgaricus* at the expense of an unknown enzyme derived from *Strep. thermophilus.*

Immunomodulating activity has been especially detected in peptides derived from milk and dairy products. These studies show that their length can be very different comprising from 2 to 64 amino acids, although those smaller than 3 KDa are the most abundant (Reyes-Díaz et al., 2016). Milk fermented by *Lb. helveticus* showed immunomodulating effect after administration to mice, whereas this bioactivity was not observed when mice were treated with milk fermented by a non-proteolytic variant of the same *Lb. helveticus* strain (Matar et al., 2001). Additionally, the possible impact of milk fermented by *Lb. helveticus* on sleep efficiency in elderly subjects was demonstrated in a clinical study (Yamamura et al., 2009). A beneficial effect of fermented milk on bone metabolism was observed in animal or human clinical studies, probably because of the high calcium concentration and bioavailability in these products (Rizzoli & Biver, 2017). A previous work by Kim and coworkers (2009) showed a positive influence on bone metabolism in ovariectomized rats exerted by milk fermented with *Lb. casei* 393. Seemingly, Narva et al., (2004 a,b) observed improvements in bone metabolism of rats fed with milk fermented by *Lb. helveticus* strains, and even on calcium metabolism in postmenopausal women.

CONCLUSION

Despite the relatively inexpensive production cost of BP by LAB species compared to the use of purified enzymes, there are still some challenges for further industrial exploitation of these bacteria; mainly the poor characterization of CEPs from different species and the low yields of peptide production by fermentation. Future and deeper studies employing LAB as BP factories have the potential to generate new peptide sequences and, therefore, new biopeptides with specific biological properties. According to Raveschot et al., (2018), metabolic engineering of the proteolytic system of LAB could overcome these challenges.

A few functional, peptide-based, products have already been launched on the market. Most of these were derived from milk, plant and marine sources. This trend is likely to continue as knowledge about the bioactivities and properties of peptides increases (Lafarga & Hayes, 2014). However, a word of caution is to be addressed on BP as a whole, since there are yet no general rules for determining the best hydrolysate with the most important activity. Each raw material, each enzyme and each hydrolysis conditions will give different final peptides. Hence, health claims and beneficial effects for the general population must be considered as 'potential advantages' more than 'effective health promoters.'

REFERENCES

Aguilar-Toalá J., Santiago-López L., Peres C. M., Peres C., Garcia H., Vallejo- Cordoba B., Hernández-Mendoza A. (2017). Assessment of multifunctional activity of bioactive peptides derived from fermented milk by specific *Lactobacillus plantarum* strains. *Journal of Dairy Science*, 100(1), 65–75.

Ahn J., Park S., Atwal A., Gibbs B., Lee B. (2009). Angiotensin I-converting enzyme (ACE) inhibitory peptides from whey fermented by *Lactobacillus* species. *Journal of Food Biochemistry*, 33, 587–602.

Aluko R. E. (2015). Antihypertensive peptides from food proteins. *Annual Review of Food Science and Technology*, 6, 235–262.

Arroume N., Froidevaux R., Kapel R., Cudennec B., Ravallec R., Flahaut C., et al., (2016). Food peptides: Purification, identification and role in the metabolism. *Current Opinion in Food Science, 7*, 101–107.

Asledottir T., Le T. T., Petrat-Melin B., Devold T. G., Larsen L. B., Vegarud G. E. (2017). Identification of bioactive peptides and quantification of b-casomorphin-7 from bovine b-casein A1, A2 and I after ex vivo gastrointestinal digestion. *International Dairy Journal*, 71, 98–106.

Ayyash M., Al-Nuaimi A., Al-Mahadin S., Liu S. (2017). In vitro investigation of anticancer and ACE-inhibiting activity, α-amylase and α -glucosidase inhibition, and antioxidant activity of camel milk fermented with camel milk probiotic: a comparative study with fermented bovine milk. *Food Chemestry*, 239, 588–597.

Ayyash M., Liu S., Al Mheiri A., Aldhaheri M., Raeisi B., Al-Nabulsi A., Osaili T., Olaimat A. (2019a). In vitro investigation of health-promoting benefits of fermented camel sausage by novel probiotic *Lactobacillus plantarum*: A comparative study with beef sausages. *LWT - Food Science and Technology*, 99, 346–354.

Ayyash M., Johnson S., Liu S., Mesmari N., Dahmani S., Al Dhaheri A., Kizhakkayil J. (2019b). In vitro investigation of bioactivities of solid-state fermented lupin, quinoa and wheat using *Lactobacillus*spp. *Food Chemistry*, 275, 50–58.

Birkemo G. A., O'Sullivan O., Ross R. P., Hill C. (2009). Antimicrobial activity of two peptides casecidin 15 and 17, found naturally in bovine colostrum. *Journal Applied Microbiology*, 106(1), 233–40.

Boutrou R., Gaudichon C., Dupont D., Jardin J., Airinei., G., Marsset-Baglieri,, M. Benamouzig R., Tomé, D., Leonil J. (2013). Sequential Release of Milk Protein-Derived Bioactive Peptides in the Jejunum in Healthy Humans. *American Journal of Clinical Nutrition*, 97, 1314–1323.

Brown, L. Vera Pingitore E., Mozzi F., Saavedra L., Villegas J., Hebert E. (2017). Lactic Acid Bacteria as Cell Factories for the Generation of Bioactive Peptides. *Protein & Peptide Letters*, 2017, 24, 1-10.

Chen Y., Liu W., Xue J., Yang J., Chen X., Shao Y., Kwok L. Y., Bilige M., Mang L., Zhang H. (2014). Angiotensin-converting enzyme inhibitory activity of *Lactobacillus helveticus* strains from traditional fermented dairy foods and antihypertensive effect of fermented milk of strain H9. *Journal of Dairy Science*, 97, 6680–6692.

Coda R., Rizzello C., Pinto D., Gobbetti M. (2012). Selected lactic acid bacteria synthesize antioxidant peptides during sourdough fermentation of cereal flours. *Applied and Environmental Microbiology*, 78(4), 1087–1096.

Daliri E. B.-M., Lee B. H., Park M. H., Kim J-H, Oh D-H. (2018). Novel angiotensin I-converting enzyme inhibitory peptides from soybean protein isolates fermented by *Pediococcuspentosaceus*SDL1409. *LWT - Food Science and Technology,* 93, 88–93.

De Noni I., Stuknytė M., Cattaneo S. (2015). Identification of β-casomorphins 3 to 7 in cheeses and in their in vitro gastrointestinal digestates. *LWT -Food Science and Technology,* 63, 550–555.

Di Bernardini R., Harnedy P. Bolton D., Kerry J., O'Neill E., Mullen A. M., Hayes M. (2011). Antioxidant and antimicrobial peptidic hydrolysates from muscle protein sources and by-products. *Food Chemistry*, 124(4), 1296–1307.

Ebner J., AşçiArslan A., Fedorova M., Hoffmann R., Küçükçetin A., Pischetsrieder M. (2015). Peptide profiling of bovine kefir reveals 236 unique peptides released from caseins during its production by starter culture or kefir grains. *Journal of Proteomics*, 117, 41–57.

Eisele, T. Stressler, T. Kranz, B. Fischer, L. (2013). Bioactive peptides generated in an enzyme membrane reactor using *Bacillus lentus* alkaline peptidase. *European Food Research Technology*, 236(3), 483–90.

Elfahri K., Vasiljevic T., Yeager T., Donkor O. (2016). Anti-colon cancer and antioxidant activities of bovine skim milk fermented by selected *Lactobacillus helveticus* strains. *Journal of Dairy Science*, 99 (1), 31–40.

El-Fattah A, Sakr S., El-Dieb S., Elkashef H. (2016). Angiotensin-converting enzyme inhibition and antioxidant activity of commercial dairy starter cultures. *Food Science Biotechnology*, 25, 1745–1751.

Escudero E., Mora L., Fraser P. D., Aristoy M. C., Arihara K., Toldrá F. (2013). Purification and identification of antihypertensive peptides in Spanish dry-cured ham. *Journal of Proteomics*, 78, 499–507.

Fekete Á., Givens D., Lovegrove J. (2015). Casein-derived lactotripeptides reduce systolic and diastolic blood pressure in a meta-analysis of randomised clinical trials. *Nutrients*, 7(1), 659–681.

Fernandez-Orozco R., Frias J., Muñoz R., Zielinski H., Piskula M., Kozlowska H., Vidal-Valverde C. (2008). Effect of fermentation conditions on the antioxidant compounds and antioxidant capacity of Lupinusangustifolius cv. Zapaton. *European Food Research and Technology*, 227(4), 979–988.

Flores M., Toldra F.(2011). Microbial enzymatic activities for improved fermented meats. *Trends in Food Science & Technology*, 22, 81-90.

Gerez, C. Cuezzo S., Rollán G. Font de Valdez G. (2008). *Lactobacillus reuteri* CRL 1100 as starter culture for wheat dough fermentation. *Food Microbiology*, 25, 253–259.

Gómez-Ruiz J., Ramos M., Recio I. (2002). Angiotensin-converting enzyme-inhibitory peptides in Manchego cheeses manufactured with different starter cultures. *International Dairy Journal*, 12(8):697–706.

Hafeez Z., Cakir-Kiefer C., Roux E., Perrin C., Miclo L., Dary-Mourot A. (2014). Strategies of producing bioactive peptides from milk proteins to functionalize fermented milk products. *Food Reserch International*, 63, 71–80.

Hayes M., Stanton C., Fitzgerald G., Ross R. (2007). Putting microbes to work: dairy fermentation, cell factories and bioactive peptides. Part II: bioactive peptide functions. *Biotechnology Journal*, 2(4), 435–449.

Hoie L., Guldstrand M., Sjoholm A., Graubaum H., Gruenwald J., Zunft H., et al., (2007). Cholesterol-lowering effects of a new isolated soy protein with high levels of non denaturated protein in hypercholesterolemic patients. *Advances in Therapy*, 24(2), 439–44.

Jin, Y., Yu Y., Qi Y., Wang F., Yan J., Zou H. (2016). Peptide profiling and the bioactivity character of yogurt in the simulated gastrointestinal digestion. *Journal of Proteomics*, 141, 24–46.

Keska, P., Stadnik, J. (2018). Stability of Antiradical Activity of Protein Extracts and Hydrolysates from Dry-Cured Pork Loins with Probiotic Strains of LAB. *Nutrients*, 10, 521. doi:10.3390/nu10040521.

Kuba M., Tanaka K., Tawata S., Takeda Y., Yasuda M. (2003). Angiotensin I-converting enzyme inhibitory peptides isolated from tofuyo fermented soybean food. *Bioscience, Biotechnology, and Biochemistry*, 67(6), 1278–1283.

Lafarga T., Hayes M. (2014). Bioactive peptides from meat muscle and by-products: generation, functionality and application as functional ingredients. *Meat Science*, 98, 227–239.

Lin L., Lv S., Li B. (2012). Angiotensin-I-converting enzyme (ACE)-inhibitory and antihypertensive properties of squid skin gelatin hydrolysates. *Food Chemistry*, 131, 225-230.

Mejri L., Vásquez-Villanueva R., Hassouna M., Marina M., García M. (2017). Identification of peptides with antioxidant and antihypertensive capacities by RP-HPLC-Q-TOF-MS in dry fermented camel sausages inoculated with different starter cultures and ripening times. *Food Research International*, 100, 708–716.

Mohanty D., Mohapatra S., Misra S., Sahu P. (2016). Milk derived bioactive peptides and their impact on human health – A review. *Saudi Journal of Biological Sciences* 23, 577–583.

Mora L., Escudero E., Aristoy M. C., Toldrá F. (2015). A peptidomic approach to study the contribution of added casein proteins to the peptide profile in Spanish dry-fermented sausages. *International Journal of Food Microbiology*, 212, 41–48.

Morimatsu F., Ito M., Budijanto S., Watanabe I., Furukawa Y., Kimura S. (1996). Plasma cholesterol-suppressing effect of papain-hydrolyzed pork meat in rats fed hypercholesterolemic diet. *Journal of Nutrition Science and Vitaminology*, 42(2), 145–153.

Nagata Y., Sonoda T., Mori M., Miyanaga N., Okumura K., Goto K., et al., (2007). Dietary isoflavones may protect against prostate cancer in Japanese men. *Journal of Nutrition*, 137(8), 1974–1979.

Nguyen D. D., Busetti F., Johnson S. K., Solah V. A. (2015). Identification and quantification of endogenous beta-casomorphins in Australian milk by LC-MS/MS and LC-HRMS. *Journal of Food Composition and Analysis*, 44, 102–110.

Nguyen D. D., Busetti F., Johnson S. K., Solah V. A. (2018). Degradation of β-casomorphins and identification of degradation products during yoghurt processing using liquid chromatography coupled with high resolution mass spectrometry. *Food Research International,* 106, 98–104.

Ondetti M., Cushman D. (1977). Design of specific inhibitors of angiotensin converting enzyme: New class of orally active antihypertensive agents. *Science,* 196, 441-444.

Pihlanto A. (2013). "Lactic fermentation and bioactive peptides," in *Lactic Acid Bacteria – R & D for Food, Health and Livestock Purposes*, M. Kongo (Ed.), Rijeka: InTech Prepress, 310–331.

Quirós A., Hernández-Ledesma B., Ramos M., Amigo L., Recio I. (2005). Angiotensin-converting enzyme inhibitory activity of peptides derived from caprine kefir. *Journal Dairy Science*, 88(10), 3480–3487.

Rajapakse N., Mendis E., Jung W., Je J. Kim S. (2005). Purification of a radical scavenging peptide from fermented mussel sauce and its antioxidant properties. *Food Research International,* 38, 175–182.

Ranathunga S., Rajapakse N., Kim S. (2006). Purification and characterization of antioxidative peptide derived from muscle of conger el (Conger myriaster). *European Food Research and Technology*, 222, 310–315.

Raveschot C., Cudennec B., Coutte F., Flahaut C., Fremont M., Drider D., Dhulster P. (2018). Production of Bioactive Peptides by *Lactobacillus Species*: From Gene to Application. *Frontiers in Microbiology*, 9, 2354. doi: 10.3389/fmicb.2018.02354.

Rizzello C., Lorusso A., Russo V., Pinto D. Marzani B., Gobbetti M. (2017). Improving the antioxidant properties of quinoa flour through fermentation with selected autochthonous lactic acid bacteria. *International Journal of Food Microbiology*, 241, 252–261.

Rojas-Ronquillo R., Cruz-Guerrero A., Flores-Nájera A., Rodríguez-Serrano G., Gómez-Ruiz L., Reyes-Grajeda J., Jiménez-Guzmán J., García-Garibay M. (2012).

Antithrombotic and angiotensin-converting enzyme inhibitory properties of peptides released from bovine casein by *Lactobacillus casei* Shirota. *International Dairy Journal*, 26(2), 147–154.

Rutella G., Solieri L., Martini S., Tagliazucchi D. (2016). Release of the antihypertensive tripeptides valine-proline-proline and isoleucine-prolineproline from bovine milk caseins during in vitro gastrointestinal digestion. *Journal of Agriculture and Food Chemistry*, 64, 8509–8515.

Seppo L., Jauhiainen T., Poussa T., Korpela R. (2003). A fermented milk high in bioactive peptides has a blood pressure lowering effect in hypertensive subjects. *American Journal of Clinical Nutrition*, 77, 326-330.

Shimizu M. Sawashita N., Morimatsu F., Ichikawa J., Taguchi Y., Ijiri Y., Yamamoto J. (2009). Antithrombotic papain-hydrolyzed peptides isolated from pork meat. *Thrombosis Research*, 123(5), 753–757.

Soleymanzadeh N., Mirdamadi S., Kianirad M. (2016). Antioxidant activity of camel and bovine milk fermented by lactic acid bacteria isolated from traditional fermented camel milk (Chal*). Dairy Science Technology*, 96, 443–457.

Solieri L., Rutella G., Tagliazucchi D. (2015). Impact of non-starter lactobacilli on release of peptides with angiotensin-converting enzyme inhibitory and antioxidant activities during bovine milk fermentation. *Food Microbiology*, 51, 108–116.

Stadnik J., Kęska P. (2015). Meat and fermented meat products as a source of bioactive peptides. *Acta ScientiarumPolonorumTechnologia Alimentaria*, 14(3), 181–190.

Suetsuna K., Ukeda H., Ochi H. (2000). Isolation and characterization of free radical scavenging activities peptides derived from casein. *Journal of Nutritional Biochemistry*, 11, 128–131.

Toldrá F., Mora L., Reig M. (2016). New insights into meat by-product utilization. Meat *Science*, 120, 54–59.

Toldrá F., Reig M., Aristoy M., Mora L. (2018). Generation of bioactive peptides during food processing. *Food Chemistry*, 267, 395–404.

Vallabha V. S., Tiku P. K. (2014). Antihypertensive peptides derived from soy protein by fermentation. *International Journal of Peptide Research and Therapeutics*, 20(2), 161–168.

Van Dorpe S., Bronselaer A., Nielandt J., Stalmans S., Wynendaele E., Audenaert K., Van De Wiele C., Burvenich C., Peremans K., Hsuchou H., De Tré G., De Spiegeleer B. (2012) Brainpeps: the blood-brain barrier peptide database. *Brain Structure & Function*, 217, 687-718.

Xiao C. W. (2008). Health effects of soy protein and isoflavones in humans. *Journal of Nutrition*, 138(6), 1244S–1249S.

In: The Many Benefits of Lactic Acid Bacteria
Editors: J. G. LeBlanc and A. de Moreno

ISBN: 978-1-53615-388-0
© 2019 Nova Science Publishers, Inc.

Chapter 7

ANTIMICROBIAL BACTERIOCINS AND PEPTIDOGLYCAN HYDROLASES: BENEFICIAL METABOLITES PRODUCED BY LACTIC ACID BACTERIA

Adriana López-Arvizu[1], Israel García-Cano[2],
María de Lourdes Pérez-Chabela[1] and Edith Ponce-Alquicira[1,]*

[1]Departamento de Biotecnología, Universidad Autónoma Metropolitana -
Unidad Iztapalapa, Ciudad de México, Mexico
[2]Department of Food Science and Technology,
The Ohio State University, Columbus, US

ABSTRACT

Lactic acid bacteria (LAB) involve a diverse group of gram-positive bacteria that primarily produce lactic acid. For centuries LAB have been recognized as safe and employed for food processing due to their ability to control the adventurous microbiota in the elaboration of dairy products, vegetables, meats, coffee, cocoa, silages and fermented beverages; but also because of their contribution to the taste, flavor and texture. Moreover, probiotic LAB may offer health benefits by acting in the modulation of the immune system, helping to manage allergies and lactose intolerance; in addition to the prevention of gastrointestinal and urinary infections by reduction of pathogens. The ability of LAB to inhibit spoilage and pathogenic microorganisms is based on several mechanisms that include acidification, production of antimicrobials and competition for nutrients and adhesion niches. Antimicrobials include the lactic and acetic acids, but also a complex range of metabolites like hydrogen peroxide, diacetyl, antifungal short-chain

---

* Corresponding Author's E-mail: pae@xanum.uam.mx.

fatty acids, pyroglutamic acid, as well as, antimicrobial peptides known as bacteriocins and antibacterial peptidoglycan hydrolases (PGHs). Bacteriocins refer to a diverse group of ribosomal antimicrobial peptides mainly active against closely-related and sensitive bacterial strains by formation of pores in the cytoplasmic membrane causing reduction of microbial competitors. Among them, nisin and other bacteriocins have been applied in the food and pharmaceutical industry to control *Listeria monocytogenes* and *Staphylococcus aureus*. On the other hand, PGH are proteins capable to cleave the peptidoglycan cell wall of gram-positive and gram-negative bacteria, examples include lysostaphin and lysozyme, which are highly active against *Staphylococcus aureus*. The objective of this chapter is to offer an overview of the characteristics and applications of bacteriocins and peptidoglycan hydrolases as part of the beneficial metabolites produced by LAB.

INTRODUCTION

Lactic acid bacteria can be found in several habitats such as the gastrointestinal tract of animals and humans, as well as in the microbiota in several foods. This microbial group comprises gram-positive, facultative anaerobic, non–spore forming, non-motile and acid tolerant bacteria from the genus *Aerococcus*, *Alloiococcus*, *Carnobacterium*, *Enterococcus*, *Lactobacillus*, *Leuconostoc*, *Oenococcus*, *Pediococcus*, *Streptococcus*, *Tetragenococcus*, *Vagococcus* and *Weissella*. Those are low G+C (31 - 49%) and belong to the *Lactobacillales* order within the *Firmicutes* phyla. Besides, the *Bifidobacterium* genus with a high G+C content (58 - 61%) belongs to the *Actinobacteria* phylum (Khalid 2011; Giraffa, 2014; Fraqueza, 2015). LAB are also classified into homofermentative and heterofermentative according to the end products derived from the glucose metabolism into lactic acid, carbon dioxide, ethanol or acetic acid.

LAB have been safely used for centuries and therefore recognized as safe (GRAS and QPS qualified presumption of safety) by the FDA and EFSA authorities. They are widely employed as starter cultures in the food industry to accelerate ripening or to control de adventitious microbiota for elaboration and preservation of several fermented foods. Applications include the processing of fermented milks, chesses, vegetables, meats, coffee, cocoa, silages, sourdough bread and wine, among others. But, LAB also contribute to the taste, flavor and texture as a result of several reactions, including lipolysis, proteolysis, conversion of lactose in citrate and various aromatic compounds, such as diacetyl, acetoin, acetaldehyde and acetic acid. In addition, their proteolytic activity induces the accumulation of small peptides and free amino acids that are further transformed into alcohols, aldehydes, acids, and esters responsible for the flavor profile and organoleptic characteristics of fermented foods. Some LAB can produce a range from 10 to > 2,000 kDa exopolysaccharides that not only confer protection to the cell producer, but may improve texture and mouthfeel of yogurt and other low-fat milk products. Moreover, LAB are industrially involved in the production of some macromolecules and enzymes (Gad, Abdel-Hamid & Farag, 2014; Özogul & Hamed, 2016).

A number of LAB strains have been used as probiotics and considered desirable members of the intestinal microbiota, including *Lactobacillus*, *Bifidobacterium*, *Propionibacterium* and *Enterococcus* strains present in foods and in dietary supplements. Other reported health benefits associated to LAB probiotics include an antihypertensive effect, reduction in the cholesterol level, antioxidant effect, relief of the allergy symptoms, decrease in dental caries and protection against colon cancer. Such, as the production of the antihypertensive angiotensin-converting enzyme produced through the proteolytic system of *Lb. helveticus*, *Lb. acidophilus* and *Lb. delbrueckii* (Giraffa, 2014; Sharma et al., 2014). Therefore, LAB are also recognized to improve healthiness due to their influence on the immune system and reduction of pathogens, in addition to their positive role in the control of some infections during pregnancy, antibiotic-derived diarrheas and intestinal inflammation, or to alleviate some allergies, as well as, to deliver vaccines and other metabolites directly in the gastrointestinal tract (Sharma et al., 2014).

Among all these beneficial metabolites produced by LAB, a major attribute is their ability to inhibit spoilage and pathogenic microorganisms based on the competition, their great acid tolerance and production of several antimicrobials (Khalid, 2011; Giraffa, 2014). These antimicrobial metabolites include lactic and acetic acids, ethanol, hydrogen peroxide, diacetyl, antifungals (short chain fatty acids derived from lipolysis reactions), and antimicrobial peptides known as bacteriocins and other antibacterial proteins like peptidoglycan hydrolases (PGHs) (Khalid, 2011; Giraffa, 2014; Sharma et al., 2014). The last two metabolites, bacteriocins and PGHs constitute an important and viable alternative to antibiotics and food grade preservatives due to their antimicrobial action against several pathogens including multi-antibiotic-resistant strains (Álvarez-Cisneros & Ponce-Alquicira 2018).

BACTERIOCINS

Most bacteria may produce antimicrobial proteins, which can be ribosomally synthetized or non-ribosomally synthetized (Chikindas et al., 2018). Bacteriocins produced by LAB constitute a heterogeneous group of ribosomal antimicrobial peptides, varying in size, structure and specificity. They are active against closely related species found in the same environmental niche where the producer strain is protected by the presence of specific immune proteins (O'Connor et al., 2015). Genes responsible for their production and immunity are usually grouped in the same cluster (Perez, Zendo and Sonomoto 2014; O'Connor et al., 2015; Chikindas et al., 2018; Field, Ross & Hill, 2018). Several studies have demonstrated the potential of LAB bacteriocins to be applied for food preservation, or an alternative antibiotics in the clinical area, mainly due to their tolerance to high temperatures and their activity over a wide pH range against spoilage microorganisms and pathogens such Bacillus, Clostridia, *Listeria (L.) monocytogenes* and methicillin-resistant *Staphylococcus (S.) aureus* and vancomycin-resistant *Enterococci* at concentrations lower than that for conventional antibiotics (Scheneider et al., 2006; Alvarez-Cisneros et al., 2010; Martin-Visscher et al., 2011; Perez, Zendo & Sonomoto, 2014). Generally, LAB bacteriocins are considered food-grade because the GRAS or QPS status of most bacteriocinogenic strains (Perez, Zendo & Sonomoto, 2014). Bacteriocins do not impart color or flavor and are easily degraded by the proteolytic enzymes, especially those formed in the pancreas (trypsin and chymotrypsin) and gastric tract (pepsin) that makes them safe for human consumption, thus minimizing the opportunity of target strains to interact and develop resistance (Zacharof & Lovitt, 2012). However, bacteriocins may have some limiting factors such as a restricted inhibitory spectrum, lack of purification protocols and/or acceptable recovery yields, a producer strain classified with non GRAS recognition, and/or the development of bacteriocin-resistant bacteria. Nonetheless, the ribosomal origins of bacteriocins enable their bioengineering either to modulate their physicochemical properties and improve their stability in food systems, or to enhance the inhibiting activity and spectra; in adidition, their heterologous expression in other hosts could be a valid alternative to increase yield or induce multiple bacteriocin production (Alvarez-Cisneros et al., 2011; Field, Ross & Hill, 2018).

Bacteriocins are diverse, some authors such as Chikindas et al., (2018) point that the primary action of bacteriocins are for signaling and repelling at their low concentration that occurs in natural environments; for instance, nisin (produced by *Lactococcus lactis)* and subtilin (from *Bacillus subtilis*) participate as modulators for their own production and transport (Balciumas et al., 2013). But, mutacin V produced by *Streptococcus (Strep.) mutants* can act as lytic agent against the producer cell to facilitate the exchange of genetic material with non-lysed competent cells. While, other bacteriocins can inhibit the quorum sensing at sub-MBC (minimal bactericidal concentration) to prevent the settle down and biofilm formation of competitors. In contrast, when bacteriocins are employed at MBC or at higher levels they can induce permeation and formation of pores in the cytoplasmic membrane and/or interfere with the cell division process and with other biological functions thus inactivating sensitive cells (Chikindas et al., (2018).

A restriction in the use of bacteriocins may be the presence of innate or acquired resistance observed in some susceptible strains. In particular, *L. monocytogenes* has shown resistance when exposed to high levels of nisin (> 10 fold MBC), as well as resistance to class IIa bacteriocins due to changes in the cell membrane properties. Also, *Bacillus (B.) cereus* can produce the enzyme nisinase capable to degrade the bacteriocins nisin and subtilin. However, no significant data is available for the resistance mechanism and the resistance can be solved by changing the pH, the type of carbohydrates o the salt concentration of the environmental growth (Ahmad et al., 2017; Balay, Gänzle & McMullen, 2018).

The number of reported bacteriocins is continuously growing, but to date the BACTIBASE, an open-access database (http://bactibase.pfba-lab-tun.org), contains calculated or predicted physicochemical and structural properties of 230 bacteriocins produced by 206 gram-positive bacteria and 19 gram-negative, being LAB the predominant bacteriocin producers. The database allows a better comprehension of the bacteriocins functionality towards the development of applications and classification by the use of tools such as similarity, sequence alignment, physicochemical profile, hidden Markov Models and structure prediction. In addition, the BAGEL4 webserver (http://bagel4.molgenrug.nl/) allows the meta-genomic DNA search for bacteriocins and post-translationally modified peptides (RIPPs) providing databases and BLAST (basic local alignment search tool) for the core peptide databases.

Bacteriocin Classification

There are several classifications of bacteriocins, mainly based in their post-translationally modification through biosynthesis and transport to the extracellular media, as well as, on their mode of action and structure. Most bacteriocins are synthetized as an inactive precursor peptide precursor having an N-terminal leader peptide, attached to the C-terminal core peptide that is removed by proteases to yield the mature bacteriocin. The leader peptide usually serves as a recognition sequence for biosynthetic enzymes; also serves to modulate the enzyme–substrate interaction for the post-translational modification (PTM) of the inactive pre-peptide that is later transported to the cell surface and transformed into the active peptide. The genetic determinants for bacteriocins production consist of less than 21 genes and their biosynthesis greatly differs, as some bacterial strains may produce more than one bacteriocin in addition to the production of PGH's (Acedo, et al., 2018).

The regulation system for the bacteriocin production is composed of an inducing peptide or pheromone–activating factor, the transmembrane histidine kinase and a response regulator. The inducer peptide is synthetized in the ribosome at low levels as a pre-peptide, which is then cleaved and secreted extracellularly by a carrier. At certain threshold concentrations the response phosphorylation receptor further activates the bacteriocin transcription. Lantibiotics (Class I) bacteriocins such nisin act as a pheromones or signal-involved membrane proteins that regulate their own production and transport; while accessory proteins induce the translocation for Class II bacteriocins (Balciunas et al., 2013).

The first bacteriocin classification scheme was proposed by Klaenhammer (1993) who defined four classes:

- Class I Lantibiotics (< 5kDa membrane-active peptides containing unusual amino acids as lanthionine, β-methyl-lanthionine and dehydrated residues);
- Class II [< 10 kDa heat-stable non-lanthionine containing membrane-active peptides with three sub-groups: IIa Listeria active peptides (with a consensus sequence in the N-terminal of Tyr-Gly-Asn-Gly-Val-Xaa-Cys- and a leader peptide GG), IIb complexes consisting of two leader GG peptides and IIc thiol-activated peptides];
- Class III for large heat-labile proteins and,
- Class IV for complex bacteriocins with a lipid or carbohydrate moiety.

Later, Cotter, Ross and Hill (2005) proposed the elimination of larger proteins from the bacteriocin classification arguing that bacteriocins should be referred as antimicrobial peptides, proposing the following classification:

- Class I for those bacteriocins subjected to an enzymatic PTM that was sub grouped according to their structure into twelve groups (MccC7-C57, lasso peptides, linear-azole, lantibiotics, linaridins, proteusins, sactibiotics, patellamide-like cyanobactins, anacyclamide-like cyanobactins, thiopeptides, bottromycins and glycocins).
- Class II included the unmodified or cyclic bacteriocins with five subgroups (IIa pediocin PA-1-like, IIb, IIc, IId and IIe peptides).
- Class III or "bacteriolysines" for thermo-labile high molecular weight peptides.

Table 1. Classification for bacteriocins from LAB (modified from Alvarez-Sieiro et al., 2016; Acedo et al., 2018)

Class	Subclasses
Class I Heat stable < 10 kDa Posttranscriptional modified peptides (RIPPs)	Class Ia. Lanthipeptides (types I, II, III [1], IV [1])
	Class Ib. Head-to-tail cyclized or circular peptides
	Class Ic. Sactibiotics[2]
	Class Id. Linear azo(in)e-containing peptides (LAPs)
	Class Ie. Glycocins
	Class If. Lasso peptides[2]
Class II Heat stable < 10 kDa Unmodified peptides Leader peptide GG	Class IIa. Pediocin-like linear (YGNGVXCXXXXCXV conserved at N-terminus)
	Class IIb. Two peptides
	Class IIc. Leaderless
	Class IId. Non pediocin-like, single-peptide
Class III Thermo-labile >10kDa	Bacteriolysins
	Non-lytic Tailocins (phage tail-like)[2]

[1]Types III and IV do not have known antimicrobial activity.
[2]None reported LAB bacteriocins, but putative clusters have been identified *in silico*.

Recently, novel bacteriocins have been reported with the emerging of genome tools applied for the screening of published LAB genomes, resulting in 785 putative bacteriocin gene clusters, which also contain new RIPPs different from the previously known bacteriocins. Therefore, a modification in the classification for most LAB bacteriocins was proposed by Alvarez-Sieiro et al., (2016). Class I resulted more diverse with six RIPP subclasses, Class II comprises four subclasses and Class III for unmodified bacteriocins > 10 kDa with bacteriolytic and non-bacteriolytic activity (Table 1). Furthermore, Acedo et al., (2018) presented an overview of the structural variety of bacteriocins and their interactions with receptors and possible modes of action.

However, many other bacteriocins are under study and some of the criteria used to classify bacteriocins are not fully applicable, such as the enterocins because they have common characteristics to more than one of the currently described bacteriocins classes or subclasses (Alvarez-Cisneros et al., 2011). For example, hemolysin from *E. faecalis* belongs to the lanthibiotic group (Class I), but its activity depends on the joint action of two peptides (Class IIb). Also enterocin P could be included in Class IIc because it is secreted by a pre-peptide translocase, but it has the YGNGV consensus sequence typical of Class IIa. Therefore, Franz et al., (2007) proposed a specific classification for enterocins that includes four main classes: lanthibiotic enterocins (Cass I), non-lanthibiotic enterocins (Class II), cyclic enterocins (Class III) and large proteins (Class IV). Class II was subdivided into three subclasses: the pediocin family of enterocins (IIa), enterocins synthesized without a leader peptide (IIb) and linear enterocins that do not belong to the pediocin family (IIc). Likewise, Ahmad et al. (2017) presented two schemes of classification, one based on the bacteriocin physicochemical properties, and other based on the presence of consensus sequence motifs claiming to include > 70% of the known bacteriocins.

Class I Bacteriocins

Class Ia of lanthipeptides or lantibiotics are small heat stable peptides that undergoes PTMs resulting in the formation of lanthionine (Lan), β-methyl-lanthionine (MeLan) and other unusual dehydrated amino residues. Serine and threonine residues go a two-step process that involve an enzymatic dehydration forming dehydroalanine (Dha) and dehydrobutyrine (Dhb), followed by a Michael-type condensation with the thiol-group of a neighboring cysteine residue to the β-carbon of Dha or Dhb forming a Lan or MeLan internal rings, that impart resistance to proteases (Acedo et al., 2018). Then the modified pre-peptide is transported and activated to the extracellular environment. The leader peptide for lantibiotics have a conserved FNLD box with helical structure essential for the ring formation trough the action of a cyclase enzyme, such as the NisC cyclase for nisin (Abts et al., 2013). Nisin A produced by *Lactococcus (Lc.) lactis ssp lactis*, lacticin 3147-A1 and lacticin 3147-A2 from *Lc. lactis* are the best studied subtype I and II from the Class Ia. The mechanism of action for most lantibiotics relies on the binding to the lipid II, a universal receptor and the main subunit transporter of peptidoglycan from the cytoplasm to the cell wall. Thereby, bacteriocins Ia block the proper cell wall synthesis that leads to the cell death. In particular, nisin A binds to Lipid II by its N-terminal rings forming a complex of eight lantibiotic bacteriocins molecules with four lipid II molecules with two dependent-concentration modes of action. At high levels, nisin A triggers the process of membrane insertion and pore formation, which leads to rapid cell death; but at low concentration, cell wall synthesis is repressed (Cotter, Hill & Ross, 2005,; Breukink & De Kruijff, 2006; Alvarez-Cisneros et al., 2011).

The head-to-tail circular, Class Ib bacteriocins, includes a group of RiPPs whose N- and C-termini are linked by a peptide bond. This group has recently attracted the attention because of their wide inhibitory action and their remarkable stability to proteases and heat treatments (Borrero et al., 2011; Grande-Burgos et al., 2014; O'Connor et al., 2015; Perez, Zendo & Sonomoto, 2018). Up to date, fourteen circular bacteriocins have been described and sub-divided in two groups according to their biochemical characteristics (Table 2).

The biosynthesis of circular bacteriocins involves three stages: the cleavage of the leader peptide, circularization and extra-cellular secretion. The biosynthetic gen cluster includes a series of five to seven genes encoding for the precursor peptide, a cationic and hydrophobic immunity protein, an ABC transporter (composed by an ATPase and a membrane protein involved in the protein secretion that also confers partial immunity), and a membrane protein that participates in the circularization. The leader peptides for these bacteriocins are heterogeneous in size and sequence (ranging from 2-48 residues) with a non-clear role in the biosynthesis process but essential for the precursor peptide folding and interaction with processing enzymes. The dehydration and cyclization are catalyzed by the bifunctional enzyme LanM, which has an N-terminal dehydratase domain and a C-terminal cyclase domain. LanM phosphorylates Ser/Thr, the resulting phosphate ester is eliminated to form Dha/Dhb (Acedo et al., 2018). Although the mode of action of most circular bacteriocins has not yet been elucidated, their antimicrobial action relies on their high net charge that facilitates their interaction with the negatively charged bacterial cell membrane causing non-selective ion channels, leaking of ions, dissipation of the membrane potential and eventually the cell death (Cebrián et al., 2012; Perez, Zendo & Sonomoto, 2018). Recent reports for the Enterocin AS-48 suggest a two stage mode of action, based on an electrostatically approach of the inactive water-soluble dimmer towards the cell membrane followed by dissociation and insertion into the lipid bilayer and establishment of electrostatic and hydrophobic interactions with the monomeric bacteriocin (Cebrián et al., 2015; Perez, Zendo & Sonomoto, 2018).

Table 2. Classification and examples of circular bacteriocins (modified from Grande-Burgos et al., 2014; Perez, Zendo and Sonomoto 2018)

Group	Name	Producer strain
Group i (Highly cationic peptides with predicted pI > 9)	Carnocyclin A	*Carnobacterium maltaromaticum* UAL307
	Garvicin ML	*Lc. garvieae* DCC43
	Leucocyclin Q	*Leuconostoc (Leuc.)mesenteroides* TK41401
	Lactocyclin Q	*Lc.sp* QU12
	Enterocin AS-48	*Ec. faecalis* UGRA10
	Uberolysin	*Strep. uberis 42*
	Amylocyclin	*B. amyloliquefaciens*
	Circularin A	*Cl. beijerinckii ATCC 25752*
	Enterocin NKR-5-3B	*Ec. faecium* NKR-5-3
	Pumilarin	*B. pumilus* B4107
Group ii (Hydrophobic cationic peptides with predicted pI < 7)	Gassericin A	*Lbgaseeri*
	Butyrivibriocin AR10*	*Butyrivibrio fibrisolvens*
	Acidocin B	*Lb. acidophilus* M46
	Plantaricyclin A	*Lb. plantarum* N326

*negatively charged, predicted pI 3.88.

Other Class I are the sactipeptides (Class Ic) that are described as sulfur-to-α-carbon-containing peptides. Examples include the subtilosin A produced by *B. subtilis* with high activity against *Clostridium* (*Cl.*) *difficile* and thurincin CD produced by *B. thuringiensis*, but only putative clusters have been identified in silico for LAB. Besides, the linear azol(in)e-containing peptides (Class Id) are composed by various heterocyclic rings of thiazole and methyl-oxazole derived from cysteine, serine and threonine residues. Being the streptolysing S produced by *Strep. pyogenes* the most representative, but also described as to potent cytolytic toxin and virulence factor (Wessel, 2005). Moreover, the Glycocins (Class Ie) are bacteriocins that contain glycosylated residues such as Glycicin F produced by *Lb. plantarum*, or the enterocin F4-9 from *Enterococcus* (*Ec.*) *faecium*. Finally, the lasso peptides (Class If) are RiPPs with an amide bond between the first residue in the core peptide chain and a negative charged residue in positions 7 to 9 generating a ring that includes de C-terminal. These bacteriocins display antimicrobial, antiviral and anti-carcinogenic activities as the Microcin J25 produced by *Eacherichia* (*E.*). *coli*, but non-reported bacteriocins from this type are produced by LAB (Alvarez-Sieiro et al., 2016).

Class II Bacteriocins

Class IIa pediocin-like bacteriocins are heat stable (< 10 kDa) and particularly active against Listeria. Their structure is characterized by a cationic and heat stable peptides lengthen from 37 to 48 amino acid residues, with isoelectric points (pI) varying from 8.3-

10. The N-terminal contains two cysteine residues joined by disulfide bridges and a conserved YGNGVXCXXXXCXV motif. The C-terminus is less conserved and participates in the target cell specificity. The gen cluster is plasmid encoded and contains four genes such as the structural pedA, the immunity *pedB*, the *pedC* and *pedD*, which encode an ABC transporter and accessory proteins for pediocin PA-1. Other IIa bacteriocins include the enterocins such as the *Ec. faecium* MXVK29 that shows accessory genes (*entK*, *entR* and *entT*) adjacent to the structural and immunity and inducer genes. The leader peptide from most Class II bacteriocins poses a highly conserved double glycine motif that serves as recognition signal for the bacteriocin production and secretion; it can also be constitutively produced or regulated by a quorum sensing system. The mode of action comprises three steps: (1) link with the mannose phosphotransferase system (Man-PTS) receptors, (2) insertion into the cytoplasmic membrane and (3) hydrophilic pore formation that induces dissipation of proton motive force, the subsequent acceleration of ATP consumption to restore the proton motive force that leads to the cell death. However, the target membrane composition plays a role for the proper bacteriocin insertion and the final pore formation and allows the exit of ions, amino acids and other essential molecules. Pediocin PA-1 (pediocin AcH) is the most studied IIa bacteriocins produced by *Pedicoccus (Ped.) acidilatici* and *Lb. plantarum*, respectively; other examples include leucocin A, acidocin A, mesenterin and sakacin P. Those bacteriocins show antagonistic activity against several spoilage and pathogen microorganisms, including *L. monocytogenes*, *S. aureus* and *Cl. perfringens* (Alvarez-Cisneros et al., 2011; Balciunas et al., 2013; Alvarez-Sieiro et al., 2016; Escamilla-Martínez et al., 2017; Ríos-Colombo, Chalón & Navarro, 2018).

Class IIb bacteriocins require the presence of two complementary peptides, at about equal quantities, that work together for their antimicrobial action forming a membrane-penetrating helix-helix structure causing membrane leakage in sensitive bacteria, such as the lactococcin G produced by *L. lactis*, the lactacin F produced by *Lb. johnsonni* and plantaricins S and T produced by *Lb. plantarum*. Although, some individual peptides can present antimicrobial activity with synergistic action like for thermophilin 13 produced by *Strep. thermophilus*, as well as, plantaricins A, E/F, J/K and W produced by *Lb. plantarum* strains. Their synthesis is regulated by at least five genes, organized in one or two operons, containing two structural genes codifying for the pre-bacteriocins, in addition to immunity, ABC transporter and accessory proteins. Their synthesis is regulated by a quorum sensing system that involves an induction factor, a membrane protein histidine kinase and response regulators. However, both two-peptide-lantibiotic and non-lantibiotic have been reported, such as the enterocin W a two-peptide-lantibiotic produced by *Ec. faecalis* NKR4-1 (Barbosa et al., 2016; O'Connor et al., 2015).

There are several reports for leaderless IIc bacteriocins conformed by one, two or more peptides, in general they are cationic with high pI values. Examples include the single peptide LsbB from *Lc. lactis* BGMN1, enterocin Q form *Ec. faecium* L50, Enterocin K1 from *E. faecium* EnGen0026, Aureocin A53 from *S. aureus* A53, Lacticin Q from *Lc. lactis* QU5, weissellicin Y and M both from *Weissella hellenica* QU13, among others. In addition to the two-peptide enterocin L50 produced by *Ec. faecium* L50 and the multi-peptide garvicin KS from *S. aureus* A70. The IIc leaderless bacteriocins are synthetized without an N-terminal leader peptide; they do not undergo any PTMs and become active after translation, but self-immunity and the secretion mechanisms remain unclear for these bacteriocins. However, Miljkovic et al., (2016) revealed that expression of LsbB requires the presence of a complete transcription terminator located downstream of the *lsbB* gene. On the other hand, the production of weissellicin Y and M has a nutrition adapted control; while the production of enterocin L50 seems to be regulated by temperature whithin the rage of 16 - 32°C. In relation to their mode of action, they do not require receptor molecules to form huge-tiroidal-pores (HTP) that cause linkage of large intracellular components lately inducing cell death (Alvarez-Sieiro et al., 2016; Perez, Zendo & Sonomoto, 2018)

Class IId or non-pediocin-like bacteriocins refer to unrelated single linear bacteriocins with different structures and mechanisms of action. Examples include lactococcin 972, lactococcin A and enterocin B, among others (Alvarez-Sieiro et al., 2016).

Class III Bacteriocins

Class III includes heat labile antimicrobial proteins with high molecular weight (> 10 kDa) with bacteriolytic and non-bacteriolytic activity. Most LAB bacteriolysins manifest their antimicrobial activity by cleaving the peptidoglycan cross-links of the target cell wall and fall into the peptidoglucano hydrolases (PGHs) group described in another section of this chapter. On the other hand, the non-lytic bacteriocins execute their antimicrobial action interfering with some biological functions that finally leads to the inactivation of sensitive cells. Examples include the dysgalacticin produced by *S. pyogenes* that inhibits the sugar uptake by interfering with the phosphotransferase systems (PTS) glucose and mannose causing membrane leakage of small molecules. As well as, caseicin from *Lb. casei* that inhibits the biosynthesis of DNA and proteins (Alvarez-Sieiro et al., 2016). Tailocins comprise large multiprotein complexes that take the shape of nanotubes larger than one million Da, resembling the tail structure of bacteriophages and causes cell death by inserting themselves into target membranes; examples include the pyocin produced by *Pseudomonads aeuroginosa* and diffocin from *C. difficile*, but no tailocins produced by LAB have been reported (Acedo et al., 2018).

Bacteriocin Applications as LAB Beneficial Metabolites

There is a worldwide demand for natural and safe products, with use of natural preservatives in combination of clean and environmentally-friendly processing technologies. From this point of view, LAB bacteriocins offer an important area of interest as they are recognized as natural and safe antimicrobials, especially those capable of inhibiting the growth of spoilage and foodborne pathogens, as well as methicillin-resistant *S. aureus* and vancomycin-resistant *Enterococci* at concentrations lower than that for conventional antibiotics. Bacteriocins differ from antibiotics in several aspects, first they are primary metabolites with a less complex biosynthetic mechanisms in comparison to the secondary metabolites antibiotics; most bacteriocins possess a narrow inhibitory activity by the action at the membrane and/or cell wall, they are active at nano- or micro-molar range, are sensible to proteases and temperature, do not alter color, taste or odor and present relatively low toxicity for eukaryotic cells, therefore they can be used for food preservation and therapeutics. In contrast, antibiotics are only approved for therapeutically use, have a broad spectrum, with activity at micro- to milli-molar range either inhibiting the synthesis of the cell wall or the cytoplasmic membrane, blocking the protein synthesis or the DNA copying processes or altering the metabolism that also may affect eukaryotic cells (Perez, Zendo & Sonomoto, 2014; Alvarez-Cisneros & Ponce-Alqucira, 2018).

López-Cuellar, Rodríguez-Hernández & Chavarría-Hernandez (2016) published a ten years review of the LAB bacteriocin application, reporting a database of 429 papers and 245 patents between years 2004 to 2015 around the world. Articles were further classified by application topics, 29% were related to food preservation; 37% focused on biomedical applications for alternative treatment of systemic, oral, stomach, vaginal and respiratory infections, contraception, cancer, skin and personal care; 25% for bio-nanomaterials and packaging materials and 9% for their veterinary use. While patents where classified by the same application topics including also their production, purification and molecular modifications; 31% patents focused on biomedical uses, 29% on food preservation, 5% on veterinary applications, and 29% for production-purification processes and recombinant molecules. The application of bacteriocins include several strategies, such *in situ* by the direct inoculation of a LAB bactericinogenic strains as starters in fermented products, or *ex situ* by the addition of a purified or semi-purified bacteriocin as an additive, as well as, the incorporation of a previously fermented product that contain bacteriocin-producer strain as an ingredient and immobilized bacteriocins into packaging film surfaces. But most industrially produced and commercially available purified or semi-purified bacteriocin preparations include nisin, pediocins, and carnocyclin A (Silva, Silva & Ribeiro, 2018).

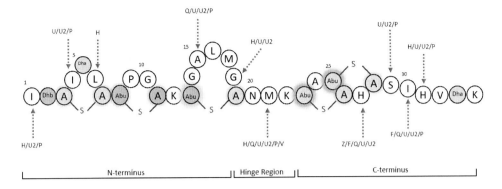

Figure 1. Representation of the nisin A structure with the Lan (A-S-A) and MeLan rings (Abu-S-A), dehydroalanine (Dha) and dehydrobutyrine (Dhb). Arrows indicate the amino acid substitution sites for natural variants (F, Q, Z, U, U2, P and H) and the semi-synthetic nisin V (Modified from Shin et al., 2016 and Campion et al., 2013).

Nisin

The Class Ia bacteriocin nisin is the most studied bacteriocin; it has been recognized as safe and approved as natural food preservative by the Joint Food and Agricultural Organization/World Health Organization (FAO/WHO) since 1969. The United States Food and Drug Administration (FDA) also approved nisin in 1988 as well as the European Food Safety Authority (EFSA) with the number E234. It has been commercially used in several countries with differences in their level and application to inhibit the growth spoilage gram-positive bacteria, especially Bacillus, Clostridium and Listeria on meat, dairy and vegetables. The most known brand is Nisaplin™ (2.5% Nisin A), but some other commercial presentations, include nisin alone or in combination with other bacteriocins and/or other natural antimicrobials that can be found from Danisco and Chr Hansen, among other companies. There are eight natural nisin variants (A, F, Q and Z produced by *Lactococcus lactis* strains, as well as the U, U2, P and H variants produced by *Streptococcus* strains) along with several semi-synthetic nisin variants (Shin et al., 2016). Nisin A contains 34 amino acid residues with five internal ring structures generated by PTMs. The N-terminal domain contains one Lan and two MeLan rings linked through an intermediate flexible hinge region to the C-terminal containing two intertwined rings (Figure 1). The N-terminal domain binds to the pyrophosphate moiety of lipid II and facilitates the pore formation by the C-terminal domain, thus interfering with the cell wall biosynthesis and the cell membrane permeation (Campion et al., 2013). Nisin Z differs from nisin A in one amino acid residue at position 27 with an asparagine instead a histidine and possess higher solubility. Nisin F has two amino acid substitutions at positions 27 and 30; while nisin Q has four substitutions at positions 15, 21, 27 and 30. All of them are highly active against *S. aureus* and *L. monocytogenes* and other gram positive foodborne bacteria. Variants U and U2 are produced by *Strep. uberis*, in their structure contain nine and ten amino acid substitutions in comparison to nisin A. Nisins H and P were detected from *Strep. hyointestinalis* and *Strep. gallolyticus* subsp *pasteurianus*, respectively.

The combination of nisin with other antimicrobials and the semi-synthetic variants have demonstrated an increased inhibitory activity against other non-food bacteria such as multidrug resistant pathogens such methicillin-resistant *S. aureus* (MRSA), vancomycin intermediate and heterogeneous *S. aureus*. However, semi-synthetic o bioengineered variants such as nisin V have shown an enhanced activity with therapeutic potential and great thermal stability and solubility, active against *Shigella*, *Pseudomonas* and *Salmonella* species (Campion et al., 2013; Shin et al., 2016).

Pediocin PA-1

Among the IIa bacteriocins, pediocin PA-A is the most studied due to its great ability to permeabilize the membrane of *L. monocytogens*, *S. aureus*, as well as, *Pseudomonas* and *E. coli*. It is thermaly stable in a wide pH range, produced by *Pedicocus* spp and resistant to heating and freezing (Schneider et al., 2006; Silva, Silva & Ribeiro, 2018). Commercial pediocin presentations include ALTA® (Kerry Bioscience), MicroGard™, Fargo 23® (Quest International) and Bactoferm F-LC® (combination of pediocin and sakacin by Chr. Hansen) mostly recommended for preservation of dairy and meat products (López-Cuellar, Rodríguez-Hernández & Chavarría-Hernandez, 2016).

Carnocyclin A

Carnobacterium maltaromaticum UAL307 that was isolated from fresh pork, produces the class Ib bacteriocin Carnocyclin A, a circular 60 residues peptide that forms anion selective channels in the lipid membrane. This bacteriocin with the commercial brand Micocin® (Griffith Laboratories) has been approved for vacuum packaged meat products in United States, Canada Costa Rica, Colombia and Mexico against *Listeria monocytogenes*; but carnocyclin can also inhibit *E. coli*, *Pseudomona aeruginosa* and *Samonella* Typhinurium (Martin-Visscher et al., 2008; 2011; Marketwire, 2011; Koné et al., 2018).

Application of Bacteriocins to Improve Food Preservation

There are several reports in relation to the use of bacteriocins for food preservation as these antimicrobial peptides can prevent the growth and transmission of spoilage and food borne pathogens by preventing the biofilm (surface-associated microbial communities) formation on surfaces, retarding the microbial growth and the biogenic amines formation during refrigeration, thus increasing the shelf life of foods while mantaining their nutritional and sensory qualities (Galié et al., 2018; Özogul & Hamed, 2018). Bacteriocins can be used in several forms, either produced *in situ* by the incorporation of LAB bacteriocinogenic strains in fermented and non-fermented foods, or produced *ex situ* by addition of a previously purified or concentrated bacteriocin in a food grade substrate. The use of bacteriocinogenic LAB for food preservation is attractive due to their GRAS status that fulfill the consumer's demands for minimally processed foods and ready-to-eat (RTE) products free from artificial additives. The producer strain can serve as a starter or be added as an additional protective culture. Protective bacteriocinogenic strains may grow and produce bacteriocins during refrigeration storage in which may be part of the predominant microbiota, thus inhibiting the proliferation of pathogenic bacteria especially during temperature abuse conditions in meat and dairy products. The use of multiple bacteriocins or multibacteriocinogenic strains seems to be more effective for food preservation; as reported by Mills et al. (2017) that generated a double bacteriocin producer (*Lc lactis* CKS2775) via conjugation capable to produce nisin and lacticin providing a multihurdle approach to control pathogens in cheese.

Other food applications include immobilized preparations applied on the surface of the food product or included in antimicrobial active packaging, coatings and nanocomposite films. For instance, Martinez-Murillo et al., (2013) incorporate nisin into whey protein isolate film with antibacterial activity against *Listeria innocua* (ATCC 33090) and *Brochothix thermosphacta* (NCIB10018) (Khaneghah et al., 2018).

Immobilized bacteriocins are more stable as the carrier protects them from inactivation due to variations in the pH, prevention of their interaction with enzymes and other food components (including lipids and other proteins). Furthermore, immobilization can regulate the bacteriocin release into the food system (Balcinas et al., 2013). Bacteriocins have been encapsulated using several methodologies including liposome formation, coextrusion, spray coating, fluid-bed coating, complex coacervation and formation of silver nanoparticles, alone or in combination with other antimicrobials, preserving their antimicrobial action (Calderón-Oliver, et al., 2017).

The thermal stability of bacterial spores is a major problem in the food industry; LAB bacteriocins may be involved as part of a hurdle processing due to their anti-spore activity against dormant spores and/or germinated spores. The bacteriocins nisin, enterocin-AS48 (produced by *Ec. faecalis* A-48-32), lacticin 3147 (produced by *L. lactis* subsp. lactis DPC3147), plantaricin TF711 (produced by *Lb. plantarum* TF711), as well as, the bacteriolysins pentocin L and pentocin S (produced by *Ped. pentosus* L and S) have shown inhibitory action against spores of *Cl. perfringens*, *Cl. sporogenes*, *Cl. botulinum*, *Cl. difficile* and *B. cereus* with differences in their mode of action. In particular nisin reduces the thermal stability of bacterial spores and prevents the outgrowth of survival spores in heat threated milk and apple juice. Similar data was observed in food models subjected to non-thermal food preservation processes such as high-pressure processing and pulsed electric field, or in systems of bacteriocins combined with plant extracts, sucrose fatty esters and potassium sorbate (Egan et al., 2016).

Bacteriocins in Human Health and Skin Care

Penicillin resistant *S. pneumonia*, vancomycin resistant enterococci (VRE), MRSA, intermediate and heterogeneous *S. aureus*, *Strep. pneumoniae* and *Cl. difficile*, among other bacteria represent a major medical risk all around the world, for their incidence in bacteremias, pneumonia and postsurgical infections, especially for immunocompromised patients. Those bacteria can be adhered or implanted on the surface of medical devices or damaged tissues in the form of a biofilms, thus causing chronical infections. Bacteriocins show a promising alternative to traditional antibiotics, their primary difference is their activity against closely related bacteria. In particular, Nisin A in combination with conventional antibiotics, lysozyme and lactoferrin can promote a synergistic effect against MRSA and reduce the biofilm formation. Therefore, this bacteriocin and other natural and bioengineered lantibiotics such as NVB302 (semi-synthetic derivate from actagardine), microbisporicin, mutacin 1140 (produced by *Strep. mutants*) and the lantibiotic lacticin are examples of bacteriocins under clinical trial to treat hospital related infections, including drug-resistant infections, as an option to substitute or decrease the use of antibiotics.

Shin et al., (2016) present a review of the biomedical applications of nisin as a promising tool to improve human health, including the treatment of oral infections and cancer. For instance, the development of an antimicrobial nanofiber wound dressing containing nisin was used to reduce the colonization of MRSA and accelerate healing with non-adverse histological effects (Heunis, Smith & Dicks, 2013). Other applications include the topical nisin application to treat mastitis and reduce the staphylococcal count in the breast milk of nisin treated woman. Moreover, this bacteriocin has a synergistic action with poly-lysine and sodium fluoride for the prevention of planktonic and biofilm forms of the cariogenic bacteria *Strep. mutants*, and other gram-positive and gram-negative oral colonizing bacteria, as well as, an anti-biofilm effect on saliva without causing cytotoxicity to human oral cells. In addition, nisin has a potential to reduce periodontal disease by inhibiting *Ec. faecalis* (frequently present in infected teeth root canals) and for the prevention of oral candidiasis caused by *Candida albicans*, a prevalent pathogen that cause mucosal fungal infections.

Cancer is characterized by the dysregulated growth of abnormal cells, usually threated by chemo- or radiotherapy and surgery with the concomitant detrimental of normal cells and severe affectation to the patients. Nisin and other bacteriocins such pyocin, colicin S4, defensing, pediocin, microcin may play a role for the treatment of cancer because of their activity against various types of cancer (Shin et al., 2016; Baindara, Korpole & Grover, 2018). In particular the use of nisin A in low concentrations (2.5%) showed an antitumor potential in head and neck squamous cell carcinoma (HNSCC) *in vitro* and *in vivo*, but higher concentrations of a combination of nisin Z and P (95%) had greatest level of apoptosis in HNSCC cells and decreased their proliferation demonstrating the potential of nisin as an anticancer agent (Kamarajan et al., 2015). Also the combination of nisin with conventional cancer drugs such doxorubicin potentiated the effectiveness against skin carcinogenesis (Preet et al., 2015). Recent data demonstrates the anti-carcinogenic effect of nisin due to the selective toxicity on melanoma cells, as well as, on gastrointestinal, hepatic and blood cancer cell lines (Lewies et al., 2018; Goudarzi et al., 2018).

Bacteriocins such as lactacin and fermenticin HV6b are potential spermicidal agents that induce reduction in the motility of human spermatozoa or sperm tail damage when used at high levels. In addition, other bacteriocins produced by *Ped. pentosaseus* SB83, *Ec. faecium* and *Lb. rhamnosus* can inhibit pathogenic vaginal bacteria such *Gardnerella vaginalis*, *Prevotella bivia*, *Bateroides*, *Peptostreptococcus* and *Mobiluncus* spp (López-Cuellar, Rodríguez-Hernández & Chavarría-Hernandez, 2016).

There are several products containing bacteriocins from probiotic LAB bacteria to prevent and treat skin diseases, including external signs of aging, acne, rosacea, bacterial and yeast infections, psoriasis and dermatitis. All these applications derive from the bacteriocin contribution to the modulation of the skin microflora, and their action in the immune system and reduction of inflammatory processes. Examples include saivaricin, nisin, mersacidin, lacticin and leucocin A (López-Cuellar, Rodríguez-Hernández & Chavarría-Hernandez, 2016).

Bacteriocins in Veterinary and Agriculture

Antibiotics are commonly used in livestock and aquaculture for the treatment of bacterial infections as a prophylactic method to prevent bacterial diseases in healthy animals; they are also used as growth promoters to improve feed conversion and body-weight gain. This issue has become one of the main safety problems worldwide and several organizations such as the WHO, the FDA and EFSA, have raise awareness to stop using antibiotics in healthy animals, in order to reduce the emergence of antibiotic-resistance bacteria strains that could be potentially transmittable to humans and to preserve the effectiveness of antibiotics. The antibiotic-resistance can take place when microorganisms are continuously exposed to sublethal doses of antibiotics as a result of an adaptation process (Álvarez-Cisneros & Ponce-Alquicira, 2018). Therefore, there is an urgent need for novel antimicrobials that can be used for animal production with low development of resistance, where bacteriocins are of great interest mainly because of their antagonism against *Cl. perfringens*, enterotoxigenic *E. coli*, *Campylobacte jejuni* and *Salmonella enterica* that can affect both animal and human health, as well as, against Streptococcus *suis* a zoonotic agent that cause economic losses for swine production. In addition the administration of bacteriocinogenic strains, such as *Lb. salivarus* can help to modulate the gastrointestinal microbiota of healthy pigs reducing the levels gram-negatives and the presence of opportunistic pathogens thus improving the meat microbial quality (Lagha, Hass, Gottschalk & Greiner, 2017).

PEPTIDOGLYCAN HYDROLASES (PGHS)

Peptidoglycan Structure

The peptidoglycan (PG) is a hetero-polymer of glycan backbones crosslinked by peptides forming a mesh-like layer in the bacterial cell wall. It is responsible for the rigidity and shape of bacterial cells; it is also involved in the bacterial resistance mechanisms to antibiotics, immune responses, phage susceptibility, serological behavior and protects against osmotic lysis, among others. The cell wall of gram-negative bacteria contains less than 10% PG. But for gram positive the PG content varies from 30-70% showing structural variations that may be considered a criterion for bacteria differentiation. In addition, the PG layer is highly dynamic showing structural changes during the cell growth and cell division processes (Meroueh et al., 2006; Bourhis & Werts, 2007; Wang & Shaevitz, 2013; Sharma et al., 2016).

The glycan backbone is made of alternating unites of *N*-acetylglucosamine (NAG) and *N*-acetylmuramic acid (NAM) linked by β-1,4 glycosidic bonds. The peptide subunits contain four and occasionally five alternating L- and D- amino acids residues joined to the lactyl group of the NAM units. In general, L-alanine is directly bound to the NAM, followed by two amino acid residues with three functional groups (D-glutamic acid or threo-3-hydroxy-glutamic [Hyg] at position 2) attached to an L-diamino acid at position 3, and a final D-alanine residue at position 4. But in some cases an additional L-alanine is also present at the C-terminus forming a penta-peptide subunit. The peptide subunit for most Gram-positive bacteria presents a L-Lysine residue at position 3, while a meso-diaminopimelate (mDAP) occurs for gram-negative and some gram-positive such Listeria. Structural modifications give rise to more than 100 different PG variants arranged into types A and B according on their peptide cross-linkage. The free amino group of the L-diamino acid residue (at position 3) directly participates in the cross-linkage to the D-alanine carboxyl group at (position 4) of an adjacent peptide subunit for the PG type A. In the case of PG type B the cross-linkage involves an inter-peptide bridge consisting of 1-7 glycine residues (Schumann, 2011).

Characteristics of Peptidoglycan Hydrolases

The enzymes capable of hydrolyze the PG cell wall promoting cell lysis, are known as peptidoglycan hydrolases (PGHs) and found in several organisms, including animals, plants, bacteria and viruses. Their synthesis is highly regulated, from their transcription to the posttranslational level in order to prevent the cell lysis (Kelly & Bayles 2003; Vollmer et al., 2008). PGHs have an important role as they participate in different phenomena of lysis, either to lyse host cells or to re-model the cell wall during growth and division (Bustamente, et al., 2012). Bacterial division results from the combination of membrane constriction with expansion and remodeling of the cell wall. For many bacteria, the cell lysis is a natural part of the life cycle and, in some cases, is useful to eliminate damaged cells within the population (Regulski et al., 2012; Höltje & Tuomanen, 1991; Smith, Blackman & Foster, 2000). PGHs represent a novel class of antimicrobials, as they can act selectively in the rapid killing of pathogenic bacteria without affecting the normal microbiota; also present low toxicity, moderate inhibition by the host immune response and low probability of developing resistance. All these features result of great interest for their biotechnological and pharmacological exploitation for food preservation and for the treatment of drug resistant infections (Bustamente, et al., 2012).

The activity of the PGHs also generates rotational products that serve as signaling molecules for the recognition of bacteria by other organisms and, in some bacteria, for the induction of β-lactamase that may provide resistance to β-lactam antibiotics. In addition, these hydrolases are also responsible for the autolysis under certain conditions. Most PGHs present a specific activity for each glycoside and amide bridges in the peptidoglycan (Vollmer et al., 2008). Although the PGHs physiological functions include the regulation of cell wall growth, the replacement of peptidoglycan during growth, the separation of daughter cells during cell division, it is difficult to assign a function other than a PGHs, because many bacteria have a high number of hydrolases with redundant roles, as they are able to break covalent bonds in sacs of peptidoglycan or their fragments (Höltje & Tuomanen, 1991; Smith, Blackman & Foster, 2000; Heidrich et al., 2001). Some specialized hydrolases enlarge the pores in the peptidoglycan in a temporal and controlled way, allowing the insertion of multiprotein complexes in the cellular sheath (pili, flagella, secretion systems), or split the PG during sporulation or germination of spores (Vollmer et al., 2008; van Heijenoort 2011; Höltje 2013; Maxime et al., 2018).

It has been proposed that some PGHs cooperate with other enzymes splitting the existing PG polymer allowing the insertion of new material, but the direct evidence for such a mechanism has not yet been proven. The role of PGHs involved in the cell separation and autolysis has been reported. Little is known about the cellular function of those PGHs involved in the maturation of PG, which includes the regulation of the length of the chains of glycans and peptides, as well as the degree of cross-linking. In a very general way, PGHs functions include regulating the availability of PG precursors, PG maturation, septum splitting, cell separation and autolysis (Höltje & Tuomanen, 1991; Höltje, 1998; Smith, Blackman & Foster, 2000; Vollmer et al., 2008; van Heijenoort, 2011; Maxime et al., 2018).

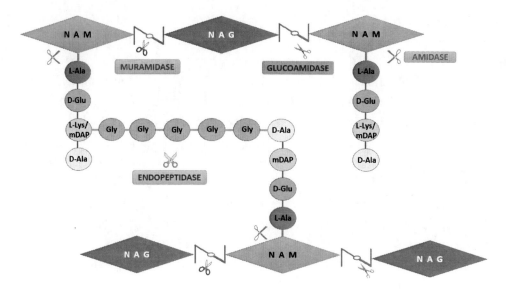

Figure 2. Representation of the of peptidoglycan (PG) structure showing the N-acetylglucosamine (NAG) and N-acetylmuramic acid (NAM) subunits linked by β-1,4 glycosidic bonds and the glycin inter-peptide bridge for PG type B. Also the sites of peptidoglycan hydrolases (PGHs) action are presented for muramidases, glucoamidases, amidases and endopeptidases (Modified from Sharma et al., 2016,; García-Cano, 2013).

Classification of Peptidoglycan Hydrolases

The PGHs are widespread in bacteria and also known as autolysins classified according to the cleavage site. The three main classes include: (1) the glucosidases that fragment the glycan backbones; (2) the amidases or peptidoglycan amidases (*N*-acetylmuramyl-*L*-alanine amidases) that hydrolyze the amide bond between NAM and the N-terminal L-Alanine residue, thus separating the peptide subunits from the glycan backbone; and (3) the peptidases (endopeptidases and carboxypeptidases) that act within the PG peptide side chains (Figure 2).

Specifically, the glucosidases consist of *N*-acetyl-β-D-glucosaminidases, lysozymes and lytic transglycosylases. The first is an endo-glucoamidase that catalyzes the hydrolysis of the glycosidic bond between NAG and adjacent monosaccharides in different oligosaccharides such as PG, chitin and other N-glycans. These enzymes have a catalytic domain and one or more associated substrate-binding domains (CBDs), such as one to six LysM motifs generally located at one end of the protein, that bind directly to the PG thus affecting the substrate binding and the hydrolysis efficiency. The latter two glucosidases are known as *N*-acetyl-β-D-muramidases or muramidases that hydrolyze the β-1,4-glycosidic bonds between the NAM and NAG residues of PG. The resulting product from the lysozyme action is a terminal reducing NAM residue; while the lytic action of transglycosylases induces a transglycosylation reaction resulting in the formation of 1,6-anhydro ring in the NAM residue.

The PG amidases or *N*-acetylmuramyl-*L*-alanine amidases hydrolyze the amide bond separating the peptide subunits from the glycan backbone. They are members of the bacterial autolytic systems and carry a signal peptide in their N-terminal site that allows their transport across the cytoplasmic membrane; while bacteriophage PG amidases are endolysins with no signal peptides (Vollmer et al., 2008). On the other hand, the peptidases that hydrolyze the amide bonds between amino acids in the PG, according to their specificity are endopeptidases (cleaving bonds within the peptide chain) or carboxypeptidases (removing the C-terminal amino acid). These PGHs are also named DD-peptidases, when cleaving between two D-amino acid residues, and LD- or DL-peptidases if the hydrolysis takes place between an L- and D- amino acid residues (Ghuysen, Tipper & Strominger, 1966; Layec, Decaris & Leblond-Bourget, 2008; Vollmer et al., 2008; Reith & Mayer, 2011; Johnson, Fisher & Mobashery, 2013; Johnson & Klaenhammer, 2016; Sharma et al., 2016).

Applications of Peptidoglycan Hydrolases as LAB Beneficial Metabolites

There is an increased interest in the study of natural antimicrobials that help to fight bacterial infections and the presence of opportunistic fungi. As they are associated with high mortality rates due to the emergence of antibiotic resistance. In this sense, PGHs are considered valuable alternatives or complements of the classic antifungal agents to fight infection by fungi (Delattin et al., 2017; Owen et al., 2018). On the other hand, multidrug-resistant strains are becoming more frequent and represent one of the main causes of morbidity and mortality. For this reason, new strategies are urgently needed to combat multi-drug-resistant pathogens. In this scenario, PGHs represent a valuable alternative as antibacterial agents due to its unique property of breaking the cell wall peptidoglycan of fungi, as well as, gram-positive and gram-negative bacteria with a low incidence to induce resistance; therefore they may replace or be used in combination with antibiotics (Sharma et al., 2016; Kashani et al., 2017).

PGHs may inhibit spoilage bacteria and food pathogens. In recent years lysozyme has been used as a natural antimicrobial, approved as a preservative in foods in USA and other countries. Its main applications have been in the prevention of gram-positive contamination in cheese and control of bacterial activity in wine. Another example is the lysostaphin active against *Staphylococcus aureus*, important pathogens in the food industry. But there is a wide range of PGHs which are being studied and that arise as an important alternative to combat a broad spectrum of bacteria, both gram-positive and gram-negative bacteria, which could be used as natural agents to fight against food-borne diseases (Callewaert et al., 2011; Carrillo et al., 2014).

Food Applications of Peptidoglycan Hydrolases

The two most studied PGHs in the food industry are lysozyme and lysostaphin. Lysozyme is a naturally occurring N-acetilmuramidase in both the animal and plant kingdoms that plays an important role in the natural defense mechanism and discovered by Alexander Fleming in 1922. Many researchers have described various biological activities of this protein, including its antibacterial activity against gram-positive bacteria and harmless to human cells. However, their antibacterial spectrum has reached some gram-negative bacteria if their outer membrane is disrupted prior to exposure through thermal, chemical or enzymatic modifications, genetic mutations and synergistic effects with other compounds. It has been used by plants and animals as a natural defense mechanism against pathogenic bacteria and it is allowed as a food additive. For instance lysozyme is used in casings and cooked RTE meat and poultry products with the GRAS Notice No. 000064 given by the USDA.

Lysostaphin (also classified as a Class III bacteriocin) has been highlighted as a next-generation agent for the treatment against *S. aureus*, one of the main pathogens transmitted by food and nosocomial infections around the world, that causes diseases such as endocarditis, pneumonia and toxic-shock syndrome, among others (Kadariya, Smith & Thapaliya, 2014). Lysostaphin is a Zn^{2+} dependent endopeptidase with a molecular mass of ~ 27 kDa, a PI of 9.5 and a pH optimum of 7.5, also classified as bacteriocin. It consists of two domains (Figure 3), the N-terminal responsible for the catalytic activity of Gly-Gly bonds, and a C-targeting domain (CWT) that specifically binds to the pentaglycine bridge of the PG cell wall of *S. aureus* and *S. epidermidis* (Ojha et al., 2018; Tossavainen et al., 2018).

This PGH is produced by *S. simulans*, which represents a main concern for their use in the food industry. Therefore, lysostaphin has been cloned and expressed in various heterologous systems including BAL. For instance, food-grade lysostaphin has been produced in *Lc. lactis* and *Ped. acidilatici* using the NICE (Nisin-controlled gene Expression) system at industrial scale (Mierau et al., 2005; García-Cano et al., 2015).

PGHs have been proposed as a tool to prevent bacterial diseases transmitted by foods, which are associated with high morbidity, death and economic losses (Fischetti 2010). These hydrolytic enzymes could inhibit the growth of *E. coli* (including the pathogenic serotypes), *L. monocytogenes* (a ubiquitous bacteria that can cause abortion in pregnant women and other serious complications in children and elderly), *Salmonella enterica* serovar Typhimurium (which can cause massive outbreaks and even death in children and elderly), and *S. aureus* (Galié et al., 2018). Recently, several PGHs have been evaluated during the cheese manufacture, where the presence of non-starter BAL contribute to the development of desirable flavor characteristics and to the quality and safety of the final product due to the production of autolysins (Cibik & Chapot-Chartier, 2004; Aldrete-Tapia et al., 2018). Although few autolysins produced by the Lactobacillus genus have been characterized; the autolysin AclB produced by *Lb. casei* was purified and fully characterized. AcIB was recognized as a species–specific cell separating enzyme responsible for the cell separation of *Lb. casei* after cell division, which helps to accelerate cheese ripening (Xu et al., 2015), in advantage with other autolysins such as the LytF produced *B. subtilis* as that only contributes synergistically to the cell separation (Margot, Pagni & Karamata, 1999; Xu et al., 2015). Some other autolysins could be applied in a similar way to improve food fermentations, such as the autolysin Acm2, required for cell separation in *Lb. plantarum* and *Lc. lactis* (Palumbo et al., 2006). There are also reports on the PGH endopeptidase activity for the surface layer (S-layer) of *Lb. acidophilus* against the cell wall of some bacteria. The S-layer is a bifunctional array of proteins or glycoproteins that overlay the cell surface of several *Eubacteria* and *Archaea*. The S-layer proteins are directly in contact with the bacterial environment and involved in surface phenomena such as bacterial co-aggregation, adhesion, protection and recognition or as carriers of virulence factors (Gerbino et al., 2015). The S-layer reported for *Lb. acidophilus* ATCC 4356 is composed of a single protein (protein S_A) of ~ 45 kDa, with endopeptidase activity against the cell wall of *Salmonella enterica*, *E. coli* and *Micrococcus luteus*. Therefore this S-layer proteins may in theory be use to prevent the growth of food spoilage microorganisms (Prado-Acosta et al., 2008; Hynönen & Palva 2013; Meng et al., 2015). Other similar PGHs like the Msp1/p75 produced by the probiotic *Lb. rhamnosus* GG that was initially reported as an anti-apoptotic and growth-promoting of intestinal cells, also shows PGH activity with D-glutamul-L-lysyl endopeptidase specificity for daughter cell separation of *Lb. rhamnosus*. In addition thirteen PGHs where detected in *Lb. casei* BL23 where the Lc-p75 was identified; this protein is the homolog of the previously mentioned Msp1/p75. Both PGHs are attached to the cell wall and secreted in the culture supernatant, which also have been confirmed being responsible for the cell wall lysis of different food deteriorating microorganisms (Claes et al., 2012; Regulski et al., 2012).

Another interesting example is the first bifunctional PGH reported, produced by *Ped. acidilactici* ATCC 8042 that possess two identified lytic sites: N-acetylglucosaminidase and N-acetilmuramoil-L-alanine amidase activities (Figure 3). These results could help in the design of new antibacterial agents of natural origin that can combat food-borne diseases and improve hygienic practices in the industrial sector (Carrillo et al., 2014; Serrano-Maldonado et al., 2018). Moreover, the search for new and natural antimicrobials includes the PGH produced by *Lb. sakei* active against various food-spoilage microorganisms. As shown in Figure 3, this PGH possess a lytic domain with *N*-acetyl-muramoil-L-alanine amidase activity joined to a LysM region (small globular domains ranging in length from 44 to 65 amino acids that bind to various PG types) (Mesnage et al., 2014; López-Arvizu, García-Cano & Ponce-Alquicira, 2017). In similar studies from the twelve predicted PGH-encoding genes in the genome of *Lb. plantarum*, two PGHs were identified, the N-acetylglucosaminidase Acm2 and NpIC/P60 D,L-endopeptidase LytA. Both enzymes revealed a potential hydrolytic activities being the

Acm2 the mayor autolysin produced by *Lb. plantarum* (Rolain et al., 2012). *Lb. plantarum* extracts with N-acetylglucosaminidase activity where capable to inhibit several microorganisms such as *Listeria innocua, Staphylococcus aureus, Weissella viridescens* and *Leuconostoc (Leuc.) mesenteroides* and where been applied to increase the shelf life of type Vienna sausage (García-Cano, 2013; Roa, 2017).

Clinical Applications of Peptidoglycan Hydrolases

It has been claimed that PGHs can help modulate the bacterial colonization of the gastrointestinal mucosa and the control of pathogenic bacteria that would be beneficial for human and animal health. The effectiveness of antibiotics against bacterial infections is decreasing due to the development of resistance in bacteria. Therefore, there is an increasing need for potential alternatives to the use of antibiotics. PGHs can be used as alternative antibacterial agents by its ability to hydrolyze the bacterial and fungi cell wall (Sharma et al., 2016). The medical applications of the recombinant lysostaphin are moving and some have entered into clinical trials (Nelson et al., 2012). The CWT domain of lysostaphin has also been considered as a promising source for alternative chemotherapy (Schmelcher et al., 2012; Osipovitch and Griswold 2014; Jagielska, Chojnacka, & Sabała, 2016) and has been redesigned to suppress its immunogenic potential (Blazanovic et al., 2015; Tossavainen et al., 2018).

As previously mentioned, lysostaphin is active against the cell wall of *S. aureus*, this microorganism has an annual incidence rate of infections from 20 to 50 cases/100,000 habitants, with 10 to 30% mortality especially for the presence of resistant strains in elderly patients (van Hal et al., 2012). In addition to the health problem, productivity losses and the economic impact involved for the control of *S. aureus* represents a burden to society (van Hal et al., 2012; Lee et al., 2012; Kashani et al., 2017). Studies of peptidoglycan hydrolase like lysozyme and lysostapnhin have been used successfully to control pathogenic bacteria resistant to antibiotics in animal models (Nakimbugwe et al., 2006; García-Cano et al., 2011).

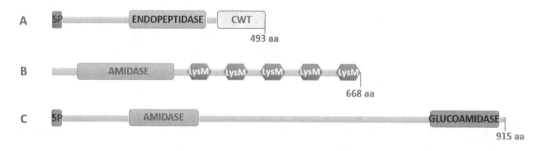

Figure 3. Domain organization of (A) lysostaphin, B) N-acetylmuramoyl-l-alanine amidase produced by *Lb. sakei* and C) Bifunctional PGH produced by *Ped. acidilactici ATCC 8042* (Modified from Sabala et al., 2014; García-Cano et al., 2015; López-Arvizu, García-Cano & Ponce-Alquicira, 2017).

The N-acetilmuramidases Lb GH25B and Lb GH25N produced by *Lb. buchneri* CD034 and NRRL B-30929 were identified as cell wall associated proteins with hydrolyzing activity. The stability of these proteins was determined through studies of thermal denaturation getting an assessment of the conditions that maintain biological activity with values of T_m around 49°C, reflecting the thermal stability of enzymes as needed for biotechnology and drug exploitation (Anzengruber et al., 2014). These values are consistent with those obtained for the glycanhydrolase Cpl-7 of *Strep. pneumoniae* bacteriophage CpI7 recently studied as an antibacterial agent (Bustamante et al., 2012). N-acetilmuramidases from the *Lb.* genus contribute to the of cell wall hydrolases (CWHs), whose development as novel therapeutic proteins represent some advantages over conventional antibiotics, mainly due to their low probability for developing resistance (Parisien et al., 2008; Anzengruber et al., 2014). Surprisingly, LbGH25B and Lb GH25N do not contain any of the typical domains that generally exhibit the PGH; a multi-domain structure that includes a catalytic domain and multiple copies of a cell wall binding domain. The binding domains that are often found in this type of proteins are LysM, SH3, PG-binding, CW-binding, or Cpl-7 known to be present in the cell wall-associated proteins (Layec, Decaris and Leblond-Bourget 2008; Anzengruber et al., 2014).

It is interesting to mention that these domains give better anchoring properties to the cell wall of sensitive microorganisms and are also present in many PGHs; Alternatively, these structural domains, can be used for the identification of foreign peptides in the cell surface or for improvement of adhesion of host bacteria; as it has been used in the administration therapeutic proteins and vaccines to the mucosa (Berlec, Ravnikar & Strukelj, 2012). The oral administration of the traditional intravenous or intramuscular vaccine minimizes possible side effects while increasing the specificity of the mucosa chronic diseases and infections (Bermudez-Humarán et al., 2011; Wyszyńska et al., 2015; Hynönen & Palva, 2013).

As already mentioned, activities of endopeptidases present in the S-layer of *Lactobacillus* also show potential for use as carriers of antigens in the design of live oral vaccines because of its adhesive and immunomodulatory properties, that can also be used in the mediation of bacterial attachment to host cells (Prado-Acosta et al., 2008; Hynönen & Palva, 2013). Bacteria treated with this type of proteins, examined by electron microscopy reveal that they exert their lethal effects forming holes in the cell wall through the digestion of PG. The high internal pressure of bacterial cells (approximately 3 to 5 atmospheres) is controlled by the highly cross-linked cell wall. Any deterioration in the integrity of the wall will result in the extrusion of the cytoplasmic membrane and the final hypotonic lysis (Loessner et al., 2002).

Currently, with the exception of the polysporin and mupirocin, which are commonly used topically, no other antimicrobials may control pathogenic bacteria colonizing the mucous membranes. The selectively for their inhibitory action against pathogenic bacteria without affecting the normal mucosa microbiota and the low likelihood of bacterial resistance, improves the potential of PGHs as a novel antimicrobials (34). Another example already mentioned is the application of the Lc-p75 (γ-D-glutamil-L-lisil-endopeptidasa) produced by the probiotic *Lactobacillus casei* BL23, which presents an anti-inflammatory activity (Regulski et al., 2012).

CONCLUSION

Bacteriocins and peptidoglycan hydrolases are beneficial metabolites produced by lactic acid bacteria. LAB bacteriocins constitute a diverse and heterogeneous group of ribosomal antimicrobial peptides, varying in size, structure and specificity, active against closely related species. These peptides are considered food-grade because the GRAS or QPS status of most bacteriocinogenic strains, do not impart color or flavor and are easily degraded by the proteolytic enzymes, that makes them safe for human consumption, minimizing the opportunity of target strains to interact and develop resistance. Therefore, bacteriocins can be applied for food preservation, as well as, an alternative antibiotics in the clinical area, mainly due to their tolerance to high temperatures and their activity over a wide pH range against spoilage microorganisms and pathogens such *Bacillus*, *Clostridia*, *L. monocytogenes* and methicillin-resistant *S. aureus* and vancomycin-resistant *Enterococci* at concentrations lower than that for conventional antibiotics. Nevertheless, more research is needed in order to expand their inhibitory spectrum, stability and increase the recovery yields; where the bioengineering offers a possibility to enhance the inhibiting activity and spectra, as well as their heterologous expression in other hosts.

There is still few data in relation to PGHs produced by lactic acid bacteria in relation to their function and characterization. But PGHs offer a wide opportunity as alternative antimicrobials from GRAS microbial sources, with great prospects for new food applications as natural antimicrobial agents.

REFERENCES

Abts, A., Montalban-Lopez, M., Kuipers, O. P., Smits, S. H. & Schmitt, L. (2013). NisC binds the FxLx motif of the nisin leader peptide. *Biochemistry, 52* (32) 5387-5395. http://doi.org/10.1021/bi4008116.

Acedo, J. Z., Chiorean, S., Vederas, J. C. & van Belkum, M. J. (2018). The expanding structural variety among bacteriocins from Gram-positive bacteria. *FEMS Microbial Reviews, 42*, 805-828. htpp://www. doi.org/10.1093/femsre/fuy033.

Ahmad, V., Khan, M. S., Jamal, Q. M. S., Alzohairy, M. A., Al Karaawi, M. A. & Siddiqui, M. U. (2017). Antimicrobial potential of bacteriocins: in therapy, agriculture and food preservation. *International Journal of Antimicrobial Agents*, *49*(1), 1-11. https://doi.org/10.1016/j.ijantimicag.2016.08.016.

Aldrete-Tapia, A., Escobar-Ramírez, C. M., Tamplin, M. L. & Hernández-Iturriaga, M. (2018). Characterization of bacterial communities in mexican artisanal raw milk bola de ocosingo cheese by high-throughput sequencing. *Frontiers in Microbiology*, *9*, 2598, 1-7. http://doi.org/10.3389/fmicb.2018.02598.

Alvarez-Cisneros, Y. M., Fernández, F. J., Wacher-Rodarte, C., Aguilar, M. B., Sáinz, Espuñes T. del R. & Ponce-Alquicira, E. (2010). Biochemical characterization of a bacteriocin-like inhibitory substance produced by *Enterococcus faecium* MXVK29 isolated from Mexican traditional sausage. *Journal of the Science and Food Agriculture*, *90*, 2475-2481. https://doi.org/10.1002/jsfa.4109.

Álvarez-Cisneros, Ye. M., Fernández, J. F., Wacher-Rodarte, C., Sáinz-Espuñes, T. & Ponce-Alquicira, E. (2011). Enterocins: Bacteriocins with applications in the food industry In: Science against microbial pathogens: communicating current research and technological advances. A. Méndez-Vilas (Ed), *Microbiology Series N° 3*, Formatex, 1130-1341.

Álvarez-Cisneros, Y. M. & Ponce-Alquicira, E. (2018). Antibiotic Resistance in Lactic Acid Bacteria. In. Antimicrobial Resistance - A Global Threat, Dr. Yashwant Kumar (Ed). *IntechOpen London UK*, 1-33. DOI: 10.5772/intechopen.80624

Anzengruber, J., Courtin, P., Claes, I. J., Debreczeny, M., Hofbauer, S., Obinger, C., Chapot-Chartier, M. P., Vanderleyden, J., Messner, P. & Schäffer, C. (2014). Biochemical characterization of the major N-acetylmuramidase from *Lactobacillus buchneri*. *Microbiology*, *160*(8), 1807-1819. http://doi.org:10.1099/mic.0.078162-0.

BACTIBASE database: http://bactibase.hammamilab.org/#. Accessed January 2019.

BAGEL4 webserver. http://bagel4.molgenrug.nl/. Aaccessed January 2019.

Baindara, P., Korpole, S. & Grove, V. (2018). Bacteriocins: perspective for the development of novel anticancer drugs. *Applied Microbiology and Biotechnology*, *102*(24), 10393-10408. https://doi.org/10.1007/s00253-018-9420-8.

Balay, D. R., Gänzle, M. G. & McMullen, L. M. (2018). The Effect of Carbohydrates and Bacteriocins on the Growth Kinetics and Resistance of *Listeria monocytogenes*. *Frontiers in Microbiology*, *9*(347), 1-12. https://doi.org/10.3389/fmicb.2018.00347.

Balciunas, E. M., Castillo Martinez, F. A., Todorov, S. D., Gombossy de Melo Franco, B. D., Converti, A. & Pinheiro de Souza, Oliveira R. (2013). Novel biotechnological applications of bacteriocins: A review. *Food Control*, *32*(1), 134-142. https://doi.org/ 10.1016/j.foodcont.2012.11.025.

Barbosa, M. S., Todorov, S. D., Ivanova, I. V., Belguesmia, Y., Choiset, Y., Rabesona, H. Chobert, J. M., Haertle, T. & Franco, B. D. G. M. (2016). Characterization of a two-peptide plantaricin produced by Lactobacillus plantarum MBSa4 isolated from

Brazilian salami. *Food Control*, *60*, 103-112. https://doi.org/10.1016/j.foodcont. 2015.07.029.

Berlec, A., Ravnikar, M. & Strukelj, B. (2012). Lactic acid bacteria as oral delivery systems for biomolecules. *Pharmazie*, *67*, 891–898. PMID 23210237 http://doi.org/ 10.1691/ph.2012.1705.

Bermudez-Humarán, L. G., Kharrat, P., Chatel, J. M. & Langella, P. (2011). Lactococci and lactobacilli as mucosal delivery vectors for therapeutic proteins and DNA vaccines. *Microbial Cell Factories*, *10*(1), S4. http://doi.org/10.1186/1475-2859-10-S1-S4.

Blazanovic, K., Zhao, H., Choi, Y., Li, W., Salvat, R. S., Osipovitch, D. C., Fields, J., Moise, L., Berwin, B. L., Fiering, S. N., Bailey-Kellogg, C. & Griswold, K. E. (2015). Structure-based redesign of lysostaphin yields potent antistaphylococcal enzymes that evade immune cell surveillance. *Molecular therapy. Methods & clinical development*, *2*, 15021, 1-10. http://doi.org/10.1038/mtm.2015.21.

Borrero, J. 1., Brede, D. A., Skaugen, M., Diep, D. B., Herranz, C., Nes, I. F., Cintas, L. M. & Hernández, P. E.. (2011). Characterization of garvicin ML, a novel circular bacteriocin produced by *Lactococcus garvieae* DCC43, isolated from *Mallard ducks* (Anas platyrhynchos). *Applied and Environmental Microbiology*, *77*(1), 369-373. http://doi.org/10.1128/AEM.01173-10.

Bourhis, L. L. & Werts, C. (2007). Role of Nods in bacterial infection. *Microbes and Infection*, *9*(5), 629–636, http://doi.org/10.1016/j.micinf.2007.01.014.

Breukink, Eefjan. & Ben, De Kruijff. (2006). Lipid II as a target for antibiotics. *Nature Reviews Drug Discovery*, *5*, 321-323. http://doi.org/10.1038/nrd2004.

Bustamante, N., Rico-Lastres, P., García, E., García, P. & Menéndez, M. (2012). Thermal stability of Cpl-7 endolysin from the Streptococcus pneumoniae bacteriophage Cp-7; cell wall-targeting of its CW_7 motifs. *PLoS One*, *7*(10), e46654, 1-11. http:// doi.org/10.1371/journal.pone.0046654.

Calderón-Oliver, Pedroza-Islas R., Escalona-Buendía, H. B., Pedraza-Chaverri, J. & Ponce-Alquicira, E. (2017). Comparative study of the microencapsulation by complex coacervation of nisin in combination with an avocado antioxidant extract. *Food Hydrocolloids*, *62*, 49-67. https://doi.org/10.1016/j.foodhyd.2016.07.028.

Callewaert, L., Walmagh, M., Michiels, C. W. & Lavigne, R. (2011). Food applications of bacterial cell wall hydrolases. *Current Opinion in Biotechnology*, *22*, 164-171. http://doi.org/10.1016/j.copbio.2010.10.012.

Campion, A., Casey, P. G., Des Field, Cotter P. D., Hill, C. & Ross, R. P. (2013). *In vivo* activity of Nisin A and Nisin V against *Listeria monocytogenes* in mice. *BMC microbiology*, *13*, 23, 1-8. http://doi.org/10.1186/1471-2180-13-23.

Carrillo, W., García-Ruiz, A., Recio, I. & Moreno-Arribas, M. V. (2014). Antibacterial activity of hen egg white lysozyme modified by heat and enzymatic treatments against oenological lactic acid bacteria and acetic acid bacteria. *Journal of Food Protection*, *77*(10), 1732-1739. https://doi.org/10.4315/0362-028X.JFP-14-009.

Cebrián, R., Baños, A., Valdivia, E., Pérez-Pulido, R., Martínez-Bueno, M. & Maqueda, M. (2012). Characterization of functional, safety, and probiotic properties of *Enterococcus faecalis* UGRA10, a new AS-48-producer strain. *Food Microbiology*, *30*(1), 59-67. https://doi.org/10.1016/j.fm.2011.12.002.

Cebrián, R., Martínez-Bueno, M., Valdivia, E., Albert, A., Maqueda, M. & Sánchez-Barrena, M. J. (2015). The bacteriocin AS-48 requires dimer dissociation followed by hydrophobic interactions with the membrane for antibacterial activity. *Journal of Structural Biology*, *190*(2), 162-172. https://doi.org/10.1016/j.jsb.2015.03.006.

Cebrián, R., Arévalo, S., Rubiño, S., Arias-Santiago, S., Rojo, M. D., Montalbán-López, M., Martínez-Bueno, M., Valdivia, E. & Maqueda, M. (2018). Control of Propionibacterium acnes by natural antimicrobial substances: Role of the bacteriocin AS-48 and lysozyme. *Scientific Reports*, *8*, 11766, 1-11. https://doi.org/10.1038/s41598-018-29580-7.

Chandra Ojha, S., Imtong, C., Meetum, K., Sakdee, S., Katzenmeier, G. & Angsuthanasombat, C. (2018). Purification and characterization of the antibacterial peptidase lysostaphin from *Staphylococcus simulans*: Adverse influence of Zn^{2+} on bacteriolytic activity. *Protein Expression and Purification*, *151*, 106-112. http://doi.org/10.1016/j.pep.2018.06.013.

Chikindas, M. L., Weeks, R., Drider, D., Chistyakov, V. A. & Dicks, L. M. (2018). Functions and emerging applications of bacteriocins. *Current Opinion in Biotechnology*, *49*, 23-28. https://doi.org/10.1016/j.copbio.2017.07.011.

Cibik, R. & Chapot-Chartier, M. P. (2004). Characterization of autolytic enzymes in *Lactobacillus pentosus*. *Letters in Applied Microbiology*, *38*, 459–463. http://doi.org/10.1111/j.1472-765X.2004.01516.x.

Claes, I. J., Schoofs, G., Regulski, K., Courtin, P., Chapot-Chartier, M. P., Rolain, T., Hols, P., von Ossowski, I., Reunanen, J., de Vos, W. M., Palva, A., Vanderleyden, J., De Keersmaecker, S. C. & Lebeer, S. (2012). Genetic and biochemical characterization of the cell wall hydrolase activity of the major secreted protein of *Lactobacillus rhamnosus* GG. *PloS One*, *7*(2), e31588, 1-8. http://doi.org/10.1371/journal.pone.0031588.

Cotter, P. D., Hill, C. & Ross, R. P. (2005). Bacteriocins: developing innate immunity for food. *Nature Reviews Microbiology*, *3*, 777-788. https://doi.org/10.1038/nrmicro1273.

Cotter, P. D., Ross, R. P. & Hill, C. (2013). Bacteriocins -a viable alternative to antibiotics?. *Nature Reviews Microbiology*, *11*, 95-105. https://doi.org/10.1038/nrmicro2937.

Delattin, N., Brucker, K., Cremer, K., Cammue, B. P. & Thevissen, K. (2017). Antimicrobial peptides as a strategy to combat fungal biofilms. *Current Topics in*

Medicinal Chemistry, 17, 604-612. https://doi.org/10.2174/15680266166661607
13142228.

Egan, K., Des Field, Rea M. C., Ross, R. P., Hill, C. & Cotter, P. D. (2016). Bacteriocins: novel solutions to age old spore-related problems?. Frontiers in Microbiology, 7(462), 1-21. https://doi.org/10.3389/fmicb.2016.00461.

Escamilla-Martínez, E. E., Cisneros, Y. M. Á., Fernández, F. J., Quirasco-Baruch, M. & Ponce-Alquicira, E. (2017). Identification of structural and immunity genes of a class iib bacteriocin encoded in the enterocin a operon of Enterococcus faecium Strain MXVK29. Journal of Food Protection, 80(11), 1851-1856. https://doi.org/10.4315/0362-028X.JFP-17-039.

Field, D., Ross, R. P. & Hill, C. (2018). Developing bacteriocins of lactic acid bacteria into next generation biopreservatives. Current Opinion in Food Science, 20, 1-6. https://doi.org/10.1016/j.cofs.2018.02.004.

Fischetti, V. A. (2010). Bacteriophage endolysins: a novel anti-infective to control Gram-positive pathogens. International Journal of Medical Microbiology, 300(6), 357-62. http://doi.org/10.1016/j.ijmm.2010.04.002.

Franz, C. M., van Belkum, M. J., Holzapfel, W. H. Abriouel, H. & Gálvez, A. (2007). Diversity of enterococcal bacteriocins and their grouping in a new classification scheme. FEMS Microbiology Reviews, 31(3), 293–310. https://doi.org/10.1111/j.1574-6976.2007.00064.x.

Fraqueza, M. J. (2015). Antibiotic resistance of lactic acid bacteria isolated from dry-fermented sausages. International Journal of Food Microbiology, 212, 76-88. https://doi.org/10.1016/j.ijfoodmicro.2015.04.035.

Gad, G. F. M., Abdel-Hamid, A. M. & Farag, Z. S. H. (2014). Antibiotic resistance in lactic acid bacteria isolated from some pharmaceutical and dairy products. Brazilian Journal of Microbiology, 45(1), 25-33. ISSN 1678-4405.

Galié, S., García-Gutiérrez, C., Miguélez, E. M., Villar, J. & Lombó, F. (2018). Biofilms in the food industry: Health aspects and control methods. Frontiers in Microbiology, 9(898), 1-18. http://doi.org/10.3389/fmicb.2018.00898.

García-Cano, I. (2013). Peptidoglucano hidrolasa de Pediococcus acidilactici ATCC 8042: Detección, localización celular y evaluación de su potencial como antibacteriano [Peptidoglycan Hydrolase from Pediococcus acidilactici ATCC 8042: Detection, cell localization and evaluation of its potential as antibacterial]. PhD thesis. Universidad Nacional Autónoma de México-Facultad de Química.

García-Cano, I., Velasco-Pérez, L., Rodríguez-Sanoja, R., Sánchez, S., Mendoza-Hernández, G., Llorente-Bousquets, A. & Farrés, A. (2011). Detection, cellular localization and antibacterial activity of two lytic enzymes of Pediococcus acidilactici ATCC 8042. Journal Applied Microbiology, 111, 607-615. http://doi.org/10.1111/j.1365-2672.2011.0588.x.

García-Cano, I., Campos-Gómez, M., Contreras-Cruz, M., Serrano-Maldonado, C. E., González-Canto, A., Peña-Montes, C., Rodríguez-Sanoja, R., Sánchez, S. & Farrés, A. (2015). Expression, purification, and characterization of a bifunctional 99-kDa peptidoglycan hydrolase from Pediococcus acidilactici ATCC 8042. *Applied Microbiology and Biotechnology*, *99*, 8563-8573. http://doi.org:10.1007/s00253-015-6593-2.

Gerbino, E., Carasi, P., Mobili, P., Serradell, M. A. & Gómez-Zavaglia, A. (2015). Role of S-layer proteins in bacteria. *World Journal Microbiol Biotechnol*, *31*, 1877. https://doi.org/10.1007/s11274-015-1952-9.

Ghuysen J. M., Tipper D. J. & Strominger J. L. (1966). Enzymes that degrade bacterial cell walls. *Methods in Enzymology*, *8*, 685–699. http://doi.org/10.1016/0076-6879(66) 08124-2.

Giraffa, Giorgio. (2014). Overview of the ecology and biodiversity of the LAB. In: *In Lactic Acid Bacteria: Biodiversity and Taxonomy*, Wilhelm H. Hotzapfel and Brian J. B. Wood (Ed.). John Wiley, 45-54 https://doi.org/10.1002/9781118655252.

Goudarzi, F., Asadi, A., Afsharpour, M. & Jamadi, R. H. (2018). *In vitro* characterization and evaluation of the cytotoxicity effects of nisin and nisin-loaded PLA-PEG-PLA nanoparticles on gastrointestinal (AGS and KYSE-30), hepatic (HepG2) and blood (K562) cancer cell lines. *American Association of Pharmaceutical Scientists*, *Pharm Sci Tech*, *19*, 1554-1566. https://doi.org/10.1208/s12249-018-0969-4.

Haddad Kashani, H., Schmelcher, M., Sabzalipoor, H., Seyed Hosseini, E. & Moniri, R. (2017). Recombinant endolysins as potential therapeutics against antibiotic-resistant *Staphylococcus aureus*: Current status of research and novel delivery strategies. *Clinical Microbiology Reviews*, *31*(1), e00071-17, 1-26. http://doi.org/10.1128/CMR.00071-17.

Heidrich, C., Templin, M. F., Ursinus, A., Merdanovic, M., Berger, J., Schwarz, H., De Pedro, M. A. & Höltje, J. V. (2001). Involvement of N-acetylmuramyl-l-alanine amidases in cell separation and antibiotic-induced autolysis of Escherichia coli. *Molecular Microbiology*, *41*, 167–178. http://doi.org/10.1046/j.1365-2958.2001.02499.x.

Heunis, T. D. J., Smith, C. & Dicks, L. M. T. (2013). *Evaluation of a nisin-eluting nanofiber scaffold to treat Staphylococcus aureus-induced skin infections in mice Antimicrobial agents and chemotherapy*, *57*(8), 3928-3935. http://doi.org/10.1128/AAC.00622-13.

Höltje, J. V. & Tuomanen, E. I. (1991). The murein hydrolases of Escherichia coli: properties, functions and impact on the course of infections *in vivo*. *Journal of General Microbiology*, *137*, 441–454. http://doi.org/10.1099/00221287-137-3-441.

Höltje, J. V. (1998). Growth of the stress-bearing and shape-maintaining murein sacculus of Escherichia coli. *Microbiology and Molecular Biology Reviews*, *62*(1), 181–203. PMID 9529891.

Hynönen, U. & Palva, A. (2013). Lactobacillus surface layer proteins: structure, function and applications. *Applied Microbiology and Biotechnology*, *97*(12), 5225-5243. http://doi.org:10.1007/s00253-013-4962-2.

Jacobs, C., Frère, J. M. & Normark, S. (1997). Cytosolic intermediates for cell wall biosynthesis and degradation control inducible beta-lactam resistance in gram-negative bacteria. *Cell*, *88*, 823–832. https://doi.org/10.1016/S0092-8674(00)81928-5.

Jacq, M., Arthaud, C., Manuse, S., Mercy, C., Bellard, L., Peters, K., Gallet, B., Galindo, J., Doan, T., Vollmer, W., Brun, Y. V., VanNieuwenhze, M. S., Di Guilmi, A. M., Vernet, T., Grangeasse, C. & Morlot, C. (2018). The cell wall hydrolase Pmp23 is important for assembly and stability of the division ring in *Streptococcus pneumoniae*. *Scientific Reports*, *8*(1), 7591, 1-14. http://doi.org/10.1038/s41598-018-25882-y.

Jagielska, E., Chojnacka, O. & Sabała, I. (2016). LytM fusion with SH3b-like domain expands its activity to physiological conditions. *Microbial Drug Resistance*, *22*(6), 461-469. http://doi.org/10.1089/mdr.2016.0053.

Johnson, B. t. R. & Klaenhammer, T. R. (2016). AcmB is an S-layer-associated β-N-acetylglucosaminidase and functional autolysin in lactobacillus acidophilus NCFM. *Applied Environmental Microbiology*, *82*(18), 5687-97. http://doi.org/10.1128/AEM.02025-16.

Johnson, J. W., Fisher, J. F. & Mobashery, S. (2013). Bacterial cell-wall recycling. *Annals of the New York Academy of Sciences*, *1277*(1), 54–75. http://doi.org/10.1111/j.1749-6632.2012.06813.x.

Kadariya, J., Smith, T. C. & Thapaliya, D. (2014). *Staphylococcus aureus* and staphylococcal food-borne disease: An ongoing challenge in public health. *Biomed Research International*, 2014, (ID 827965.), 1-9. http://doi.org/10.1155/2014/827965.

Kamarajan, P., Hayami, T., Matte, B., Liu, Y., Danciu, T., Ramamoorthy, A., Worden, F., Kapila, S. & Kapila, Y. (2015). Nisin ZP, a bacteriocin and food preservative, inhibits head and neck cancer tumorigenesis and prolongs survival. *Plos One*, *10*(7) e0131008, 1-20. https://doi.org/10.1371/journal.pone.0131008.

Khalid, K. (2011). An overview of lactic acid bacteria. *International Journal of Biosciences*, *1*(3), 1-13. ISSN: 2220-6655 http://www.innspub.net.

Klaenhammer, T. R. (1993). *Genetics of bacteriocins produced by lactic acid bacteria FEMS Microbiology Reviews*, *12*, 39–85. https://doi.org/10.1111/j.1574-6976.1993.tb00012.x.

Koné, A. P., Zea, J. M. V., Gagné, D., Cinq-Mars, D., Guay, F. & Saucier, L. (2018). Application of *Carnobacterium maltaromaticum* as a feed additive for weaned rabbits to improve meat microbial quality and safety. *Meat Science*, *135*, 174-188. http://dx.doi.org/10.1016/j.meatsci.2017.09.017.

Lagha, A. Ben., Haas, B., Gottschalk, M. & Grenier, D. (2017). Antimicrobial potential of bacteriocins in poultry and swine production. *Veterinary Research*, *48*(22), 1-12. https://doi.org/10.1186/s13567-017-0425-6.

Layec, S., Decaris, B. & Leblond-Bourget, N. (2008). Diversity of Firmicutes peptidoglycan hydrolases and specificities of those involved in daughter cell separation. *Research in Microbiology*, *159*, 507-515. http://doi.org/10.1016/j.resmic.2008.06.008.

Lee, B. Y., Singh, A., David, M. Z., Bartsch, S. M., Slayton, R. B., Huang, S. S., Zimmer, S. M., Potter, M. A., Macal, C. M., Lauderdale, D. S., Miller, L. G. & Daum, R. S. (2012). The economic burden of community-associated methicillin-resistant *Staphylococcus aureus* (CA-MRSA). *Clinical microbiology and infection: the official publication of the European Society of Clinical Microbiology and Infectious Diseases*, *19*(6), 528-536. http://doi.org:10.1111/j.1469-0691.2012.03914.x.

Lewies, A., Wentzel, J. F., Miller, H. C. & Du Plessis, L. H. (2018). The antimicrobial peptide nisin Z induces selective toxicity and apoptotic cell death in cultured melanoma cells. *Biochimie*, *144*, 28-40. https://doi.org/10.1016/j.biochi.2017.10.009.

Loessner, M. J., Kramer, K., Ebel, F. & Scherer, S. (2002). C-terminal domains of *Listeria monocytogenes* bacteriophage murein hydrolases determine specific recognition and high-affinity binding to bacterial cell wall carbohydrates. *Molecular Microbiology*, *44*, 335–349. http://doi.org/10.1046/j.1365-2958.2002.02889.x.

López-Arvizu, L. A., García-Cano, I. & Ponce-Alquicira, E. (2017). Clonación y expresión de la peptidoglucano hidrolasa producida por *Lactobacillus sakei* [Cloning and expression of peptidoglycan hydrolase produced by *Lactobacillus sakei*]. In *XVII Congreso Nacional de Biotecnología y Bioingeniería SMBB* [*XVII National Congress of Biotechnology and Bioengineering SMBB*]. Puerto Vallarta, Jalisco Mexico.

López-Cuellar, M. R., Rodríguez-Hernández, A. I. & Chavarría-Hernandez, N. (2016). LAB bacteriocin applications in the last decade. *Biotechnology & Biotechnological Equipment*, *30*(6), 1039-1050. https://doi.org/10.1080/13102818.2016.1232605.

Margot, P., Pagni, M. & Karamata, D. (1999). *Bacillus subtilis* 168 gene lytF encodes a gammaD-glutamate-meso-diaminopimelate muropeptidase expressed by the alternative vegetative sigma factor, sigmaD. *Microbiology*, *145* (Pt 1), 57–65. http://doi.org/10.1099/13500872-145-1-57.

Marketwire. (2011). *Revolutionary new food ingredient, Micocin®, fights Listeria*. http://www.marketwired.com/press-release/revolutionary-new-food-ingredient-micocinfights-listeria-1391478.htm.

Martin-Visscher, L. A., van Belkum, M. J., Garneau-Tsodikova, S., Whittal, R. M., Zheng, J., McMullen, L. M. & Vederas, J. C. (2008). Isolation and characterization of Carnocyclin A, a novel circular bacteriocin produced by *Carnobacterium maltaromaticum* UAL307. *Applied and Environmental Microbiology*, *74*(15), 4756-63. http://doi.org/10.1128/AEM.00817-08.

Martin-Visscher, L. A., Yoganathan, S., Sit, C. S., Lohans, C. T. & Vederas, J. C. (2011). The activity of bacteriocins from *Carnobacterium maltaromaticum* UAL307 against Gram-negative bacteria in combination with EDTA treatment. *FEMS Microbiology Letters*, *317*(2), 152–159. https://doi.org/10.1111/j.1574-6968.2011.02223.x.

Meng, J., Gao, S. M., Zhang, Q. X. & Lu, R. R. (2015). Murein hydrolase activity of surface layer proteins from *Lactobacillus acidophilus* against *Escherichia coli*. *International Journal of Biological Macromolecules*, *79*, 527-532. http://doi.org/10.1016/j.ijbiomac.2015.03.057.

Meroueh, S. O., Bencze, K. Z., Hesek, D., Lee, M., Fisher, J. F., Stemmler, T. L. & Mobashery, S. (2006). Three-dimensional structure of the bacterial cell wall peptidoglycan. *Proceedings of the National Academy of Sciences*, *103*(12), 4404–4409. http:// doi.org/10.1073/pnas.0510182103.

Mesnage, S., Dellarole, M., Baxter, N. J., Rouget, J. B., Dimitrov, J. D., Wang, N., Fujimoto, Y., Hounslow, A. M., Lacroix-Desmazes, S., Fukase, K., Foster, S. J. & Williamson, M. P. (2014). Molecular basis for bacterial peptidoglycan recognition by LysM domains. *Nature Communications*, *5*, 4269, 1-11. http://doi.org/10.1038/ncomms 5269.

Mierau, I., Leij, P., van Swam, I., Blommestein, B., Floris, E., Mond, J. & Smid, E. J. (2005). Industrial-scale production and purification of a heterologous protein in Lactococcus lactis using the nisin-controlled gene expression system NICE: The case of lysostaphin. *Microbial Cell Factories*, *4*(15), 1-9. http://doi.org/ 0.1186/1475-2859-4-15.

Miljkovic, M., Uzelac, G., Mirkovic, N., Devescovi, G., Diep, D. B., Venturi, V. & Kojic, M. (2016). LsbB bacteriocin interacts with the third transmembrane domain of the YvjB receptor. *Applied and Environmental Microbiology*, *82*(17), 5364- 537. https://doi.org/10.1128/AEM.01293-16.

Mills, S., Griffin, C., O'Connor, P. M., Serrano, L. M., Meijer, W. C., Hill, C. & Ross, R. P. (2017). A multibacteriocin cheese starter system, comprising nisin and lacticin 3147 in *Lactococcus lactis*, in combination with plantaricin from *Lactobacillus plantarum*. *Applied and Environmental Microbiology*, *83*(14), 799-17. http://doi.org// 10.1128/AEM.00799-17.

Murillo-Martínez, M. M., Tello-Solís, S. R., García-Sánchez, M. A. & Ponce-Alquicira, E. (2013). Antimicrobial activity and hydrophobicity of edible whey protein isolate films formulated with nisin and/or glucose oxidase. *Journal of Food Science*, *78*(4), M560-M566. http://www.doi.org/10.1111/1750-3841.12078.

Nakimbugwe, D., Masschalck, B., Deckers, D., Callewaert, L., Aertsen, A. & Michiels, C. W. (2006). Cell wall substrate specificity of six different lysozymes and lysozyme inhibitory activity of bacterial extracts. *FEMS Microbiology Letters*, *259*, 41–46. http://doi.org/10.1111/j.1574-6968.2006.00240.x.

Nelson, D. C., Schmelcher, M., Rodriguez-Rubio, L., Klumpp, J., Pritchard, D. G., Dong, S. & Donovan, D. M. (2012). Endolysins as antimicrobials. *Advances in Virus Research*, *83*, 299–365. http://doi.org/10.1016/B978-0-12-394438-2.00007-4.

O'Connor, P. M., Ross, R. P., Hill, C. & Cotter, P. D. (2015). Antimicrobial antagonists against food pathogens: a bacteriocin perspective. *Current Opinion in Food Science*, *2*, 51–57. http://dx.doi.org/10.1016/j.cofs.2015.01.004.

Osipovitch, D. C. & Griswold, K. E. (2014). Fusion with a cell wall binding domain renders autolysin LytM a potent anti-*Staphylococcus aureus* agent. *FEMS Microbiology Letters*, *362*(2), 1-7. http://doi.org/10.1093/femsle/fnu035.

Owen, R. A., Fyfe, P. K., Lodge, A., Biboy, J., Vollmer, W., Hunter, W. N. & Sargent, F. (2018). Structure and activity of ChiX: a peptidoglycan hydrolase required for chitinase secretion by *Serratia marcescens*. *The Biochemical Journal*, *475*(2), 415-428. http://doi.org/10.1042/BCJ20170633.

Özogul, F. & Hamed, I. (2018). The importance of lactic acid bacteria for the prevention of bacterial growth and their biogenic amines formation: A review. *Critical reviews in food science and nutrition*, *58*(10), 1660-1670. https://doi.org/10.1080/10408398.2016.1277972.

Palumbo, E., Deghorain, M., Cocconcelli, P. S., Kleerebezem, M., Geyer, A., Hartung, T., Morath, S. & Hols, P. (2006). D-alanyl ester depletion of teichoic acids in *Lactobacillus plantarum* results in a major modification of lipoteichoic acid composition and cell wall perforations at the septum mediated by the Acm2 autolysin. *Journal of Bacteriology*, *188*(10), 3709-3715. http://doi.org/10.1128/JB.188.10.3709-3715.2006.

Parisien, M. A., Allain, B., Zhang, J., Mandeville, R. & Lan, C. Q. (2008). Novel alternatives to antibiotics: bacteriophages, bacterial cell wall hydrolases, and antimicrobial peptides. *Journal Applied Microbiology*, *104*, 1–13. http://doi.org/10.1111/j.1365-2672.2007.03498.x.

Perez, R. H., Zendo, T. & Sonomoto, K. (2014). Novel bacteriocins from lactic acid bacteria (LAB): various structures and applications. *Microbial Cell Factories*, *13*, 1-13. https://doi.org/10.1186/1475-2859-13-S1-S3.

Perez, R. H., Zendo, T. & Sonomoto, K. (2018). Circular and leaderless bacteriocins: Biosynthesis, mode of action, applications, and prospects. *Frontiers in Microbiology*, *9*(2085), 1-18. https://doi.org/10.3389/fmicb.2018.02085.

Prado Acosta, M., Mercedes Palomino, M., Allievi, M. C., Sanchez Rivas, C. & Ruzal, S. M. (2008). Murein hydrolase activity in the surface layer of *Lactobacillus acidophilus* ATCC 4356. *Applied & Environmental Microbiology*, *74*(24), 7824-7827. http:// doi.org/10.1128/AEM.01712-08.

Preet, S., Bharati, S., Panjeta, A., Tewari, R. & Rishi, P. (2015). Effect of nisin and doxorubicin on DMBA-induced skin carcinogenesis—a possible adjunct therapy. *Tumor Biology, 36*, 8301- 8308. https://doi.org/10.1007/s13277-015-3571-3.

Regulski, K., Courtin, P., Meyrand, M., Claes, I. J., Lebeer, S., Vanderleyden, J., Hols, P., Guillot, A. & Chapot-Chartier, M. P. (2012). Analysis of the peptidoglycan hydrolase complement of *Lactobacillus casei* and characterization of the major γ-D-glutamyl-L-lysyl-endopeptidase. *PLoS One, 7*(2), e32301, 1-11. http://doi.org/ 10.1371/journal.pone.0032301.

Reith, J. & Mayer, C. (2011). Peptidoglycan turnover and recycling in Gram-positive bacteria. *Applied Microbiology and Biotechnology, 92*(1), 1–11. http://doi.org/ 10.1007/s00253-011-3486-x.

Rice, K. C. & Bayles, K. W. (2003). Death's toolbox: examining the molecular components of bacterial programmed cell death. *Molecular Microbiology, 50*, 729–738. http:// doi.org/10.1046/j.1365-2958.2003.t01-1-03720.x.

Ríos-Colombo, N. S., Chalón, M. C. & Navarro, S. A. (2018). Pediocin-like bacteriocins: new perspectives on mechanism of action and immunity. *Current Genetics, 64*, 345-351. http://www.doi.org/10.1007/s00294-017-0757-9.

Roa, K. (2017). *Evaluación de la vida de anaquel de una salchicha tipo Viena con la adición de metabolitos antimicrobianos producidos por bacterias ácido lácticas* [*Evaluation of the shelf life of a Vienna type sausage with the addition of antimicrobial metabolites produced by lactic acid bacteria*] (BSc Thesis). Universidad Nacional Autónoma de México-Facultad de Química.

Rolain, T., Bernard, E., Courtin, P., Bron, P. A., Kleerebezem, M., Chapot-Chartier, M. P. & Hols, P. (2012). Identification of key peptidoglycan hydrolases for morphogenesis, autolysis, and peptidoglycan composition of *Lactobacillus plantarum* WCFS1. *Microbial Cell Factories, 11*, 137, 1-15. http://doi.org/10.1186/1475-2859-11-137.

Sabala, I., Jagielska, E., Bardelang, P. T., Czapinska, H., Dahms, S. O., Sharpe, J. A., James, R., Than, M. E., Thomas, N. R. & Bochtler, M. (2014). Crystal structure of the antimicrobial peptidase lysostaphin from *Staphylococcus simulans. The FEBS Journal, 281*(18), 4112-4222. http://doi.org/10.1111/febs.12929.

Schmelcher, M., Powell, A. M., Becker, S. C., Camp, M. J. & Donovan, D. M. (2012). Chimeric phage lysins act synergistically with lysostaphin to kill mastitis-causing *Staph. aureus* in murine mammary glands. *Applied Environmental Microbiology, 78*(7), 2297-2305. http://doi.org/10.1128/AEM.07050-11.

Schneider, R., Fernández, J. F., Aguilar, M. B., Guerrero-Legarreta, I., 1 Alpuche-Solis, A. & Ponce-Alquicira, E. (2006). Partial characterization of a class IIa pediocin produced by *Pediococcus parvulus* 133 strain isolated from meat (Mexican chorizo). *Food Control, 17*, 909-915. https://doi.org/10.1016/j.foodcont. 2005.06. 010.

Schumann, P. (2011). Chapter 5 peptidoglycan structure. In: Methods in Microbiology, Academic Press. *Fred Rainey and Aharon Oren* (Eds), Vol. *38*, 101-129. http://doi.org/10.1016/B978-0-12-387730-7.00005-X.

Serrano-Maldonado, C., García-Cano, I., González-Canto, A., Ruiz-May, E., Elizalde-Contreras, J. M. & Quirasco, M. C. (2017). Cloning and Characterization of a Novel N-acetylglucoaminidase (AtlD) from *Enterococcus faecalis*. *Journal of Molecular Microbiology and Biotechnology*, *28*(1), 14-27. https://doi.org/10.1159/000486757.

Sharma, A. K., Kumar, S., Harish, K., Dhakan, D. B. & Sharma, V. K. (2016). Prediction of peptidoglycan hydrolases- a new class of antibacterial proteins. *BMC Genomics*, *17*(411), 2-12 http://doi.org/10.1186/s12864-016-2753-8.

Sharma, P., Tomar, S. K., Goswami, P., Sangwan, V. & Singh, R. (2014). Antibiotic resistance among commercially available probiotics. *Food Research International*, *57*, 176-195. https://doi.org/10.1016/j.foodres.2014.01.025.

Shin, J. M., Gwak, J. W., Kamarajan, P., Fenno, J., Rickard, A. H. & Kapila, Y. L. (2016). Biomedical applications of nisin. *Journal of Applied Microbiology*, *120*, 1449-1465. http://doi.org/10.1111/jam.13033.

Silva, C. C. G., Silva, S. P. M. & Ribeiro, S. C. (2018). Application of bacteriocins and protective cultures in dairy food preservation. *Frontiers in Microbiology*, *9*(594), 1-15. https://doi.org/10.3389/fmicb.2018.00594.

Smith, T. J., Blackman, S. & Foster, S. J. (2000). Autolysins of *Bacillus subtilis*: multiple enzymes with multiple functions. *Microbiology*, *146*, 249–262. http://doi.org/10.1099/00221287-146-2-249.

Tossavainen, H., Raulinaitis, V., Kauppinen, L., Pentikäinen, U., Maaheimo, H. & Permi, P. (2018). Structural and functional insights into lysostaphin-substrate interaction. *Frontiers in Molecular Biosciences*, *5*(60), 1-14. http://doi.org/10.3389/fmolb.2018.00060.

van Hal, S. J., Jensen, S. O., Vaska, V. L., Espedido, B. A., Paterson, D. L. & Gosbell, I. B. (2012). Predictors of mortality in *Staphylococcus aureus* Bacteremia. *Clinical Microbiology Reviews*, *25*(2), 362-386. http://doi.org/10.1128/CMR.05022-11.

van Heijenoort, J. (2011). Peptidoglycan hydrolases of *Escherichia coli*. *Microbiology & Molecular Biology Reviews*, *75*(4), 636-663. http://doi.org/10.1128/MMBR.00022-11.

Vollmer, W., Joris, B., Charlier, P. & Foster, S. (2008). Bacterial peptidoglycan (murein) hydrolases. *FEMS Microbiology Reviews*, *32*, 259–286. http://doi.org/10.1111/j.1574-6976.2007.00099.x.

Wang, S. & Shaevitz, J. W. (2013). The mechanics of shape in prokaryotes. *Frontiers in Bioscience*, *5*, 564–574. http://doi.org/10.2741/S390.

Wessels, M. R. (2005). Streptolysin S. *The Journal of Infectious Diseases*, *192*(1), 13–15. *https://doi.org/10.1086/430625.*

Wyszyńska, A., Kobierecka, P., Bardowski, J. & Jagusztyn-Krynicka, E. K. (2015). Lactic acid bacteria--20 years exploring their potential as live vectors for mucosal vaccination. *Applied Microbiology and Biotechnology*, *99*(7), 2967-2977. http://doi.org:10.1007/s00253-015-6498-0.

Xu, Yi., Wang, T., Kong, J. & Wang, H. L. (2015). Identification and functional characterization of AcIB, a novel cell-separating enzyme from *Lactobacillus casei*. *International Journal of Food Microbiology*, *203*, 93-100. http://doi.org/10.1016/j.ijfoodmicro.2015.03.011.

Zacharof, M. P. & Lovitt, R. W. (2012). *Bacteriocins produced by lactic acid bacteria APCBEE Procedia*, *2*, 50-56. https://doi.org/10.1016/j.apcbee.2012.06.010.

In: The Many Benefits of Lactic Acid Bacteria
Editors: J. G. LeBlanc and A. de Moreno

ISBN: 978-1-53615-388-0
© 2019 Nova Science Publishers, Inc.

Chapter 8

ANTIMICROBIAL PRODUCTS: AN IMPORTANT FEATURE OF LAB AND THEIR APPLICATIONS

Sujitra Techo[1], and Somboon Tanasupawat[2],†*

[1]Mahidol University, Nakhonsawan Campus, Nakhonsawan, Thailand
[2]Department of Biochemistry and Microbiology, Faculty of Pharmaceutical Sciences,
Chulalongkorn University, Bangkok, Thailand

ABSTRACT

Lactic acid bacteria (LAB) are known to be present in diverse origins which are rich in nutrients and energy sources, for instance, milk, meat, plants and gastrointestinal tract of humans and animals. LAB that have extensively been found to belong to the genera *Lactobacillus* (*Lb*), *Streptococcus* (*Strep*), *Enterococcus* (*Ec*), *Pediococcus* (*Ped*), *Lactococcus* (*Lc*), *Leuconostoc* (*Leuc*), *Aerococcus* (*A*), *Oenococcus* (*O*), *Carnobacterium* (*Cb*), *Vagococcus* (*V*), *Tetragenococcus* (*T*), *Weisella* (*W*) and *Fructobacillus* (*F*) (a genus of fructophilic lactic acid bacteria). They can produce lactic acid as the primary metabolic end product via carbohydrate metabolisms. Therefore, various fermented foods are developed by the action of these bacteria. Antimicrobial compounds secreted by LAB have long been recognized to prolong the shelf-life and improve the safety of foods. LAB produce a variety of compounds, including lactic acid, acetic acid, hydrogen peroxide (H_2O_2), ethanol, formic acid, fatty acids, diacetyl, acetaldehyde, acetoin, reuterin, reutericyclin, bacteriocins, bacteriocin–like compounds and other low molecular mass compounds which exhibit the antimicrobial activity against spoilage and pathogenic microorganisms and germination of spore as well. LAB cultures are marketed today as protective cultures against common food spoilage bacteria and pathogens. Consequently, the application of such cultures may lead to improvements in food quality and sensory attributes by controlling spoilage microbiota. The use of

---

* Corresponding Author's Email: Sujitra.tac@mahidol.ac.th.
† Corresponding Author's Email: Somboon.T@chula.ac.th.

antimicrobials from LAB as food preservatives has become a popular practice. Moreover, the addition of LAB as probiotics into food and medical products has also been reported.

INTRODUCTION

The interest in naturally- or bio-preservation has been increasing during recent years. For this reason, many researchers have been paying attention on antagonistic microorganisms and their antimicrobial metabolites for application in the food industry (Reis et al., 2012). Among the microorganisms, lactic acid bacteria (LAB) are fascinating since ancient time; many kinds of food are fermented naturally by autochthonous LAB, which form the desirable characteristics of the products. The action of these bacteria can improve texture, flavor, odor and taste of foods. These phenomena are responsible from the generation of numerous metabolites such as lactic acid, acetic acid, ethanol, diacetyl, acetone, exopolysaccharide, specific proteases and bacteriocin by fermentation (Porto et al., 2017). The reduction of pH and conversion of sugars to organic acids are the primary preserving actions that these bacteria provide to fermented food. Therefore, shelf-life of foods is extended and the safety for consumption is ensured. In addition to organic acids, other antimicrobials produced by LAB; H_2O_2, ethanol, formic acid, fatty acids, diacetyl, acetaldehyde, acetoin, reuterin, reutericyclin, bacteriocins, bacteriocin–like substances (BLIS) also exhibit antagonistic activity against undesirable microorganisms (Šušković et al., 2010).

In regard to antagonistic property, LAB are used as protective cultures and preservatives in food-based products because of their bioactive compound production. Furthermore, they are also developed as the pure and mixed culture for the alleviation of human suffering from the pathogenic infection. In this chapter, we focus on the diversity of LAB distributed in natural materials, their antimicrobials, antimicrobial spectrum and application of LAB in food products and clinical alleviation.

LAB AND HABITATS

LAB comprise a clade of Gram-positive bacteria integrated by a cluster of morphological, metabolic, and physiological characteristics (Axelsson, 2004). They are catalase-negative, acid tolerant, non-sporulating, non-respiring rods or cocci that produce lactic acid as the major metabolic end-product of carbohydrates fermentation (Salminen et al., 1996; Klaenhammer & de Vos, 2011). They have less than 55% mol G+C content in their DNA (Stiles & Holzapfel, 1997). LAB consisted of several genera including *Lactobacillus (Lb)*, *Streptococcus (Strep)*, *Enterococcus (Ec)*, *Pediococcus (Ped)*, *Lactococcus (Lc)*, *Leuconostoc (Leuc)*, *Aerococcus (A)*, *Oenococcus (O)*,

Carnobacterium (*Cb*), *Vagococcus* (*V*), *Tetragenococcus* (*T*), *Weisella* (*W*) and *Fructobacillus* (*F*) (Axelsson, 2004). Among these genera, *Lactobacillus* is the largest genus (Parte, 2018). LAB can be divided into two groups based on their glucose metabolisms including homofermentative LAB which converted glucose almost quantitatively to lactic acid (more than 85%) through Embden-Meyerhof-Parnas or glycolysis pathway and heterofermentative LAB which fermented glucose to lactic acid (50%), ethanol/acetic acid, and CO_2 through the phosphoketolase pathway (Reddy et al., 2008).

LAB require numerous essential nutrition for their growth; hence, distribution of LAB is associated with the sources containing the high concentration of carbohydrates, amino acid, vitamins, purines and pyrimidines (Saguir et al., 2007). Consequently, LAB are typically detected in a large number of natural sources that are rich-nutrients such as dairy, meat, cereal, plants, and their fermentative products, and functioned as natural inhabitants gastro-intestinal (GI) tract of human and animals. These bacteria produce many metabolic products especially, lactic acid that provide a unique characteristic of each fermented product. *Lactobacillus* seems to be the most widespread LAB which distributed in these environments. *Lb. plantarum*, *Lb. reuteri*, *Lb. delbrueckii* subsp. *lactis*, *Lb. delbrueckii* subsp. *bulgaricus*, *Lb. paracasei*, *Lb. curvatus*, *Lb. rhamnosus*, *Lb. fructivorans*, *Lb. parabuchneri*, *Lb. brevis*, *Lb. animalis*, *Lb. curvatus*, *Lb. fermentum*, *Lb. helveticus*, *Lb. kefiri*, *Ec. faecium*, *Ec. faecalis*, *Ec. devriesei*, *Ec. durans*, *Ec. saccharominimus*, *Lc. lactis*, *Lc. garvieae*, *Leuc. pseudomesenteroides*, *Leuc. mesenteroides*, *Leuc. lactis*, *Leuc. citreum*, *W. viridescens*, *Strep. thermophilus* and *Strep. bovis* are isolated from dairy-based products including camel milk, goat milk, cow milk, Greek traditional yoghurt, fermented milk, Maroilles (French cheese) and Romanian dairy products (Zamfir et al., 2006; Tormo et al., 2015; Abushelaibi et al., 2017; Georgalaki et al., 2017; Nacef et al., 2017). Various species of LAB including *Lb. sakei*, *Lb. casei*, *Lb. plantarum*, *Lb. paraplantarum*, *Lb. paracasei*, *Lb. johnsonii*, *Lb. brevis*, *Lb. curvatus*, *Lb. fermentum*, *Lb. graminis*, *Lb. carnosus*, *Ped. pentosaceus*, *Ped. acidilactici*, *Ec. faecalis*, *Ec. faecium*, *W. cibaria*, *W. confusa*, *W. halotolerans*, *Leuc. inhae*, *Leuc. mesenteroides*, *Leuc. carnosum*, *Leuc. citreum*, *Lc. lactis* and *Carnobacterium* spp. are existed in meat-originated products such as Ciauscolo salami, dry-salted cod and albacore muscle, Harbin dry sausages, Uzicka sausages, Pastırma (traditional meat product in Turkey) and Lukanka (traditional Bulgarian fermented meat product) (Federici et al., 2014; Gómez-Sala et al., 2015; Moracanin et al., 2015; Han et al., 2017; Öz et al., 2017; Todorov et al., 2017). *Lb. plantarum*, *Ped. pentosaceus*, *Ped. acidilactici*, *Lb. brevis*, *Lb. coryniformis*, *Lb. paracasei*, *Lb. fermentum*, *Lb. rhamnosus*, *Lb. sakei*, *Lb. sanfranciscensis*, *Lb. crustorum*, *Lb. paralimentarius*, *Lb. rossiae*, *Lb. mindensis*, *Lb. helveticus*, *Lc. lactis*, *Lc. garvieae*, *Ec. casseliflavus*, *Ec. faecium*, *Ec. faecalis*, *Ec. mundtii*, *Ec. hirae*, *Ec. gallinarum*, *Ec. hermanniensis*, *Leuc. citreum*, *Leuc. mesenteroides* and *W. confusa* are involved in cereal-based products i.e., wheat, brewer's grains, quinoa grains and sourdough, Nigerian fermented cereal-based foods and Chinese traditional sourdough (Asurmendi et al., 2015; Carrizo et al., 2016; Liu et al., 2016; Alfonzo et al., 2017; Adesulu-Dahunsi et al., 2018). *Lb. plantarum, Lb. brevis, Lb. paracasei, Lb. fermentum, Lb. casei, Lc. lactis, Lc. mesenteroides, Ped. pentosaceus, Lb. kunkeei, Lb. florum, F. pseudoficulneus, F. fructosus, Ec. durans*, are found in plant-derived product for instance; fruit (pulp and byproduct of *Malphigia glabra* L. and *Mangifera indica* L., and *Annona muricata* L.), fresh flowers and fruit peels, Kahudi (traditional rapeseed fermented food) (Endo et al., 2009; Endo et al. 2010; Goswami et al., 2017; Da Costa et al., 2018). Noteworthy, the species belonged to genus *Fructobacillus* are found only in fructose-rich origins such as flower and fruit. The indigenous species to human, *Lb. paracasei, Lb. casei, Lb. gasseri, Lb. fermentum, Lb. rhamnosus, Lb. oris, Lb. vaginalis, Lb. ruminis, Lb. plantarum, Lb. pentosus, Lb. salivarius, Lb. saerimneri, Lb. johnsonii, Lb. reuteri* and *W. confusa* are presented in GI tract of human (Ladda et al., 2015; Noguchi et al., 2012; Rubio et al., 2014; Taweechotipatr et al., 2009; Spinler et al., 2008). Intestine samples of summer adult worker bees (*Apis mellifera* L.) from different hives are used for LAB isolation. *Lb. acidophilus, Lb. crispatus, Lb. johnsonii, Ec. faecium, Lactobacillus* spp. and *Enterococcus* spp. are found in the samples (Audisio et al., 2011). In marine animals, *Leuc. mesenteroides, Leuc. lactis, Lb. pentosus, W. cibaria, W. confusa, Lb. plantarum, Ped. pentosaceus* are isolated from fish gut, marine fish, shellfish and shrimp (Hongpattarakere et al., 2012; Gurovic et al. 2014). *Lb. johnsonii, Lb. mucosae, Lb. murinus, Lb. salivarius, Lb. amylovorus Lb. fermentum, Lb. perolens, Lb. reuteri, Lb. johnsonii, Lb. acidophilus, Lb. animalis, Lb. casei, Lb. farciminis, W. paramesenteroides, W. cibaria, W. hellenica* and *Ec. faecium* are existed in GI tract of calves (Maldonado et al., 2012: Sandes et al., 2017). *Ec. durans, Ec. faecium, Ec. hirae, Ped. acidilactici, Ped. pentosus, W. paramesenteroides, Lb. acidophilus, Lb. amylovorus, Lb. buchneri, Lb. mucosae* and *Lb. plantarum* are isolated from cattle faeces (Maldonado et al., 2018).

ANTIMICROBIAL PRODUCTS

LAB can produce several active metabolites against other microorganisms such as organic acids (lactic, acetic, formic, propionic and butyric acids), that intensify their action by reducing the pH of the media, and other substances, such as ethanol, fatty acids, acetoin, hydrogen peroxide, diacetyl, antifungal compounds (propionate, phenyl-lactate, hydroxyphenyl-lactate, cyclic dipeptides and 3-hydroxy fatty acids), bacteriocins (nisin, pediocin, lacticin, enterocin and others) and bacteriocin-like inhibitory substances (BLIS), reuterin and reutericyclin (Reis et al., 2012). These LAB metabolic products have a positive effect on food taste, smell, color, and texture, but they may also have the beneficial side effects of extending shelf-life and inhibiting the growth of pathogenic organisms (O'Bryan et al., 2015). Antimicrobial substances produced by LAB can be divided into two groups (Šušković et al., 2010) including:

1 (Low molecular mass substances) < 1,000 Da :(organic acids, hydrogen peroxide, diacetyl, acetaldehyde and acetoin, carbon dioxide, reuterin, reutericyclin, cyclic dipeptides, phenyllactic acid, 4-hydroxyphenyllactic acid and 3-hydroxy fatty acids) Table 1).

Table 1. Low molecular mass antimicrobial metabolites of lactic acid bacteria (Modified from Šušković et al., 2010)

Compound	Producers	Indicator strains
Lactic acid	All LAB	Yeasts, Gram-positive bacteria, Gram-negative bacteria
Acetic acid	Heterofermentative LAB	Yeasts, Gram-positive bacteria, Gram-negative bacteria
Diacetyl, acetaldehyde, acetoin	Variety of genera of LAB including: *Lactococcus, Leuconostoc, Lactobacillus* and *Pediococcus*	Yeasts, Gram-positive bacteria, Gram-negative bacteria
Hydrogen peroxide	All LAB	Yeasts, Gram-positive bacteria, Gram-negative bacteria
Carbon dioxide	Heterofermentative LAB	Most of microorganisms
Reuterin	*Lb. reuteri*	Fungi, protozoa, Gram-positive and Gram-negative bacteria
Reutericyclin	*Lb. reuteri*	Gram-positive bacteria
Cyclic dipeptides	*Lb. plantarum, Lb. pentosus*	Fungi
3-Phenyllactic acid, 4-hydroxyphenyllactic acid	*Lb. plantarum, Lb. alimentarius, Lb. rhamnosus, Lb. sanfranciscensis, Lb. hilgardii, Leuc. citreum, Lb. brevis, Lb. acidophilus, Leuc. mesenteroides*	Fungi
3-Hydroxy fatty acids	*Lb. plantarum*	Fungi
Benzoic acid, methylhydantoin, mevalonolactone	*Lb. plantarum*	Fungi, Gram-negative bacteria

2 (High molecular mass substances) > 1,000 Da :(Bacteriocins)

Bacteriocins are ribosomally-synthesized peptides or proteins with the anti-microbial activity that are produced by different groups of bacteria including many members of LAB) Gálvez et al., 2007 .(Bacteriocins released from LAB are generally recognized as safe substances) GRAS); they have attracted interest to use extensively as food preservatives .Since nisin A was recognized in 1928, diverse bacteriocins have been identified from LAB) Zendo and Sonomoto, 2011 .(Interfering of the cell wall and or the membrane of target organisms, either by inhibiting cell wall biosynthesis or causing pore formation, consequently resulting in death, are the proposed anti-microbial mechanism of bacteriocins) O'Sullivan et al., 2002 .(Bacteriocins are classified into 4 groups based on their structures and characteristics) Table.

Table 2. Suggested classification of bacteriocins from LAB
(Modified from Cotter et al., 2005; Šušković et al., 2010)

Classification	Major characteristics	Examples
Class I lantibiotics/lanthionine-containing bacteriocins subdivided into: Type A lantibiotics Type B lantibiotics	Small (< 5 kDa) membrane-active peptides containing unusual amino acids; – elongated peptides with a net positive charge – smaller globular peptide with negative or no net charge	Type A: Nisin, lacticin 481, lactocin S Type B: Mersacidin
Class II non-lanthionine-containing bacteriocins Subdivided into: Subclass IIa Subclass IIb Subclass IIc	Heterogeneous class of small (< 10 kDa) heat-stable post-translation unmodified non-lantibiotics IIa: pediocin-like IIb: two-peptide IIc: with wide range of effects on membrane permeability and cell wall formation	IIa: Pediocin PA1, sakacin A, sakacin P, leucocin A, curvacin A IIb: Lactococcin G, lactococcin M, lactacin F, plantaricin A IIc: Acidocin B, enterocin P, enterocin B, reuterin 6
Class III Bacteriolysins	Large (> 30 kDa) heat-labile antimicrobial proteins, complex proteins with domain-type structure that function through the lyses of sensitive cell by catalyzing cell wall hydrolysis	Lysostaphin, enterolysin A, helveticin J, helveticin V-1829
Class IV	Complex bacteriocins carrying lipid or carbohydrate moieties	Plantaricin S, leuconocin S, lactocin 27, pediocin SJ1

ANTIMICROBIAL SPECTRUM OF ANTIMICROBIALS AGAINST SPOILAGE AND PATHOGENIC MICROORGANISMS

Bactericidal Activity

Gram-Negative Bacteria

Helicobacter Pylori

Helicobacter (*H*) *pylori* was successfully cultivated by Warren and Marshall in 1983 who first described that this bacterium is associated with chronic gastritis and peptic ulcer Warren & Marshall, 1983 .(*H* .*pylori* commonly colonized on the mucous layer of the gastric epithelium of the human stomach .Hence, it has been isolated from materials of gastric sites of diverse populations throughout the world) Owen, 1995 .(Over the past two decades, increased attention had been paid on *H* .*pylori* since it plays an important role in the pathogenesis of chronic gastritis and peptic ulcer diseases .Both epidemiological and clinical evidence has indicated that *H.pylori* is associated with an increased risk of gastric carcinoma) Heavey & Rowland, 2004 .(The International Agency for Research into Cancer) IARC (and the World Health Organization) WHO (have classified this bacterium into Class I) definite (biological carcinogen in humans) Park et al., 2004 .(Nowadays, *H. pylori* infection remains a commonly spread disease with definite morbidity and mortality) Zullo et al., 2012 .(Moreover, the prevalence of infection in the developing countries was higher) greater than 80 (%than the developed ones) less than 40) (%Vale & Vítor, 2010)

European guidelines suggested the standard triple therapies including a proton pump inhibitor) PPI (clarithromycin) 500 mg (plus amoxicillin) 1g (or metronidazole/tinidazole) 500 mg, (all given twice daily for 7–14 days for *H.pylori* eradication) Zullo et al., 2012 . (Unfortunately, the decrease in the efficacy of this regimen occurred throughout the world during the last decade. Consequently, the sequential therapy) 10-days therapy (was proposed to get better results than triple therapies .This novel regimen consisted of a PPI plus amoxicillin 1 g) both twice daily (given for the first five days followed by a triple therapy including a PPI, clarithromycin 500 mg, and tinidazole) all drugs were given twice daily (for the remaining five days .However, antibiotic-based treatment for *H . pylori* infection had a higher effect on the composition of intestinal microbiota) Myllyluoma et al., 2007 .(Besides, antibiotic-based therapy would not be cost-effective, causes side effects and in particular, encourages the spread antibiotic resistance) Michetti, 2001 .(Therefore, an increased interest of many researchers had been paid on LAB since some strains of these bacteria had exhibited antagonistic activity against *H. pylori*. Organic acids especially, lactic acid produced by LAB has well known to exhibit antibacterial activity against *H. pylori* which are related to pH value in the medium. Lactic acid produced by *Lb. acidophilus* suppresses the growth of all seven strains of *H. pylori in vitro* (Bhatia et al., 1989). Extracellular lactic acid secreted by *Lb. acidophilus* and *Lb. rhamnosus* inhibited the growth of *H. pylori* (Midolo et al., 1995). The spent culture supernatant (SCS) of the *Lb. acidophilus* strain LB (LB-SCS) dramatically decreased the viability of *H. pylori in vitro* independent of pH and lactic acid levels (Coconnier et al., 1998). LB activity was strain-dependent and better at low pH. *Lb. paraplantarum* KNUC25 isolated from over-fermented kimchi, is effective in inhibiting the growth of *H. pylori*, which is related to pH and the adherence of cagA-positive *H. pylori* to gastric cells (Ki et al., 2010). *Lb. acidophilus* LY5 produced lactic and acetic acid that reduced urease activity and the viability of both *H. pylori* BCRC 17021 and CMU83 (Lin et al., 2011). The investigation of Zheng et al. (2016) reported that *Lb. pentosus* strain LPS16 had the broad-spectrum antibacterial activity against multidrug-resistant *H. pylori* strains, suggesting that it can be used to prevent *H. pylori* infection. *Lb .fermentum* UCO-979C significantly inhibited *H .pylori*-induced IL-8 production in AGS cells and reduced the viability of *H .pylori* (García et al., 2017). The strains of *Lb. delbrueckii* subsp. *bulgaricus* inhibited many *H. pylori* strains, including antibiotic resistant strains by secreting lactic acid, bacteriocins-like substances or other bioactive substances (Boyanova et al., 2009). In addition to lactic acid and low pH, viable and nonviable *W. confusa* strain PL9001 inhibited the binding of *H. pylori* ATCC 43504 to human gastric-cell line MKN-45 cells by more than 90%. The spent culture supernatant of PL9001 rapidly decreased the viability of *H. pylori* and ruptured the cell walls by the action of bacteriocins (Nam et al., 2002). Kim et al. (2003) reported that the bacteriocins; nisin A, lacticin A164, lacticin BH5, lacticin JW3 lacticin NK24, pediocin PO2, and leucocin K produced from *Lc. lactis* subsp. *lactis* ATCC 11454, *Lc. lactis* subsp. *lactis* A164, *Lc. lactis* BH5, *Lc. lactis* JW3, *Lc. lactis* NK24, *Ped. acidilactici* PO2, and *Leuconostoc* sp. LAB145-3A, respectively, showed the anti-*H. pylori* activity. MICs of these bacteriocins were ranged from 0.39–25, 0.097–12.5, 0.097–12.5, 0.195–25, 0.097–25, 6.25–50 and 25 mg/L, respectively, that varied among the *H. pylori* strains tested, of which strain ATCC 43504 was the most tolerant. The inhibitory activity of *Ec. faecium* TM39 against *H. pylori* could result from the inhibition on *H. pylori* urease activity and the secreted bacteriocin-like substances (Tsai et al., 2004). *Lb. delbrueckii* subsp. *bulgaricus* BB18 isolated from authentic Bulgarian dairy products produced a novel bacteriocin (bulgaricin BB18) which contained the antagonistic activity against four tested strains of *H. pylori* (Simova et al., 2009). The cell-free culture supernatants obtained from *Lb. brevis* BK11, *Lb. acidophilus* BK13, *Ped. pentosaceus* BK34, *Lb. paracasei* BK57, and *Lc. lactis* BK65 strains producing very high levels of lactic acid dramatically decreased the viability of *H. pylori*. In addition, the bactericidal activity of *Lb. plantarum* BK10, *Lb. brevis* BK11, *Lb. acidophilus* BK13, *Lb. paracasei* BK57, and *Ec. faecalis* BK61 strains was significantly correlated with their bacteriocin production (Lim, 2014). Purified bacteriocin BK11 and BK61 (MW 6.5 and 4.5 kDa) produced from *Lb. brevis* BK11 and *Ec. faecalis* BK61, exhibited anti-*H. pylori* activity and lowered the urease activity of *H. pylori* (Lim, 2015).

Other Gram-Negative Bacteria

Bacteriocins produced by LAB generally possess antibacterial activity against Gram-positive foodborne pathogens (Zendo & Sonomoto, 2011). In recent years, many bacteriocins active against Gram-negative bacteria have been identified proving their efficacy in foods and treating infections (Ghodhbane et al., 2015). Therefore, the limitation of bacteriocins has been improved. There are some reports that demonstrated that bacteriocins have the ability to inhibit Gram-negative bacteria. In addition to anti-*H. pylori* activity, bulgaricin BB18 has an effect on the growth of other Gram-negative pathogens such as *Escherichia* (*E*) *coli* WF and *Salmonella* (*S*) *Typhimurium* FVK781 as well (Simova et al., 2009). Bacteriocin 7293A and B from *W. hellenica* BCC 7293 have broad antimicrobial spectra against many strains of tested Gram-positive bacteria and exceptionally inhibited several important Gram-negative food-borne pathogens such as *Pseudomonas aeruginosa*, *Aeromonas hydrophila*, *S. Typhimurium* and *E. coli*

(Woraprayote et al., 2015). Strains of *Lc. lactis* isolated from tropical rainforest wild-type fruits produce antimicrobial substances which suppress the growth of 3 strains of *E. coli*, *S. enterica* subsp. *enterica*, *S. enterica* subsp. *Abaetetuba*, *Salmonella* sp., *Enterobacter* sp. and *Shigella sonnei* and *Shigella* sp. (Tenea et al., 2018). Active compounds in CFS from *Ec. faecalis* KT11 show antimicrobial activity against significant Gram-negative bacteria including *Klebsiella pneumoniae*, *Serratia marcescens*, *Enterobacter aerogenes*, *Pseudomonas aeruginosa* and *Kocuria rhizophila* (reclassified name of *Micrococcus luteus*) (Abanoz & Kunduhoglu, 2018).

Gram-Positive Bacteria

Listeria monocytogenes

Listeria (*L*) *monocytogenes* causes listeriosis, a rare foodborne disease with a mortality rate of 20%–30%. The elderly and immunocompromised hosts are particularly susceptible to listeriosis. *L. monocytogenes* is ubiquitous and can contaminate food-processing environments, posing a threat to the food chain. This is particularly important for ready-to-eat foods as there is no heat treatment or other antimicrobial steps between production and consumption. Thus, occurrence and control of *L. monocytogenes* are essential for industry and public health. Advances in whole-genome sequence technology are facilitating the investigation of disease outbreaks, linking sporadic cases to outbreaks, and linking outbreaks internationally. Novel control methods, such as bacteriophage and bacteriocins, can contribute to a reduction in the occurrence of *L. monocytogenes* in the food-processing environment, thereby reducing the risk of food contamination and contributing to a reduction of this public health issues (Jordan & McAuliffe, 2018).

Tomé et al. (2006) reported that lactic acid, hydrogen peroxides (H_2O_2) and bacteriocin produced by strains of *Lb. plantarum*, *Lb. brevis* and *Lc. lactis* showed inhibitory activity against *L. monocytogenes* isolated from cold smoked salmon. Two strains of *Ped. pentosaceus* isolated from Alheira, a traditional Portuguese fermented sausage demonstrated anti-*Listeria* activity of a proteinaceous nature (Albano et al., 2007). *Ec. faecium* MMZ17 isolated from Gueddid, a traditionally Tunisian fermented meat, produced enterocin MMZ17 which exhibited anti-Listeria activity. Enterocin MMZ17 can be classified as a small, heat-stable anti-*Listeria*-active peptide, presumably belonging to class IIa bacteriocins (Belgacem et al., 2008). Two bacteriocins similar to pediocin PA-1 and enterocin B secreted by *Ec. faecium* ALP7 and *Ped. pentosaceus* ALP57 (isolated from non-fermented seafood), reduced the growth of *L. monocytogenes* (Pinto et al., 2009). *Ped. pentosaceus* BCC 3772 isolated from Nham (a Thai fermented pork sausage), produced anti-*Listeria* activity during the early logarithmic growth phase with the maximum production of 520,000 AU/mL detected at 12 h of growth. The anti-*Listeria* substance was found to be a class IIa bacteriocin, subsequently identified as pediocin PA-1/AcH (Kingcha et al., 2012). Kimchi-derived *Ped. pentosaceus* showed an anti-*L. monocytogenes* activity. The active compound was identified the active moiety as a LysM domain in a 23-kDa protein that could potentially be a peptidoglycan hydrolase enzyme. This protein was stable over a wide pH range (pH 4–8) and heat treatment (80–100°C for 20 min) (Jang et al., 2014). *W. hellenica* BCC 7239, isolated from Thai fermented pork sausage called Nham, produced two putatively novel bacteriocins, 7293A and 7293B with the molecular mass 6249.302 and 6489.716 Da, respectively. Antimicrobial activity of both peptides was not inactivated by organic solvents (ethanol, isopropanol, acetone, acetonitrile) and surfactants (Tween 20, Tween 80 and Triton X-100). Both novel bacteriocins hold promise for applications in the prevention or treatment of pathogenic infections as food and feed additives to replace antibiotics for enhancing the productivity and sustainability of food production (Woraprayote et al., 2015). A bacteriocin with antagonistic activity against *L. monocytogenes* was produced by a strain of *Lc. lactis* (L3A21M1) which isolated from an Azorean cheese. This bacteriocin with a molecular mass of 2900.23 Da, similar to that of lacticin 481, was heat stable (100°C for 30 and 60 min), active across a wide pH range (2.0-12.0) that is useful for application in the fermentation process (Ribeiro et al. 2016). The two bacteriocins produced by *Ec. durans* 152 were characterized and identified as Dur 152A (an enterocin L50A derivative with two amino acid substitutions of I/M) and enterocin L50B. At 400 AU/mL, the bacteriocins prevented growth of *Listeria* in deli ham for at least 10 weeks at 8°C and at least 30 days at 15°C. For comparison, 500 ppm Nisin controlled *Listeria* growth for up to 6 weeks at 8°C and up to 18 days at 15°C (Du et al., 2017). Plantaricin LPL-1, produced by *Lb. plantarum* LPL-1, was a novel class IIa bacteriocin with bactericidal activity against *L. monocytogenes* 54002. The minimal inhibitory concentration (MIC) was confirmed as 16 µg/mL, and the time-kill kinetics showed that plantaricin LPL-1 significantly decrease the viable cell numbers of *L. monocytogenes* 54002 in a time-dependent and dose-dependent manner (Wang et al., 2019).

Other Gram-Positive Bacteria

Various organic acids, including acetic acid, lactic acid, citric acid, and malic acid have been shown to possess antibacterial effects and therefore are used as antibacterial additives. Organic acids released by *Ped. pentosaceus* strain T1 exert inhibiting effects against the significant pathogenic bacteria; *L. grayi*, *L. ivanovii* subsp. *ivanovii*, *L. innocua*, *L. seeligeri*, *Staphylococcus* (*St*) *aureus* subsp. *aureus*, *St. epidermides*, *St. saprophyticus* subsp. *saprophyticus* and *Bacillus* (*B*) *cereus* (Jang et al., 2014). Bacteriocins produced by LAB have been widely known to inhibit their related species especially to Gram-positive bacteria (Zendo & Sonomoto, 2011). Several studies have reported that bacteriocins released by LAB had an interesting feature to decrease the number of Gram-positive pathogenic bacteria in many foods and clinical trials. Bacteriocin-containing CFS collected from *Ec. faecium* strains exhibited the suppression against *L. innocua*, *L. ivanovii* and strains of LAB such as *Lc. lactis* subsp. *cremoris*, *Lb.*

casei, *Lb. curvatus*, *Lb. delbrueckii*, *Ec. faecalis*, *Ec. feacium* and *Ped. acidilactici* (Belgacem et al., 2008). The study of Woraprayote et al. (2015) indicated that many strains of Gram-positive bacteria (*B. circulans*, *B. coagulans*, *B. subtilis*, *Ec. faecalis*, *Ec. faecium*, *Lb. sakei* subsp. *sakei*, *Lc. lactis* subsp. *lactis*, *Leuc. mesenteroides* subsp. *mesenteroides*, *L. innocua*, *Ped. pentosaceus*, *St. aureus*, *St. xylosus*, *W. confusa*, *W. cibaria*, *W. paramesenteroides*, *W. hellenica*) are inhibited by the action of bacteriocin 7293A and 7293B secreted by *W. hellenica* BCC 7923. Methicillin-resistant *St. aureus* (MRSA) is a multidrug-resistant micro-organism and the important nosocomial pathogen worldwide. Antibacterial compounds produced by *Lb. acidophilus* LC1285 and *Lb. casei* LBC80R strains were active against clinical isolates of MRSA strains 43, 64, 75, 27, 61, 22, 18, 69, 80, 36 and ATCC 43300 (Karska-Wysocki et al., 2010).

Fungicidal Activity

Fresh fruits, vegetables, and also grains and nuts are rich sources of nutrition which is important for fungal growth (Bueno et al., 2004). Food spoilage by mold and the occurrence of their mycotoxins constitute a potential health hazard. Development of biological control should help improve the safety of products by controlling mycotoxin contamination (Dalié et al., 2010). LAB have a GRAS status. They have been used to ferment foods for many years, which suggest the nontoxic nature of metabolites produced by these organisms. Moreover, it is believed that the metabolic activity of LAB that may contribute in many ways to the control of bacterial pathogens, and improvement of shelf life and sensory qualities, might also have applications for preventing mold growth (Bianchini, 2015). To date, the use of LAB and their antifungal compound for controlling growth of fungi have received much attention by researchers. The growth of *Aspergillus* (*A*) *nidulans* was inhibited during co-cultivation with *Lb. plantarum* MiLAB 393 and the expression of fungal proteins was altered. Cyclo (L-Phe–L-Pro), lactic acid and 3-phenyllactic acid were the active compounds produced by this LAB strain which affected on the growth of *A. nidulans* (Ström et al. 2005). *Lb. acidophilus* LMG 9433 suppressed the growth of *Penicillium* (*Pe*) spp. The supernatant of *Lb. amylovorus* DSM 20532 was fairly inhibiting for *Cerinosterus* sp., *Cladosporium* sp., *Endomyces (Ed) fibuliger* MUCL 11443, *Penicillium* spp. A, F and G and *Rhizopus* (*R*) *oryzae* MUCL 20145. With *Lb. coryniformis* subsp. *coryniformis* LMG 9196, clear inhibition zones were detected against *Cerinosterus* sp., *Cladosporium* sp., *Penicillium* spp. F and G and *R. oryzae* MUCL 20145. The antifungal properties of supernatants collected from these LAB were due to organic acids and other pH-dependent antifungal substance (De Muynck et al., 2004). Bacteriocin-synthesizing *Lc. lactis* strains 194 and K-205 were isolated from raw cow milk. These strains had high antibiotic activity against *Eurotium (Er) repens* up to 3600 and 2700 IU/mL as compared to nisin and up to 2500-1700 IU/mL as compared to fungicidal antibiotic nystatin (Stoyanova et al., 2010). *Ec. durans* A5-11 was isolated from traditional Mongolian airag cheese. This strain inhibited the growth of several fungi including *Fusarium* (*F*) *culmorum*, *Pe. roqueforti*, *Pe. expansum*, *Debaryomyces* (*D*) *hansenii*, *Candida krusei* and *Mucor plumbeus*. It produced two bacteriocins: durancin A5-11a and durancin A5-11b which contained antagonistic activity against tested *D. hansenii* strain LMSA2.11.003 (Belguesmia et al., 2012). The antifungal properties of cell-free culture filtrate (CCF) from *Lb. casei* AST18 were detected, and the antifungal compounds of CCF were found to be lactic acid, cyclo-(Leu-Pro), 2,6-diphenyl-piperidine, and 5,10-diethoxy-2,3,7,8-tetrahydro-1*H*,6*H*-dipyrrolo[1,2-a;1′,2′-d]pyrazine. The antifungal activity against *Penicillium* sp. of *Lb. casei* AST18 was a synergistic effect of lactic acid and cyclopeptides (Li et al., 2012). An anti-*Candida* protein (ACP) produced by *Ec. faecalis* showed a broad-spectrum activity against multidrug-resistant *Candida albicans* strain MTCC 183, MTCC 7315, MTCC 3958, NCIM 3557, NCIM 3471 and DI. A molecular weight of ACP was approximately 43 kDa, stable after heat-treatment at 90°C for 20 min, maintained activity over a pH range 6–10 and remained active after treatment with α-amylase, lipase, organic solvents, and detergents (Shekh & Roy 2012). Lactic, acetic and phenyllactic acids (PLA) were responsible for antifungal activity produced by *Lb. casei* CRL 239, *Lb. plantarum* CRL 681, 788, 759, 142, *Lb. fermentum* CRL 220, 236, 251, *Lb. reuteri* CRL 1098 and *Lb. acidophillus* CRL 1070. Among these strains, *Lb. fermentum* CRL 251 produced fungus inhibitory peptide/s, smaller than 10 kDa, thermostable, active in the pH range of 4-7 and sensitive to trypsin (Gerez et al., 2013). *Lb. casei*, *Lb. rhamnosus*, *Lb. fermentum*, *Lb. acidophilus*, *Lb. plantarum*, *Lb. sakei* and *Lb. reuteri* had the capabilities to produce 3-phenyllactic acid which exhibited the inhibition of *Colletotrichum gloeosporioides*, *Botrytis cinerea*, *Pe. expansum* and *A. flavus* (Cortés-Zavaleta et al., 2014). Phenyllactic, OH-phenyllactic and benzoic acids were the active compounds produced by *Lb. amylovorus* DSM 19280 and *Lb. reuteri* R29 which were active against *F. culmorum*. *Lb. amylovorus* DSM 19280 cell-free-supernatant (CFS) inhibited *F. culmorum* spores at levels of 10^4 spores/mL, for seven days, whereas *Lb. reuteri* R29 CFS inhibited up to 10^5 spores/mL for the same time (Oliveira et al., 2015). The organic acid content of crude antifungal substances produced by *Lb. sakei* ALI033 showed high concentrations of lactic acid (502.47 mg/100 g). Antifungal substance against *Pe. brevicompactum* FI02 produced by *Lb. sakei* ALI033 is most likely due to its ability in producing organic acid (Huh & Hwang, 2016). Reuterin produced by *Lb. reuteri* ATCC 53608 showed a fungicidal activity (killed 99.9% of all tested microorganisms) against *Aspergillus* spp., *Paecilomyces* spp., *Penicillium* spp., *Candida* spp., *Aureobasidium pullulans*, *Er.rubrum/amstelodami*, *Kluyveromyces* spp., *Saccharomyces* spp., *Lachancea thermotolerance*, *Rhodotorula mucilaginosa*, *Torulospora delbrueckii*, *Wickerhamomyces anomalus*, *Zygosaccharomyces rouxii*. MICs of reuterin ranged from 1.0 to 4.8 mM (Vimont et al. 2019).

Sporicidal Activity

Control and eradication of *Bacillus* and *Clostridium* spores are one of the most challenging aspects of microbial control faced by the modern food industry. Traditionally, spores have been controlled using extreme treatments such as high heat alone or in combination with chemical additives. However, modern consumers are more conscious than previous generations of the adverse health effects associated with the consumption of certain chemical preservatives and the significant effects of heat on the nutritional value and flavor of many foods (Egan et al., 2016). Therefore, application of LAB containing sporicidal activity in medium and food model has been widely investigated. *B. cereus* cultures were inoculated into the nonfat milk medium at 0, 24, and 48 h after the growth of LAB (*Strep. lactis*, *Strep. thermophilus*, *Lb. acidophilus*, and *Lb. bulgaricus*). After 72 h of fermentation of all the tested LAB, *B. cereus* declined to less than 10 CFU/mL within 40 min. Spore germination of *B. cereus* was not affected in media that were fermented for 24 h. Germination frequencies were lowered to 33 to 43% and 23 to 27% in media fermented for 48 and 72 h, respectively (Wong et al., 1988). The bacteriocinogenic *Lb. plantarum* BN, *Lc. lactis* ATCC 11454 and *Ped. pentosaceus* ATCC 43200 and 43201 inhibited the germination of *Clostridium* (*Cl*) *botulinum* spores (10^4 spores/mL) at the incubation temperature of 4, 10, 15 and 30°C. *Lc. lactis* 11454 was the most active and was followed by *Ped. pentosaceus* 43200 and 43201 and *Lb. plantarum* BN. Increasing the bacteriocinogenic cell inoculum levels from 10^2 through 10^4 to 10^6 CFU/mL increased the radio of inhibition zones (Okereke et al., 1991). In the present work, bacteriocins produced by *Lc. lactis* subsp. *lactis* INIA 415 apparently inhibited outgrowth of *Cl. beijerinckii* spores, since spore counts remained fairly constant in cheese made from milk inoculated with this culture and clostridial spores throughout the ripening period (Garde et al., 2011). Ávila et al. (2014) reported that *Lb. reuteri* INIA P572 was used as reuterin-producing strain. Reuterin were able to inhibit the growth of vegetative cells and spores of all assayed *Clostridium* strains (*Cl. tyrobutyricum*, *Cl. butyricum*, *Cl. beijerinckii* and *Cl. sporogenes*) in milk and modified RCM broth. In milk, the MICs of reuterin suppressed the vegetative cells and spores of *Clostridium* ranged from 1.02–16.25 mM and 4.06–32.50 mM, respectively. In mRCM, the concentration of reuterin required to inhibit the growth of vegetative cells (0.51–8.13 mM) of the four tested species of *Clostridium* was lower than that required to inhibit the spores (2.03–16.25 mM). Reuterin, with a broad inhibitory activity spectrum against *Clostridium* spp. spores and vegetative cells, may be the best option to control *Clostridium* growth in dairy products and to prevent associated spoilage, such as late blowing defect of cheese. The sensitivity of *Cl. perfringens* to reuterin which is produced by *Lb. reuteri* INIA P572, was depended on strain and culture medium. In general, vegetative cells exhibited higher sensitivity than spores. Reuterin (MIC values 2.03–16.25 mM) inhibited the growth of vegetative cells and the outgrowth of spores of all tested *Cl. perfringens*, both in mRCM and milk, with higher resistance in milk (Garde et al., 2014).

USE OF LAB WITH ANTIMICROBIAL ACTIVITY AND THEIR ANTIMICROBIALS IN FOOD AND CLINICAL APPLICATIONS

In the past, fermentation was the process used to improve the quality and shelf life of food products and the secret of their extended shelf life is the presence of antimicrobial LAB, which are natural inhabitants of fermented foods (Varsha & Nampoothiri, 2016). Then, application of pure or mixed culture of LAB as starter culture has become widespread in the industrial production of fermented food products (Brashears et al., 2005). Nowadays, the employment of LAB containing antagonism effect has extended not only in fermented foods but also in fresh and minimally processed foods. Numerous studies have shown the potential use of either LAB cells or antimicrobial compounds from them as protective cultures or biopreservatives in food products for many years. Moreover, LAB have shown to be probiotic cultures based on their GRAS status and probiotic features including antagonistic activity against pathogens, acid and bile resistance, ability to colonization on the digestive tract, immunomodulatory modulation, antimutagenic and anticarcinogenic properties (Saarela et al., 2000; Saad et al., 2013).

In the past years, there has been a particular focus on the application of bacteriocins produced by LAB in controlling the growth of pathogenic bacteria in foods. In six of the CFSs, different class IIa bacteriocins, namely leucocin A and B, mundticin L, pediocin PA-1, sakacin A, and sakacin X produced by *Leuc. carnosum* IDE1105, *Enterococcus* sp. IDE0886, *Ped. acidilactici*, *Lb. plantarum* IDE0105, *Lb. sakei* IDE0216 and *Lb. curvatus* IDE0444, respectively, were identified as the major anti-*Listeria* compounds. The minimal effective concentration (MEC) of the individual CFSs to achieve a 2.3 $\log_{10}$ reduction of *L. monocytogenes* was determined in culture broth, whole milk, and ground beef at 4°C. Best results were obtained using CFSs containing pediocin PA-1, that displayed only three- and ten-times higher MECs in milk (307 AU/mL) and ground meat (1024 AU/g) compared to the broth, respectively. A twenty-fold increase in the MEC (2048 AU/mL) was observed for a mundticin L-containing fermentate, and a CFS containing leucocin A and B was inactivated more than fifty-fold (> 1280 AU/mL) in both food matrices. Remarkably, the sakacin A and sakacin X containing CFSs displayed very selective inactivation rates, in which sakacin A was only effective in meat (512 AU/g), while sakacin X was only effective in milk (2048 AU/mL) (Hartmann et al., 2011). Moreover, the effect of poly (lactic acid) (PLA)/sawdust particle (SP) biocomposite film incorporated with pediocin PA-1/AcH (Ped) on the reduction of *L. monocytogenes* in raw sliced pork was investigated. A model study of PLA/SP + Ped as a food-contact antimicrobial packaging on raw sliced pork suggests a potential inhibition of *L. monocytogenes* (99% of the total listerial population) on raw sliced pork during the chilled storage (Woraprayote et al., 2013). Pathogens were inoculated on freshly harvested spinach, followed by application of the LAB antimicrobial (LactiGuard™, commercial LAB product). Treated spinach was aerobically incubated up to 12 days at 7°C and surviving pathogens enumerated via selective/differential plating. L-lactic acid and a bacteriocin-like inhibitory substance (BLIS) were detected and quantified from cell-free fermentates obtained from LAB-inoculated liquid microbiological medium. Application of 8.0 $\log_{10}$ CFU/g LAB produced significant reductions in *E. coli* O157:H7 and *Salmonella* populations on spinach of 1.6 and 1.9 $\log_{10}$ CFU/g, respectively (Cálix-Lara et al. 2014). Using bacteriocin-producing strains in the manufacture of fresh cheese might contribute to preventing the growth of undesirable pathogenic bacteria such as *L. monocytogenes*. Fresh cheese was made from pasteurized cows' milk inoculated with bacteriocin-producing LAB and artificially contaminated with approximately 10^6 CFU/mL of *L. monocytogenes*. In comparison, an increase of 4 log CFU/mL in pathogen numbers in cheese was detected over the same period, in the absence of bacteriocin-producing LAB. The combination of two bacteriocin producing *Enterococcus* sp. optimized the reduction of *L. monocytogenes* counts in fresh cheese, reducing by approximately 5 log units after seven days (Coelho et al., 2014). The biopreservative effect of the cell-free supernatant containing bacteriocin on quality and shelf life of fresh pork stored for 16 days at 4°C were investigated. The antimicrobial substance could retain the good quality and extend the shelf life pork during storage and 1.2% antimicrobial substance could extend the shelf life to 12 days and the meat showed good characteristics (Miao et al., 2015). Outbreaks of food-borne disease associated with the consumption of fresh and minimally processed fruits and vegetables have increased dramatically over the last few years. Traditional chemical sanitizers are unable to completely eradicate or kill the microorganisms on fresh produce. These conditions have stimulated the search for alternative methods for increasing food safety. The use of protective cultures, particularly LAB, has been proposed for minimally processed products (Siroli et al., 2015). Applying the *Lb. plantarum* strains CIT3 and V7B3 to apples and lettuce, respectively, increased both the safety and shelf-life. The *Lb. plantarum* CIT3 strain achieved a significant inhibition of *E. coli* and *L. monocytogenes* in sliced apples. In fact, in the products inoculated with *Lb. plantarum* CIT3, the kinetics of *E. coli* death accelerated significantly and was below the limit of detection after seven days (Siroli et al., 2015). In fermented meat, *Clostridium* spp. growth is kept under control by the addition of nitrite. The growing request of consumers for safer products has led to consider alternative bio-based approaches, the use of protective cultures being one of them. When *Lb. plantarum* PCS20 (initial concentration 9 log CFU/ g) was added to the batter without nitrite treatment, a significant reduction of *Cl. perfrigens* was observed between day 2 and 5 in comparison with the batch containing only the pathogen. The results from the last sampling day showed that, after nine days of the fermentation process, *Cl. perfrigens* growth was inhibited to 1.5 log CFU/g in the batches inoculated with the protective culture (Di Gioia et al., 2016). *Vibrio (V) parahaemolyticus* is recognized as the main cause of gastroenteritis associated with the consumption of seafood. Bacteriocin-producing *Lb. plantarum* FGC-12 isolated from golden carp intestine had strong antibacterial activity toward *V. parahaemolyticus*. This bacteriocin showed MIC for *V. parahaemolyticus* at 6.0 mg/mL. The addition of the fish-borne bacteriocin to shrimps leaded *V. parahaemolyticus* to reduce 1.3 log units at 4 °C storage for six day (Lv et al., 2017). *Lb. curvatus* 54M16 produced bacteriocins *sak* X, *sak* T_α, *sak* T_β and *sak* P containing antibacterial activity against *L. monocytogenes*. In MRS broth, *Lb. curvatus* 54M16 was able to inhibit *L. monocytogenes* to undetectable levels after 48 h at 20°C or 5 days at 15°C. Anti-*Listeria* activity was lower during the production of fermented sausages with pathogen inoculation at levels of approximately 4 log CFU/g. However, total inhibition of *L. monocytogenes* native to the raw ingredients was achieved after the fermentation period (Giello et al., 2018). *Ped. acidilactici* CRL 1753 produced phenyl-lactic acid was used as a liquid bio-preserver in bread making to increase the shelf life of wheat bread. A significant increase in shelf life was observed when CRL 1753 culture (incubated for 48 h) and calcium propionate (CP) were combined into the bread formulation; at 18 days of storage no molds were observed when the lactic strain was added while 70% of bread was spoilage in the bread made with CP alone (Bustos et al., 2018).

Dietary components such as vegetable or probiotic microorganisms have been proposed as an alternative solution to decrease *H. pylori* colonization in at-risk populations. Some strains of LAB have been shown to exert bacteriostatic or bactericidal effects against *H. pylori* in *in vitro* and *in vivo* models. The study of Cruchet et al. (2003) suggested that regular ingestion of a probiotic strain, *Lb. johnsonii* La1, may interfere with *H. pylori* colonization in asymptomatic children and may be an effective alternative to modulate *H. pylori* infection and its associated gastritis in pediatric populations with high prevalences of infection by this pathogen. The patients received the probiotic therapy (*Lb. rhamnosus* GG, *Lb. rhamnosus* LC705, *Bifidobacterium* (*B*) *breve* Bb99 and *Propionibacterium* (*P*) *freudenreichii* subsp. *shermanii* JS) or a placebo during *H. pylori* eradication and for three weeks following the treatment and their daily symptoms were recorded in a standardized diary. No significant differences were found between the two groups for individual symptoms. However, the probiotic group showed less treatment-related symptoms throughout the *H. pylori* eradication therapy in contrast to the placebo group (Myllyluoma et al., 2005). In 40 *H. pylori*-positive subjects receiving *Lb. reuteri* once a day for four weeks or placebo, *Lb. reuteri* ATCC 55730 reduces *H. pylori* load as semi-quantitatively, assessed by both ^{13}C-urea breath test δ-value and *H. pylori* stool antigen quantification after four weeks of treatment ($p < 0.05$). No changes were observed in patients receiving placebo. *Lb. reuteri* administration was followed by a significant decrease in the gastrointestinal symptom as compared to pretreatment value that was not present in those receiving placebo. No difference in eradication rates was observed (Francavilla et al., 2008). In another study, people infected with *H. pylori* ingested probiotic apple snack containing 10^8 CFU of *Lb. salivarius* subsp. *salivarius* for four weeks. A decrease in the values of ^{13}C in the expired breath of subjected was observed. This probiotic strain showed the beneficial effects against *H. pylori* infection

Table 3. Use of LAB and their antimicrobials in food products

Food model	Producer	Antagonistic substance	Tested bacteria	Reduction in pathogenic population	Reference
Whole milk and Ground beef	*Enterococcus* sp. IDE 0886, *Lb. curvatus* IDE 0444, *Lb. plantarum* IDE 0105, *Lb. sakei* IDE0216, *Leuc. carnosum* IDE1105, *Ped. acidilactici*	Mundticin L, sakacin X, pediocin PA-1, sakacin A, leucocin A and B pediocin PA-1	*L. monocytogenes*	2.3 $\log_{10}$ reduction (MEC: 2048, 2048, 307, >1280, >1280 and 307 AU/mL in whole milk MEC: 2048, >10,240, 1024, 512, >1280 and 1024 AU/mL in ground beef)	Hartmann et al. (2011)
Raw slice pork	*Ped. pentosaceus* BCC 3772	Pediocin PA-1/AcH in poly (lactic acid)/sawdust particle biocomposite film	*L. monocytogenes*	2 log cycles or 99% of total listerial population	Woraprayote et al. (2013)
Fresh spinach	LactiGuard™ (commercial LAB product)	L-Lactic acid and a bacteriocin-like substance	*E. coli* O157:H7 and *S. enterica*	1.6 and 1.9 $\log_{10}$ CFU/g	Cálix-Lara et al. (2014)
Azorean artisanal cheese (Pico cheese)	*Lc. lactis* and *Ec. faecalis*	bacteriocin	*L. monocytogenes*	~ 5 $\log_{10}$ CFU/g	Coelho et al. (2014)
Fresh pork	*Lb. paracasei* subsp. *tolerans* FX-6	Bacteriocin with broad-spectrum	Spoilage bacteria (Total viable bacteria)	Maintain total viable count at 4 $\log_{10}$ CFU/g	Miao et al. (2015)
Minimally processed apples and lamb's lettuce	*Lb. plantarum* CIT3 and V7B3	-	*E. coli* and *L. monocytogenes*	below the limit of detection	Siroli et al. (2015)
Pork ground meat	*Lb. plantarum* PCS20	-	*Clostridium* spp.	1 to 1.5 $\log_{10}$ CFU/g	Di Gioia et al. (2016)
Shrimp (*Penaeus vannamei*)	*Lb. plantarum* FGC-12	Bacteriocin	*V. parahaemolyticus*	1.3 $\log_{10}$ CFU/g	Lv et al. (2017)
Fermented sausages	*Lb. curvatus* 54M16	bacteriocin	*L. monocytogenes*	<1 $\log_{10}$ CFU/g	Giello et al. (2018)
Wheat bread.	*Ped. acidilactici* CRL 1753	Phenyllactic acid + calcium propionate	*A. niger* CH2, *A. japonicus* CH5, *Pe. roqueforti* CH4, *Pe. digitatum* CH10 and *Metschnikowia pulcherrima* CH7	None mould observed	Bustos et al. (2018)

-; not determined. MEC, Minimal effective concentrations.

(Betoret et al., 2012). A total of 229 patients were randomly assigned to either a 1-week triple therapy of rabeprazole (10 mg bid), amoxicillin (750 mg bid), and clarithromycin (200 mg bid) or triple therapy plus *Lb. gasseri*-containing yogurt. In the yogurt-plus-triple therapy groups, yogurt containing *Lb. gasseri* OLL2716 (112 g) was given twice daily for four weeks (3 weeks pretreatment and also one week during eradication therapy). Overall eradication (intention to treat/per protocol) was 74.5% for the triple-only group, and 85.6% for the yogurt-plus-triple group. Eradication of primary clarithromycin-resistant strains tended to be higher for yogurt-plus-triple therapy than triple-only therapy (38.5 vs 28.0%, respectively) (Deguchi et al., 2012). A meta-analysis study of Fang et al. (2018) concluded that *Lactobacillus*, as an adjunct to triple therapy, can increase *H. pylori* eradication rates as well as reduce the incidence of therapy-related diarrhea in children. Moreover, a higher dose and a longer duration of supplementation may conduce to the positive impact of *Lactobacillus* on *H. pylori* eradication.

A total of 64 children with acute watery diarrhea were studied. The probiotic group received 10^8 CFU *Lb. reuteri* DSM 17938 for five days in addition to oral rehydration solution (ORS) and the second group was treated with ORS only. The mean duration of diarrhea was significantly reduced in the *Lb. reuteri* group compared to the control group (approximately 15 h, 60.4 ± 24.5 h vs.74.3 ± 15.3 h). The percentage of children with diarrhea was lower in the *Lb. reuteri* group (13/29; 44.8%) after 48 h than the control group (27/31; 87%) (Dinleyici et al., 2015). The intervention was a 28 day, the once-daily course of a four-strain oral probiotic capsule containing *Lb. acidophilus* NCFM, *Lb. paracasei* Lpc-37, *B. lactis* Bi-07 and *B. lactis* Bl-04. Probiotic adjunct therapy was associated with a significant improvement in diarrhoea outcomes. The primary duration of diarrhoea outcome (0.0 versus 1.0 days) and two exploratory outcomes, total diarrhoea days (3.5 versus 12.0 days) and rate of diarrhoea (0.1 versus 0.3 days of diarrhoea/stool diary days submitted), all decreased in participants with probiotic use compared with placebo. There was no significant difference in the rate of *Clostridium difficile* infection (CDI) recurrence or functional improvement over time between treatment groups (Barker et al., 2017). The oral administration of *Ped. pentosaceus* LI05 improved the survival rate and alleviated the histopathological impact of *Cl. difficile* in mice. Compared to the CDI group, the levels of inflammatory mediators in the colon, as well as inflammatory cytokines and chemokines in serum, were substantially attenuated in the LI05 group (Xu et al., 2018). Over 70% of patients are prescribed antibiotics during their intensive care unit (ICU) admission. The gut microbiome is dramatically altered early in an ICU stay, increasing the risk of antibiotic-associated diarrhea (AAD) and *Cl. difficile* infections (CDI). Evidence suggests that some probiotics are effective in the primary prevention of AAD and CDI. In a recent study, where thirty-two patients participated, there were no serious adverse events in the probiotic group, compared to three in the control group. AAD was documented in 12.5% of the probiotic group and 31.3% in the control group.

One patient in the probiotic group developed CDI compared to three in the control group (Alberda et al., 2018).

Table 4. Use of LAB containing antibacterial activity as probiotics in clinical application

Infectious disease	Products	LAB strain	Dose and duration time	Results	Reference
H. pylori infection	Probiotics	*Lb. johnsonii* La1	>10^7 CFU/mL for 4 weeks	Regular ingestion of a product containing *Lactobacillus* La1 represent an interesting alternative to modulate *H. pylori* colonization in children	Cruchet et al. (2003)
	Probiotics	*Lb. rhamnosus* GG, *Lb. rhamnosus* LC705, *B. breve* Bb99 and *Propionibacterium freudenreichii* subsp. *shermanii* JS	10^9 CFU twice a day for 13 weeks	The *H. pylori* eradication rate was non-significantly higher in the group receiving probiotic therapy compared with control group	Myllyluoma et al. (2005)
	Probiotic tablets	*Lb. reuteri* ATCC 55730	10^8 CFU for 4 weeks	*Lb. reuteri* effectively suppresses *H. pylori* infection in humans and decreases the occurrence of dyspeptic symptoms.	Francavilla et al. (2008)
	Probiotic apple snack	*Lb. salivarius* subsp. *salivarius*	10^8 CFU/d for 4 weeks	Decrease in the values of ^{13}C in the expired breath	Betoret et al. (2012)
	Probiotic yoghurt	*Lb. gasseri* OLL2716	$\geq 10^9$ CFU twice daily for 4 weeks	Improves the efficacy of triple therapy in patients with *H. pylori* infection	Deguchi et al. (2012)
	Probiotics or yoghurt supplemented with probiotics	*Lactobacillus* spp.	-	Increase *H. pylori* eradication rates as well as reduce the incidence of therapy-related diarrhea in children	Fang et al. (2018)
Clostridium difficile associated diarrhea	Probiotics	*Lb. reuteri* DSM 17938	10^8 CFU/d for 5 days	Significantly reduces the mean duration of diarrhea in children.	Dinleyici et al. (2015)
Clostridium difficile Infection (CDI)	Probiotics	*Lb. acidophilus* NCFM, *Lb. paracasei* Lpc-37, *B. lactis* Bi-07 and *B. lactis* Bl-04	1.7×10^{10} CFU/d for 4 weeks	Decrease significantly the duration of CDI diarrhea.	Barker et al. (2017)
	Probiotics	*Ped. pentosaceus* LI05	3×10^9 CFU/d for 14 days	Reduce the risk of CDI by down-regulating the degree of intestinal inflammatory response and the production of serum inflammatory mediators, protecting the intestinal barrier function.	Xu et al. (2018)
	Probiotic yoghurt drink	*Lb. casei*	10^9 CFU twice a day for 30 days	A probiotic containing drink can safely be delivered via feeding tube and should be considered as a preventative measure for AAD and CDI in ICU.	Alberda et al. (2018)

CONCLUSION

LAB are found in a variety of natural habitats, not only in dairy, meat, cereal, plants and their fermentative products but also in GI tract of human and animals. These niches are suitable sources for the isolation of LAB. *Lactobacillus* seems to be frequently detected in these origins. Antimicrobials such as organic acids, ethanol, fatty acids, acetoin, hydrogen peroxide, diacetyl, antifungal compounds, reuterin, reutericyclin and bacteriocins released by LAB, have significant effects against spoilage and pathogenic microorganisms which are fascinating for food and clinical applications. In consequence, LAB have extensively served as protective cultures or biopreservative agents in various food products instead of the use of chemical a method named "bio-control approach." Based on GRAS status, LAB have also been used as probiotics which are combined with antibiotic-based therapy for treating patients suffering from some bacterial infections.

REFERENCES

Abanoz H. S., Kunduhoglu B. (2018). Antimicrobial activity of a bacteriocin produced by *Enterococcus faecalis* KT11 against some pathogens and antibiotic-resistant bacteria. *Korean Journal for Food Science of Animal Resources,* 38(5), 1064–1079.

Abushelaibi A., Al-Mahadin S., El-Tarabily, K., Shah N. P., Ayyash M. (2017). Characterization of potential probiotic lactic acid bacteria isolated from camel milk. *LWT - Food Science and Technology,* 79, 316–325.

Adesulu-Dahunsi A. T., Jeyaram K., Sanni A. I. (2018). Probiotic and technological properties of exopolysaccharide producing lactic acid bacteria isolated from cereal-based Nigerian fermented food products. *Food Control,* 92, 225–231.

Albano H., Oliveira M., Aroso R., Cubero N., Hogg T., Teixeira P. (2007). Antilisterial activity of lactic acid bacteria isolated from "Alheiras" (traditional Portuguese fermented sausages): *In situ* assays. *Meat Science,* 76, 796–800.

Alberda C., Marcushamer S., Hewer T., Journault N., Kutsogiannis D. (2018). Feasibility of a *Lactobacillus casei* drink in the intensive care unit for prevention of antibiotic associated diarrhea and *Clostridium difficile. Nutrients,* 10(5), 539.

Alfonzo A., Miceli C., Nasca A., Franciosi E., Ventimiglia G., Gerlando R. D., Tuohy K., Francesca N., Moschetti G., Settanni L. (2017). Monitoring of wheat lactic acid bacteria from the field until the first step of dough fermentation. *Food Microbiology,* 62, 256–269.

Asurmendi P., García M. J., Pascual L., Barberis, L. (2015). Biocontrol of *Listeria monocytogenes* by lactic acid bacteria isolated from brewer's grains used as feedstuff in Argentina. *Journal of Stored Products Research,* 61, 27–31.

Axelsson L. 2004. Lactic acid bacteria: classification and physiology. In, *Lactic acid bacteria: microbiological and functional aspects 3rd edn.* S. Salminen, A. von Wright and A. C., Ouwehand (Eds.). New York: Marcel Dekker, 1–66.

Audisio M. C., Torresa M. J., Sabatéa D. C., Ibargurena C., Apella, M. C. (2011). Properties of different lactic acid bacteria isolated from *Apis mellifera* L. bee-gut. *Microbiological Research*, 166, 1–13.

Ávila M., Gómez-Torres N., Hernández M., Garde S. (2014). Inhibitory activity of reuterin, nisin, lysozyme and nitrite against vegetative cells and spores of dairy-related *Clostridium* species. *International Journal of Food Microbiology, 172, 70–75.

Barker A. K., Duster M., Valentine S., Hess T., Archbald-Pannone L., Guerrant R. Safdar N. (2017). A randomized controlled trial of probiotics for *Clostridium difficile* infection in adults (PICO). *Journal of Antimicrobial Chemotherapy, 72, 3177–3180.

Belgacem Z. B., Ferchichi M., Prévost H., Dousset, X., Manai M. (2008). Screening for anti-listerial bacteriocin-producing lactic acid bacteria from "Gueddid" a traditionally Tunisian fermented meat. *Meat Science, 78, 513–521.

Belguesmia Y., Choiset Y., Rabesona H., Baudy-Floc'h, M., Le Blay G., Haertlé T., Chobert J.-M. (2012). Antifungal properties of durancins isolated from *Enterococcus durans* A5-11 and of its synthetic fragments. *Letters in Applied Microbiology, 56, 237–244.

Betoret E., Betoret N., Arilla A., Bennár M., Barrera C., Codoñer P., Fito P. (2012). No invasive methodology to produce a probiotic low humid apple snack with potential effect against *Helicobacter pylori. Journal of Food Engineering, 110, 289–293.

Bhatia S. J., Kochar N., Abraham P., Nair N. G., Mehta A. P. (1989). *Lactobacillus acidophilus* inhibits growth of *Campylobacter pylori in vitro. Journal of Clinical Microbiology, 27(10), 2328–2330.

Bianchini A. (2015). Lactic acid bacteria as antifungal agents. In, *Advances in Fermented Foods and Beverages Improving Quality, Technologies and Health Benefits*. W. Holzapfel (Ed.), Cambridge, Woodhead Publishing in Food Science, Technology and Nutrition, UK, 333–353.

Boyanova L., Stephanova-Kondratenko M., Mitov I. (2009). Anti-*Helicobacter pylori* activity of *Lactobacillus delbrueckii* subsp. *bulgaricus* strains, preliminary report. *Letters in Applied Microbiology, 48, 579–584.

Brashears M. M., Amezquita A., Jaroni D. (2005). Lactic acid bacteria and their uses in animal feeding to improve food safety. In, *Advances in food and nutrition research*. Vol. 50, S. L. Taylor (Ed.). California, Elsevier Academic Press, USA, 1–31.

Bueno D. J., Silva, J. O., Oliver G. (2004). Fungal Isolation and Enumeration in Foods. In, *Methods in Molecular Biology, Public Health Microbiology: Methods and Protocols*. Vol. 268, J. F. T. Spencer and A. L. Ragout de Spencer (Eds.). Totowa, ©Humana Press Inc. New Jersey, 127–132.

Bustos A. Y., De Valdez G. F., Gerez C. L. (2018). Optimization of phenyllactic acid production by *Pediococcus acidilactici* CRL 1753. Application of the formulated bio-preserver culture in bread. *Biological Control*, 123, 137–143.

Cálix-Lara T. F., Rajendran M., Talcott S. T., Smith S. B., Miller R. K., Castillo A., Sturino J. M., Taylor T. M. (2014). Inhibition of *Escherichia coli* O157: H7 and *Salmonella enterica* on spinach and identification of antimicrobial substances produced by a commercial lactic acid bacteria food safety intervention. *Food Microbiology*, 38, 192–200.

Carrizo S. L., De Oca C. E. M., Laiño J. E., Suarez N. E., Vignolo G., LeBlanc J. G., Rollán G. (2016). Ancestral Andean grain quinoa as source of lactic acid bacteria capable to degrade phytate and produce B-group vitamins. *Food Research International*, 89, 488–494.

Coconnier M.-H., Lieven V., Hemery E., Servin A. L. Antagonistic activity against *Helicobacter* infection *in vitro* and *in vivo* by the human *Lactobacillus acidophilus* strain LB. *Applied and Environmental Microbiology*, 64(11), 4573–4580.

Coelho M. C., Silva C. C. G., Ribeiro S. C., Dapkevicius M. L. N. E., Rosa H. J. D. (2014). Control of *Listeria monocytogenes* in fresh cheese using protective lactic acid bacteria. *International Journal of Food Microbiology*, 191, 53–59.

Cotter P. D., Hill C., Ross R. P. (2005). Bacteriocins, developing innate immunity for food. *Nature Reviews Microbiology*, 3, 777-788.

Cortés-Zavaleta O., López-Malo A., Hernández-Mendoza A., García H. S. (2014). Antifungal activity of lactobacilli and its relationship with 3-phenyllactic acid production. *International Journal of Food Microbiology*, 173, 30–35.

Cruchet S., Obregon M. C., Salazar G., Diaz E., Gotteland M. (2003). Effect of the ingestion of a dietary product containing *Lactobacillus johnsonii* La1 on *Helicobacter pylori* colonization in children. *Nutrition*, 19(9), 716–721.

Da Costa W. K. A., De Souza G. T., Brandão L. R., De Lima R. C., Garcia E. F., Lima M. D. S., De Souza E. L., Saarela M., Magnani M. (2018). Exploiting antagonistic activity of fruit-derived *Lactobacillus* to control pathogenic bacteria in fresh cheese and chicken meat. *Food Research International*, 108, 172–182.

Dalié D. K. D., Deschamps A. M., Richard-Forget F. (2010). Lactic acid bacteria – Potential for control of mould growth and mycotoxins, A review. *Food Control*, 21, 370–380.

De Muynck C., Leroy A. I. J., De Maeseneire S., Arnaut F., Soetaert W., Vandamme E. J. (2004). Potential of selected lactic acid bacteria to produce food compatible antifungal metabolites. *Microbiological Research*, 159, 339–346.

Deguchi R., Nakaminami H., Rimbara E., Noguchi N., Sasatsu M., Suzuki T., Matsushima M., Koike J., Igarashi M., Ozawa H., Fukuda R., Takagi A. (2012). Effect of pretreatment with *Lactobacillus gasseri* OLL2716 on first-line *Helicobacter pylori* eradication therapy. *Journal of Gastroenterology and Hepatology,* 27, 888–892.

Di Gioia D., Mazzola G., Nikodinoska I., Aloisio I., Langerholc T., Rossi M., Raimondi S., Melero B., Rovira J. (2016). Lactic acid bacteria as protective cultures in fermented pork meat to prevent *Clostridium* spp. growth. *International Journal of Food Microbiology,* 235, 53–59.

Dinleyici E. C., Dalgic N., Guven S., Metin O., Yasa O., Kurugol Z., Turel O., Tanir G., Yazar A. S., Arica V., Sancar M., Karbuz A., Eren M., Ozen M., Kara A., Vandenplas Y. (2015). *Lactobacillus reuteri* DSM 17938 shortens acute infectious diarrhea in a pediatric outpatient setting. *Jornal de Pediatria,* 91(4), 392–396.

Du L., Liu F., Zhao P., Zhao T., Doyle M. P. (2017). Characterization of *Enterococcus durans* 152 bacteriocins and their inhibition of *Listeria monocytogenes* in ham. *Food Microbiology,* 68, 97–103.

Egan K., Field D., Rea M. C., Ross R. P., Hill C., Cotter P. D. Bacteriocins, novel solutions to age old spore-related problems? *Frontiers in Microbiology,* 7, 461.

Endo A., Futagawa-Endo Y., Dicks L. M. T. (2009). Isolation and characterization of fructophilic lactic acid bacteria from fructose-rich niches. *Systematic and Applied Microbiology,* 32, 593–600.

Endo A., Futagawa-Endo Y., Sakamoto M., Kitahara M., Dicks L. M. T. (2010). *Lactobacillus florum* sp. nov., a fructophilic species isolated from flowers. *International Journal of Systematic and Evolutionary Microbiology,* 60, 2478–2482.

Fang H. R., Zhang G. Q., Cheng J. Y., Li Z. Y. (2019). Efficacy of *Lactobacillus*-supplemented triple therapy for *Helicobacter pylori* infection in children, a meta-analysis of randomized controlled trials. *European Journal of Pediatrics,* 178(1), 7-16.

Federici S., Ciarrocchi F., Campana R., Ciandrini E., Blasi G., Baffone W. (2014). Identification and functional traits of lactic acid bacteria isolated from Ciauscolo salami produced in Central Italy. *Meat Science,* 98, 575–584.

Francavilla R., Lionetti E., Castellaneta S. P., Magistà A. M., Maurogiovanni G., Bucci N., De Canio A., Indrio F., Cavallo L., Ierardi E., Miniello V. L. (2008). Inhibition of *Helicobacter pylori* infection in humans by *Lactobacillus reuteri* ATCC 55730 and effect on eradication therapy a pilot study. *Helicobacter,* 13, 127–134.

Gálvez A., Abriouel H., López R. L., Ben Omar N. (2007). Bacteriocin-based strategies for food biopreservation. *International Journal of Food Microbiology,* 120, 51–70.

García A., Navarro K., Sanhueza E., Pineda S., Pastene E., Quezada M., Henríquez K., Karlyshev A., Villena J., González C. (2017). Characterization of *Lactobacillus fermentum* UCO-979C, a probiotic strain with a potent anti-*Helicobacter pylori* activity. *Electronic Journal of Biotechnology,* 25, 75–83.

Garde S., Ávila M., Arias R., Gaya P., Nuñez M. (2011). Outgrowth inhibition of *Clostridium beijerinckii* spores by a bacteriocin-producing lactic culture in ovine milk cheese. *International Journal of Food Microbiology,* 150, 59–65.

Garde S., Gómez-Torres N., Hernández M., Ávila M. (2014). Susceptibility of *Clostridium perfringens* to antimicrobials produced by lactic acid bacteria, reuterin and nisin. *Food Control,* 44, 22–25.

Georgalaki M., Zoumpopoulou G., Mavrogonatou E., Van Driessche G., Alexandraki V., Anastasiou R., Papadelli M., Kazou M., Manolopoulou E., Kletsas D., Devreese B., Papadimitriou K., Tsakalidou E. (2017). Evaluation of the antihypertensive angiotensin-converting enzyme inhibitory (ACE-I) activity and other probiotic properties of lactic acid bacteria isolated from traditional Greek dairy products. *International Dairy Journal*, 75, 10–21.

Gerez C. L., Torres M. J., de Valdez G. F., Rollán G. (2013). Control of spoilage fungi by lactic acid bacteria. *Biological Control,* 64, 231–237.

Ghodhbane H., Elaidi S., Sabatier J. -M., Achour S., Benhmida J., Regaya I. (2015). Bacteriocins active against multi-resistant Gram negative bacteria implicated in nosocomial infections. *Infectious Disorders – Drug Targets,* 15, 2–12.

Giello M., La Storia A., De Filippis F., Ercolini D., Villani F. (2018). Impact of *Lactobacillus curvatus* 54M16 on microbiota composition and growth of *Listeria monocytogenes* in fermented sausages. *Food Microbiology*, 72, 1–15.

Gómez-Sala B., Muñoz-Atienza E., Sánchez J., Basanta A., Herranz C., Hernández P. E., Cintas L. M. (2015). Bacteriocin production by lactic acid bacteria isolated from fish, seafood and fish products. *European Food Research and Technology*, 241, 341–356.

Goswami G., Bora S. S., Parveen A., Boro R. C., Barooah M. (2017). Identification and functional properties of dominant lactic acid bacteria isolated from *Kahudi*, a traditional rapeseed fermented food product of Assam, India. *Journal of Ethnic Foods*, 4, 187–197.

Gurovic M. S. V., Gentili A. R., Olivera N. L., Rodríguez M. S. (2014). Lactic acid bacteria isolated from fish gut produce conjugated linoleic acid without the addition of exogenous substrate. *Process Biochemistry*, 49, 1071–1077.

Han Q., Kong B., Chen Q., Sun F., Zhang H. (2017). *In vitro* comparison of probiotic properties of lactic acid bacteria isolated from Harbin dry sausages and selected probiotics. *Journal of Functional Foods*, 32, 391–400.

Hartmann H. A., Wilke T., Erdmann R. (2011). Efficacy of bacteriocin-containing cell-free culture supernatants from lactic acid bacteria to control *Listeria monocytogenes* in food. *International Journal of Food Microbiology,* 146, 192–199.

Heavey P. M., Rowland I. R. (2004). Gastrointestinal cancer. *Best Practice and Research Clinical Gastroenterology,* 18(2), 323–336.

Hongpattarakere T., Cherntong N., Wichienchot S., Kolida S., Rastall R. A. (2012). *In vitro* prebiotic evaluation of exopolysaccharides produced by marine isolated lactic acid bacteria. *Carbohydrate Polymers,* 87, 846–852.

Huh C. K., Hwang T. Y. (2016). Identification of antifungal substances of *Lb. sakei* subsp. ALI033 and antifungal activity against *Penicillium brevicompactum* strain FI02. *Preventive Nutrition and Food Science,* 21(1), 52–56.

Jang S., Lee J., Jung U., Choi H. -S., Suh H. J. (2014). Identification of an anti-listerial domain from *Pediococcus pentosaceus* T1 derived from Kimchi, a traditional fermented vegetable. *Food Control,* 43, 42–48.

Jordan K., McAuliffe O. (2018). *Listeria monocytogenes* in Foods. In, *Advances in Food and Nutrition Research*. D. Rodríguez-Lázaro (Ed.). Elsevier Inc., USA, 181-213.

Karska-Wysocki B., Bazoa M., Smoragiewicz W. (2010). Antibacterial activity of *Lactobacillus acidophilus* and *Lactobacillus casei* against methicillin-resistant *Staphylococcus aureus* (MRSA). *Microbiological Research,* 165, 674–686.

Ki M. R., Ghim S. Y., Hong I. H., Park J. K., Hong K. S., Ji A. R., Jeong K. S. (2010). *In vitro* inhibition of *Helicobacter pylori* growth and of adherence of cagA-positive strains to gastric epithelial cells by *Lactobacillus paraplantarum* KNUC25 isolated from Kimchi. *Journal of Medicinal Food,* 13(3), 629–634.

Kim T. S., Hur J. W., Yu M. A., Cheigh C. I., Kim K. N., Hwang J. K., Pyun Y. R. (2003). Antagonism of *Helicobacter pylori* by bacteriocins of lactic acid bacteria, *Journal of Food Protection*, 66(1), 3–12.

Kingcha Y., Tosukhowong A., Zendo T., Roytrakul S., Luxananil P., Chareonpornsook K., Valyasevi R., Sonomoto K., Visessanguan, W. (2012). Anti-listeria activity of *Pediococcus pentosaceus* BCC 3772 and application as starter culture for Nham, a traditional fermented pork sausage. *Food Control*, 25, 190–196.

Klaenhammer T. R., de Vos W. M. (2011). An incredible scientific journey. In, *The 10th LAB symposium Thirty Years of Research on Lactic Acid Bacteria*. M. Blngham (Ed.). Rotterdam, 24 Media Labs, 1-11.

Ladda B., Theparee T., Chimchang J., Tanasupawat S., Taweechotipatr M. (2015). *In vitro* modulation of tumor necrosis factor α production in THP-1 cells by lactic acid bacteria isolated from healthy human infants. *Anaerobe,* 33, 109–116.

Li H., Liu L., Zhang S., Cui W., Lv J. (2012). Identification of antifungal compounds produced by *Lactobacillus casei* AST18. *Current Microbiology*, 65, 156–161.

Lim S. M. (2014). Anti-*Helicobacter pylori* activity of antimicrobial substances produced by lactic acid bacteria isolated from Baikkimchi, *Journal of the Korean Society for Applied Biological Chemistry,* 57(5), 621−630.

Lim E. S. (2015). Purification and characterization of two bacteriocins from *Lactobacillus brevis* BK11 and *Enterococcus faecalis* BK61 showing anti-*Helicobacter pylori* activity. *Journal of the Korean Society for Applied Biological Chemistry,* 58(5), 703–714.

Lin W. -H., Wu C. -R., Fang T. J., Guo J. -T., Huang S. -Y., Leef M. -S., Yang H. -L. (2011). Anti-*Helicobacter pylori* activity of fermented milk with lactic acid bacteria. *Journal of the Science of Food and Agriculture*, 91, 1424–1431.

Liu T., Li Y., Chen J., Sadiq F. A., Zhang G., Li Y., He G. (2016). Prevalence and diversity of lactic acid bacteria in Chinese traditional sourdough revealed by culture dependent and pyrosequencing approaches. *LWT – Food Science and Technology*, 68, 91–97.

Lv X., Du J., Jie Y., Zhang B., Bai F., Zhao H., Li J. (2017). Purification and antibacterial mechanism of fish-borne bacteriocin and its application in shrimp (*Penaeus vannamei*) for inhibiting *Vibrio parahaemolyticus*. *World Journal of Microbiology and Biotechnology*, 33, 156.

Maldonado N. C., Ficoseco C. A., Mansilla F. I., Melián C., Hébert E. M., Vignolo G. M., Nader-Macías M. E. F. (2018). Identification, characterization and selection of autochthonous lactic acid bacteria as probiotic for feedlot cattle. *Livestock Science*, 212, 99–110.

Maldonado N. C., De Ruiz C. S., Otero M. C., Sesma F., Nader-Macías M. E. (2012). Lactic acid bacteria isolated from young calves – Characterization and potential as probiotics. *Research in Veterinary Science*, 92, 342–349.

Michetti P. (2001). Lactobacilli for the management of *Helicobacter pylori*. *Nutrition*, 17(3), 268-269.

Miao J., Peng W., Liu G., Chen Y., Chen F., Cao Y. (2015). Biopreservative effect of the natural antimicrobial substance from *Lactobacillus paracasei* subsp. *tolerans* FX-6 on fresh pork during chilled storage. *Food Control*, 56, 53–56.

Midolo P. D., Lambert J. R., Hull R., Luo F., Grayson M. L. (1995). *In vitro* inhibition of *Helicobacter pylori* NCTC 11637 by organic acids and lactic acid bacteria. *Journal of Applied Microbiology*, 79, 475–479.

Moracanin S. V., Stefanovic S., Radicevic T., Borovic, B., Djukic D. (2015). Production of biogenic amines by lactic acid bacteria isolated from Uzicka sausages. *Procedia Food Science*, 5, 308–311.

Myllyluoma E., Ahlroos T., Veijola L., Rautelin H., Tynkkynen S., Korpela R. (2007). Effects of anti-*Helicobacter pylori* treatment and probiotic supplementation on intestinal microbiota. *International Journal of Antimicrobial Agents*, 29, 66–72.

Myllyluoma E., Veijola L., Ahlroos T., Tynkkynen S., KankuriE., Vapaatalo H., Rautelin H., Korpela R. (2005). Probiotic supplementation improves tolerance to *Helicobacter pylori* eradication therapy–a placebo-controlled, double-blind randomized pilot study. *Alimentary Pharmacology and Therapeutics*, 21, 1263–1272.

Nacef M., Chevalier M., Chollet S., Drider D., Flahaut C. (2017). MALDI-TOF mass spectrometry for the identification of lactic acid bacteria isolated from a French cheese, The Maroilles. *International Journal of Food Microbiology*, 247, 2–8.

Nam H., Ha M., Bae O., Lee Y. (2002). Effect of *Weissella confusa* strain PL9001 on the adherence and growth of *Helicobacter pylori*. *Applied and Environmental Microbiology,* 68(9), 4642–4645.

Noguchi S., Hattori M., Sugiyama H., Hanaoka A., Okada S., Yoshida T. (2012). *Lactobacillus plantarum* NRIC1832 enhance IL-10 production from CD4$^+$ T cells *in vitro*. *Biosciences, Biotechnology and Biochemistry,* 76(10), 1925–1931.

Oliveira P. M., Brosnan B., Furey A., Coffey A., Zannini E., Arendt E. K. (2015). Lactic acid bacteria bioprotection applied to the malting process. Part I, Strain characterization and identification of antifungal compounds. *Food Control,* 51, 433–443.

O'Bryan C. A., Crandall P. G., Ricke S. C., Ndahetuye, J. B. (2015). Lactic acid bacteria (LAB) as antimicrobials in food products, types and mechanisms of action. In, *Handbook of Natural Antimicrobials for Food Safety and Quality*. M. Taylor (Ed.). Cambridge, Woodhead Publishing, UK, 117-136.

Okereke A., Montville T. J. (1991). Bacteriocin-mediated inhibition of *Clostridium botulinum* spores by lactic acid bacteria at refrigeration and abuse temperatures. *Applied and Environmental Microbiology,* 57(12), 3423–3428.

O'Sullivan L., Ross R. P., Hill C. (2002). Potential of bacteriocin-producing lactic acid bacteria for improvements in food safety and quality. *Biochimie,* 84, 593–604.

Owen R. J. (1995). Bacteriology of *Helicobacter pylori*. *Clinical Gastroenterology*, 9(3), 415–446.

Öz E., Kaban G., Barış Ö., Kaya M. (2017). Isolation and identification of lactic acid bacteria from pastırma. *Food Control*, 77, 158–162.

Park S., Kim W. S., Choi U. J., Han S. U., Kim Y. S., Kim Y. B. (2004). Amelioration of oxidative stress with ensuing inflammation contributes to chemoprevention of *H. pylori*-associated gastric carcinogenesis. *Antioxidants and Redox Signaling,* 6, 549–560.

Parte A. C. (2018). LPSN – List of prokaryotic names with standing in bacteria integrated by a cluster of nomenclature (bacterio.net), 20 years on. *International Journal of Systematic and Evolutionary Microbiology,* 68, 1825-1829.

Pinto A. L., Fernandes M., Pinto C., Albano H., Castilho F., Teixeira P., Gibbs P. A. (2009). Characterization of anti-*Listeria* bacteriocins isolated from shellfish, Potential antimicrobials to control non-fermented seafood. *International Journal of Food Microbiology,* 129, 50–58.

Porto M. C. W., Kuniyoshi T. M., Azevedo P. O. S., Vitolo M., Oliveira R. P. S. (2017). *Pediococcus* spp., An important genus of lactic acid bacteria and pediocin producers. *Biotechnology Advances,* 35, 361–374.

Reddy G., Altaf M., Naveena B. J., Venkateshwar M., Kumar E. V. (2008). Amylolytic bacterial lactic acid fermentation–A review. *Biotechnology Advances,* 26, 22-34.

Reis J. A., Paula A. T., Casarotti S. N., Penna A. L. B. (2012). Lactic acid bacteria antimicrobial compounds, characteristics and applications. *Food Engineering Reviews,* 4, 124–140.

Ribeiro S. C., O'Connor P. M., Ross R. P., Stanton C., Silva C. C. G. (2016). An anti-listerial *Lactococcus lactis* strain isolated from Azorean Pico cheese produces lacticin 481. *International Dairy Journal,* 63, 18–28.

Rubio R., Jofré A., Martín B., Aymerich T., Garriga M. (2014). Characterization of lactic acid bacteria isolated from infant faeces as potential probiotic starter cultures for fermented sausages. *Food Microbiology,* 38, 303–311.

Saad N., Delattre C., Urdaci M., Schmitter J. M., Bressollier P. (2013). An overview of the last advances in probiotic and prebiotic field. *LWT-Food Science and Technology,* 50, 1–16.

Saarela M., Mogensen G., Fondén R., Mättö J., Mattila-Sandholm T. (2000). Probiotic bacteria, safety, functional and technological properties. *Journal of Biotechnology,* 84, 197–215.

Saguir F. M., de Nadra M. C. M. (2007). Improvement of a chemically defined medium for the sustained growth of *Lactobacillus plantarum*, nutritional requirements. *Current Microbiology,* 54, 414–418.

Salminen S., Laine M., von Wright A., Vuopio-Varkila J., Korhonen T., Mattila-Sandholm T. (1996). Development of selection criteria for probiotic strains to assess their potential in functional foods, A Nordic and European approach. *Bioscience and Microflora,* 15, 61–67.

Sandes S., Alvim L., Silva B., Acurcio L., Santos C., Campos M., Santos C., Nicoli J., Neumann E., Nunes Á. (2017). Selection of new lactic acid bacteria strains bearing probiotic features from mucosal microbiota of healthy calves; looking for immunobiotics through *in vitro* and *in vivo* approaches for immunoprophylaxis applications. *Microbiological Research,* 200, 1–13.

Shekh, R. M., Roy U. (2012). Biochemical characterization of an anti-*Candida* factor produced by *Enterococcus faecalis. BMC Microbiology,* 12, 132.

Simova E. D., Beshkova D. B., Dimitrov Z. H. P. (2009). Characterization and antimicrobial spectrum of bacteriocins produced by lactic acid bacteria isolated from traditional Bulgarian dairy products. *Journal of Applied Microbiology,* 106(2), 692–701.

Siroli L., Patrignani F., Serrazanetti D. I., Tabanelli G., Montanari C., Gardini F., Lanciotti R. (2015). Lactic acid bacteria and natural antimicrobials to improve the safety and shelf-life of minimally processed sliced apples and lamb's lettuce. *Food Microbiology*, 47, 74–84.

Spinler J. K., Taweechotipatr M., Rognerud C. L., Ou C. N., Tumwasorn S., Versalovic J. (2008). Human-derived probiotic *Lactobacillus reuteri* demonstrate antimicrobial activities targeting diverse enteric bacterial pathogens. *Anaerobe*, 14, 166–171.

Stiles M. E., Holzapfel W. H. (1997). Lactic acid bacteria of foods and their current taxonomy. *International Journal of Food Microbiology*, 36, 1-29.

Stoyanova L. G., Ustyugova E. A., Sultimova T. D., Bilanenko E. N., Fedorova G. B., Khatrukha G. S., Netrusov A. I. (2010). New antifungal bacteriocin-synthesizing strains of *Lactococcus lactis* ssp. *lactis* as the perspective biopreservatives for protection of raw smoked sausages. *American Journal of Agricultural and Biological Sciences*, 5 (4), 477–485.

Ström K., Schnürer J., Melin P. (2005). Co-cultivation of antifungal *Lactobacillus plantarum* MiLAB 393 and *Aspergillus nidulans*, evaluation of effects on fungal growth and protein expression. *FEMS Microbiology Letters*, 246, 119–124.

Šušković J., Kos B., Beganović J., Panuvc A. L., Habjanič K., Matošić S. (2010). Antimicrobial activity – The most important property of probiotic and starter lactic acid bacteria. *Food Technology and Biotechnology*, 48(3), 296–307.

Taweechotipatr M., Iyer C., Spinler J. K., Versalovic J., Tumwasorn S. (2009). *Lactobacillus saerimneri* and *Lactobacillus ruminis*, novel human derived probiotic strains with immunomodulatory activities. *FEMS Microbiology Letters*, 293, 65–72.

Tenea G. N., Hurtado P., Ortega C. (2018). Inhibitory effect of substances produced by native *Lactococcus lactis* strains of tropical fruits towards food pathogens. *Preventive Nutrition and Food Science*, 23(3), 260–268.

Todorov S. D., Stojanovski S., Iliev I., Moncheva P., Nero L. A., Ivanova I. V. (2017) Technology and safety assessment for lactic acid bacteria isolated from traditional Bulgarian fermented meat product "lukanka." *Brazilian Journal of Microbiology*, 48, 576–586.

Tomé E., Teixeira P., Gibbs P. A. (2006), Anti-listerial inhibitory lactic acid bacteria isolated from commercial cold smoked salmon. *Food Microbiology*, 23, 399–405.

Tormo H., Lekhal D. A. H., Roques C. (2015). Phenotypic and genotypic characterization of lactic acid bacteria isolated from raw goat milk and effect of farming practices on the dominant species of lactic acid bacteria. *International Journal of Food Microbiology*, 210, 9–15.

Tsai C. C., Huang L. F., Lin C. C., Tsen H. Y. (2004). Antagonistic activity against *Helicobacter pylori* infection *in vitro* by a strain of *Enterococcus faecium* TM39, *International Journal of Food Microbiology*, 96(1), 1–12.

Vale F. F., Vítor J. M. B. (2010). Transmission pathway of *Helicobacter pylori*, Does food play a role in rural and urban areas? *International Journal of Food Microbiology,* 138, 1–12.

Varsha K. K., Nampoothiri K. M. (2016). Appraisal of lactic acid bacteria as protective cultures. *Food Control*, 69, 61–64.

Vimont A., Fernandez B., Ahmed G., Fortin H.-P., Fliss I. (2019). Quantitative antifungal activity of reuterin against food isolates of yeasts and moulds and its potential application in yogurt. *International Journal of Food Microbiology,* 289, 182–188.

Wang Y., Qin Y., Zhang Y., Wu R., Li P. (2019). Antibacterial mechanism of plantaricin LPL-1, a novel class IIa bacteriocin against *Listeria monocytogenes. Food Control,* 97, 87–93.

Warren J. R., Marshall B. J. (1983). Unidentified curved bacilli on gastric epithelium in active chronic gastritis, *The Lancet,* 321(8336), 1273–1275.

Wong C. –H, Chen Y. -L. (1988). Effects of lactic acid bacteria and organic acids on growth and germination of *Bacillus cereus. Applied and Environmental Microbiology,* 54(9), 2179–2184.

Woraprayote W., Kingcha Y., Amonphanpokin P., Kruenate J., Zendo T., Sonomoto K., Benjakul S., Visessanguan W. (2013). Anti-listeria activity of poly(lactic acid)/sawdust particle biocomposite film impregnated with pediocin PA-1/AcH and its use in raw sliced pork. *International Journal of Food Microbiology*, 167, 229–235.

Woraprayote W., Pumpuang L., Tosukhowong A., Roytrakul S., Perez R. H., Zendo T., Sonomoto K., Benjakul S., Visessanguan W. (2015). Two putatively novel bacteriocins active against Gram-negative food borne pathogens produced by *Weissella hellenica* BCC 7293. *Food Control*, 55, 176–184.

Xu Q., Gu S., Chen Y., Quan J., Lv L., Chen D., Zheng B., Xu L., Li L. (2018). Protective effect of *Pediococcus pentosaceus* LI05 against *Clostridium difficile* infection in a mouse model. *Frontiers in Microbiology*, 9, 2396.

Zamfir M., Vancanneyt M., Makras L., Vaningelgem F., Lefebvre K., Pot B., Swings J., De Vuyst L. (2006). Biodiversity of lactic acid bacteria in Romanian dairy products. *Systematic and Applied Microbiology,* 29, 487–495.

Zendo T., Sonomoto K. (2011). Classification and diversity of bacteriocins. In, *Lactic Acid Bacteria and Bifidobacteria Current Progress in Advanced Research.* K. Sonomoto and Y. Atsushi (Eds.). Norfolk, Caister Academic Press, UK, 159-164.

Zheng P. X., Fang H. Y., Yang H. B., Tien N. -Y., Wang M. C., Wu J. J. (2016). *Lactobacillus pentosus* strain LPS16 produces lactic acid, inhibiting multidrug-resistant *Helicobacter pylori*. *Journal of Microbiology, Immunology and Infection*, 49, 168–174.

Zullo A., Hassan C., Ridola L., De Francesco V., Vaira D. (2012). Standard triple and sequential therapies for *Helicobacter pylori* eradication, An update. *European Journal of Internal Medicine*, 24, 16–19.

In: The Many Benefits of Lactic Acid Bacteria
Editors: J. G. LeBlanc and A. de Moreno

ISBN: 978-1-53615-388-0
© 2019 Nova Science Publishers, Inc.

Chapter 9

DIVERSITY AND PRODUCTION OF γ-AMINOBUTYRIC ACID BY LACTIC ACID BACTERIA

*Sukanya Phuengjayaem and Somboon Tanasupawat**

Department of Biochemistry and Microbiology,
Faculty of Pharmaceutical Sciences, Chulalongkorn University,
Bangkok, Thailand

ABSTRACT

Gramma-aminobutyric acid (GABA) is a non-protein amino acid, which is produced from the decarboxylation of L-glutamic acid by glutamate decarboxylase. GABA provides beneficial effects for human health, inhibiting cancer cell proliferation, helping the recovery from alcohol-related symptoms, controlling stress, improves memory and learning abilities. Therefore, it has been classified as a bioactive compound in foods and medicine.GABA was produced by various micro-organisms including fungi, yeasts and lactic acid bacteria (LAB) Currently, LAB strains are accepted for high-performance GABA production including *Lactobacillus* (*Lb*) and *Pediococcus* (*Ped*) from Nham (Thai fermented pork sausage), *Lb*. senmaizukei from traditional pickles and *Lactococcus* (*Lc*) from cheese in Japan, *Lb. paracasei* from cheese in Italy and Japanese traditional fermented fish, *Lb. brevis* from many fermented foods (Korean fermented vegetable kimchi, Chinese traditional paocai, fresh milk, alcohol distillery lees, and black raspberry juice); and *Lb. plantarum*, *Leuconostoc* (*Leuc*) and *Weissella* (*W*) from kimchi in Japan and Korea. The production of GABA by microorganisms is dependent on several factors, including pH, temperature, glutamate concentration, medium component, and cultivation time. These factors have been optimized by the one-factor-at-a-time method. The limitation of this method varies only one independent factor at a time while fixing all others, and the interactions between each factor are neglected. Eventually, response surface methodology (RSM), as a statistical method, is more rapid, reduces the number of

---

* Corresponding Author's Email: Somboon.T@chula.ac.th.

experiments and studies interactions several factors, which saves the cost of experiment and time. The statistical RSM method will be applied.

INTRODUCTION

Gamma-aminobutyric acid (GABA) is an amino acid that consists of four carbons, nine hydrogens, two oxygens, and one molecule of nitrogen. The chemical formula is $C_4H_9NO_2$ and molecular weight of 103.12 g/mole. GABA is widely distributed in nature including microorganisms, plants and animals (Ueno, 2000). Many microorganisms such as fungi, *Monascus purpureus (Su et al., 2003) and Rhizopus microsporus* (Aoki et al., 2003); and LAB strains of the genera *Lactobacillus, Pediococcus, Lactococcus, Leuconostoc* and *Weissella* are reported to produce GABA (Dhakal, BajpaiBaek, 2012). GABA has shown several important functions in humans and animals (Foster & Kemp, 2006), therefore some high GABA-producing LAB have been emphasized for GABA production and GABA-enriched in fermented food. Furthermore, GABA or GABA-enriched in fermented food has been entering the commercial process as food additives and functional food supplements. In this chapter, GABA structure and function, diversity of GABA producing LAB, the method for the determination, improvement and development process for high yield of GABA, application, and benefits of GABA have been described.

γ-AMINOBUTYRIC ACID (GABA): STRUCTURE AND FUNCTION

GABA was first synthesized in 1883 and was known only as a plant and microbial metabolic product; however, it was found to be an integral part of the mammalian central nervous system in 1950. Until 1959, it was shown to be an inhibitory synapse on crayfish muscle fibers and GABA acted to stimulate the inhibitory nerve. Both inhibitions by nerve stimulation and the application of GABA were blocked by picrotoxin (Roth, Cooper & Bloom, 2003). GABA was decarboxylated from L-glutamic acid, catalyzed by α-glutamate decarboxylase (GAD) with pyridoxal 5' phosphate (PLP: the active form of vitamin B6) as a cofactor. This process converts irreversible alpha-decarboxylation of L-glutamate (the principal excitatory neurotransmitter) into GABA (the principal inhibitory neurotransmitter) (Figure 1).

GAD, a catalyzed enzyme for GABA, is generally discovered in eukaryotes and prokaryotes such as in microorganisms, plants, and animals. Although some GABA can be found in pancreatic cells and kidney, there is no significant amount of GABA in mammalian tissues other than in the nervous system. In addition, GABA in many foods such as tea, beni-koji, and germinated brown rice also have been found (Thwe et al., 2011).

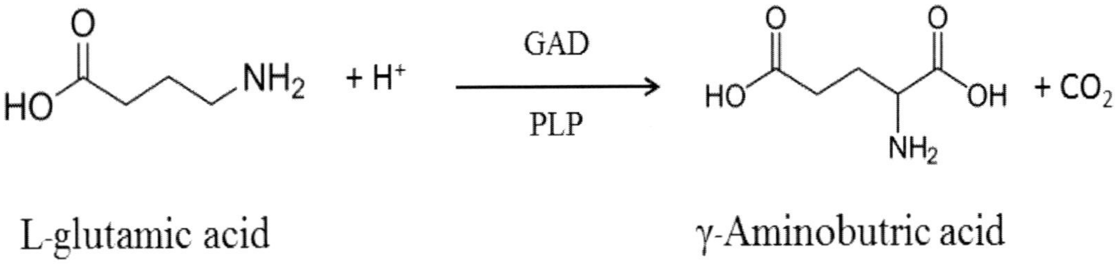

L-glutamic acid   γ-Aminobutric acid

Figure 1. The decarboxylation of L-glutamic acid to GABA GAD, Glutamate decarboxylase; PLP, Pyridoxal-5'-phosphate.

GABA is an inhibitory neurotransmitter in the central nervous system of mammals by binding to specific transmembrane receptors in the plasma membrane of both presynaptic and postsynaptic neurons (Kinnersley & Turano, 2000). GABA plays a role in regulating neuronal activation in the nervous system. This binding causes the opening of ion channels to allow the flow of either negatively-charged chloride ions into the cell or positively-charged potassium ions out of the cell. This action results in a negative change in the transmembrane potential, usually causing hyperpolarization. Three typical classes of GABA receptor are known: $GABA_A$ and $GABA_C$ ionotropic receptors, which are ion channels themselves, and $GABA_B$ metabotropic receptors, which are G protein-coupled receptors that open ion channels via intermediaries (G proteins) (Olsen & DeLorey, 1999).

GABA is found mostly as a zwitterion, which is when the carboxyl group deprotonated and the amino group protonated. Its conformation depended on the environments. In the gas form, a highly folded conformation is strongly favored due to the electrostatic attraction between the two functional groups. The stability is 50 kcal per mole according to quantum chemistry calculations. In the solid form, more extensively modeled variants are found at the end of the amino end and the concave form at the end of the carboxylate. This is due to the packing interactions with the neighboring molecules. In solution form, five different conformations, some folded and some extended are found as a result of solvation effects. The conformational flexibility of GABA is important for its biological function, it has been found to bind to different receptors with different conformations. Many GABA analogues with pharmaceutical applications have more rigid structures in order to control the binding better (Majumdar & Guha, 1988; Sapse, 2000).

DIVERSITY OF γ-AMINOBUTYRIC ACID PRODUCING LACTIC ACID BACTERIA

GABA was produced by various microorganisms based on glutamate decarboxylase (GAD). Furthermore, LAB used for the production of GABA-enriched in fermented foods has increasingly attracted great attention due to they are safe and several health-promoting properties associated with LAB themselves (Nejati et al., 2013).

Currently, GABA production using LAB has been the focus of research (Li & Cao, 2010). LAB are getting a lot of attention because they could produce bioactive peptides, amino acids, volatile fatty acids, and lactic acid throughout the fermentation process (Lahtinen, Tuomivan & Dijken, 2011). Nonetheless, to produce functional foods with the potential of modulating the blood pressure, production of GABA and L-glutamic acid in a sustained fashion is of primary importance (Li & Cao, 2010). L-Glutamic acid serves as a precursor for the formation of GABA. Addition of glutamate to the fermentation process could have an adverse effect on the accumulation of GABA (Li & Cao, 2010). Furthermore, the addition of GABA to a food matrix is recognized to be illegal (Kim et al., 2009). With such drawbacks for the development of GABA-containing foods, incorporation of L-glutamic acid-producing LAB into the fermentation process is a great advantage.

The GABA-producing ability varies widely among the strains and some GABA-producing LAB strains have shown a great promise potential in large-scale fermentation for the production of GABA (Li et al., 2010). Some fermented products enriched in GABA using LAB as starters such as in dairy products (Hayakawa et al., 2004), black raspberry juice (Kim et al., 2009), soy milk (Tsai et al., 2006), kimchi (Seok et al., 2008), and cheese (Nomura et al., 1998) have been developed.

Currently, *Lactobacillus* (*Lb*), *Pediococcus* (*Ped*), *Lactococcus* (*Lc*), *Leuconostoc* (*Leuc*), *Streptococcus* (*Strep*), and, *Weissella* (*W*) strains including *Lb. plantarum*, *Leuc. mesenteroides*, *W. cibaria*, *Lb. paracasei*, *Lb. brevis*, *Lb. casei*, *Lb. pseudomesenteroides*, *Lc. lactis*, *Lb. delbrueckii* sub sp. *bulgaricus*, *Lb. rhamnosus*, *Strep. thermophilus* and *Enterococcus* (*Ec*) *durans* has been accepted for high performance GABA production (Dhakal, BajpaiBaek, 2012). They were isolated and screened from traditional fermented food. *Lb. namurensis* NH2 and *Ped. pentosaceus* HN8 were isolated from Nham (Thai fermented pork sausage) (Ratanaburee et al., 2013). *Lb. senmaizukei* was isolated from traditional pickles kimchi (Hiraga et al., 2008). *Lb. rhamnosus* YS9 was isolated from Chinese traditional food (Lin, 2013). *Lb. paracasei* was isolated from cheese in Italy and Japanese traditional fermented fish (Komatsuzaki et al., 2005). *Lb. buchneri* and *Lb. brevis* were isolated from many fermented foods including Korean fermented vegetable kimchi, Chinese traditional paocai, fresh milk, alcohol distillery lees and black raspberry juice (Kim et al., 2009; Park & Oh 2006). The strains of *Lb. senmaizukei* from pickles (Park & Oh, 2006). *Lb. plantarum*, *Leuc. mesenteroides*, *Leuc. lactis* and *W. viridescens* were isolated from kimchi in Japan and Korea (Kim & Kim, 2012). *Lc. lactis* strains were isolated from cheese starters and cheese in Japan (Lu et al., 2008; Nomura et al., 1998). Moreover, LAB were isolated from black raspberry juice and soymilk (Tsai et al., 2006). The high GABA producing strains, *Lb. paracasei* PF6, *Lb. delbrueckii* sub sp. *bulgaricus* PR1, *Lc. lactis* PU1 and *Lb. brevis* PM17 have been isolated from Pecorino di Filiano,

Pecorino del Reatino, Pecorino Umbro, and Pecorino Marchigiano cheeses in Italy, respectively (Siragusa et al., 2007).

Multitudinous of GABA-producing LAB are important for the food industry (Komatsuzaki et al., 2005). Because of the potential for producing GABA-enriched foods, LAB are received a lot of attention. The ability of LAB to produce GABA is varied among species and the strains as shown in Table 1.

Table 1. GABA produced by lactic acid bacteria

Strain	Source	Medium	GABA Yield	Time (h)	Reference
Bifidobacterium dentium NFBC2243, DPC6333 and UCC35624	Dental carries, Infant feces and ileal-caecal region	MRS with 5% (w/v) MSG	8.63, 6.16 and 2.04 g/L	48	Barrett, 2014
Co-fermented with *Leuc. mesenteroides* SM and *Lb. plantarum* K154	Carrot juice and kimchi	MRS with 3% (w/v) MSG	10 mg/ml	30	Kwon et al., 2016
Ec. faecium NCIM 5593	Takarishta	Rice flour medium (3% w/v) with 1% (w/v) glutamate	750.55 mg/100g DM	48	Divyashri & Prapulla, 2016
Lb. brevis CECT 8183	Goat cheese	Sourdough medium	0.96 mM	24	Diana et al., 2014
Lb. brevis CRL 1942	Quinoa sourdough	MRS with 270 mM MSG	255 mM	48	Villegas et al., 2016
Lb. brevis GABA100	-	Black raspberry juice with 2% (w/v) MSG	26.5 mg/ml	12 day	Kim et al., 2009
Lb. brevis HYE1	Kimchi	MRS with 1% (w/v) MSG	18.76 mM	48	Lim et al., 2017
Lb. brevis NCL912	Paocai (Chinese fermented vegetable)	MRS with 134.45 mM glutamate	1005.81 mM	48	Li et al., 2010
Lb. brevis NPS-QW-145, NPS-QW-171, NPS-QW-177, NPS-QW-193, NPS-QW-216, NPS-QW-242, NPS-QW-255 and NPS-QW-267	Korean kimchi	MRS with 7% (w/v) MSG	25.831, 19.631, 24.098, 23.33, 21.694, 22.986, 19.072, and 24.992 g/L, respectively	72	Wu & Shah, 2015
Lb. brevis TCCC 13007	Pickled Chinese vegetables	3.6 g corn steep liquor, 22 g glucose, 15 g yeast extract, 2 g sodium acetate, 0.2 g MgSO$_4$, 0.2 g (NH$_4$)$_2$SO$_4$ with 7% (w/v) MSG	61 g/L	66	Zhang et al., 2012
Lb. buchneri WPZ001	Fermented sausages	MRS, Corncob hydrolysate with 0.1% MSG	129 g/L, 117 g/L	48	Zhao et al., 2015

Table 1. (Continued)

Strain	Source	Medium	GABA Yield	Time (h)	Reference
Lb. bulgaricus CFR 2028	CFTRI culture collection centre	MRS with 1% (w/v) MSG	22.7 mM	48	Gangaraju, Murty & Prapulla, 2014
Lb. paracasei 15C	Nostrano-cheeses	MRS	14.8 mg/kg	24	Franciosi et al., 2015
Lb. paracasei NFRI 7415	Fermented fish (funa-sushi)	MRS with 500 mM glutamate	302 mM	144	Komatsuzaki et al., 2005
Lb. plantarum MNZ	Malaysian fermented foods	MRS with 5% (w/v) MSG	115.2 mg/kg dosa	120	Zareian et al., 2015
Lb. plantarum NDC75017	Fermented dairy products in China	MRS with 80 mM MSG	314.56 mg/100 g substrate	-	Shan et al., 2015
Lb. plantarum Taj-Apis362	Honeybees	MRS with 497.97 mM of glutamate	7.15 mM	60	Tajabadi et al., 2015
Lb. rhamnosus 21D-B	Nostrano-cheeses	MRS with vancomycin 8 μg/mL	11.3 mg/kg	24	Franciosi et al., 2015
Lb. rhamnosus GG	Bioresource and Collection Center	MRS with 2.27% (w/v) MSG	1.13 mg/mL	36	Song & Yu, 2018
Lb. rhamnosus YS9	Chinese pickles	MRS with 200 mM glutamate	187 mM	84	Lin, 2013
Lb. senmaizukei L13	Japanese pickles	GYP with 5%MSG	Not report	48	Hiraga et al., 2008
Lc. lactis subsp.*lactis*	Kimchi	MRS with 1% MSG	3.68 g/L	48	Lu et al., 2008
Lc. lactis subsp. *lactis* B	Kimchi and youghurt	MRS with 1% (w/v) MSG, brown rice juice, germinated soybean juice and skim milk (33:58:9, v/v/v)	3.68 g/L, 6410 mg/l	48	Lu et al., 2008
Lc. lactis subsp. *lactis* PU1	Cheese	Wheat flour sourdough	258.71 mg/kg	48	Rizzello et al., 2008
Monascus purpureus CMU001	-	Rice	28370 mg/kg	7 days	Jannoey et al., 2010
Ped. pentosaceus NH102	Fermented pork	MRS with 0.5% (w/v) MSG	8.386 g/L	24	Ratanaburee et al., 2013)
Ped. pentosaceus HN8	Fermented beef	MRS with 0.5% (w/v) MSG	9.06 g/L	24	Ratanaburee et al., 2013
Ped. pentosaceus HN8 and *Lb. namurensis* NH2	Fermented meat (Nham)	MRS with 0.5% (w/v) MSG	4051 mg/kg	48	Ratanaburee et al., 2013
Strep. thermophilus 84C	Nostrano-cheeses	M17	80.0 mg/kg	24	Franciosi et al., 2015
Strep. thermophilus ST110	Yogurt starter	Tryptone, yeast extract-lactose (TYL) with 100 mM MSG	655 mg/L	16	Somkuti et al., 2012

MSG, Monosodium glutamate; MRS, de Man, Rogosa and Sharpe.

Although there have been many studies on GABA production by LAB, the isolation of GABA-producing LAB and the development of their GABA production need to be further explored since fermented foods as the isolation sources for GABA producers have various kinds and diverse LAB (Lim et al., 2017).

METHODS FOR THE DETERMINATION OF γ-AMINOBUTYRIC ACID

The monitoring of GABA production can be determined using several techniques, including qualitative and quantitative analysis. Thin layer chromatography (TLC) is the most primary screening method for GABA qualitative analysis. This method is simple, rapid and dose not require the derivatization process for GABA, but TLC method does not provide enough accuracy for quantifying GABA. Therefore, it is only an initial analysis and requires other techniques to confirm the results. Following derivatization, GABA is measured using High-performance liquid chromatography (HPLC). This technique is extremely rapid, efficient and can be analyzed both qualitatively and quantitatively. However, it was necessary to derivative the sample before being analyzed using HPLC. Many studies usually combine TLC and HPLC methods together to evaluate and quantify GABA. The detailed analysis is summarized in Table 2.

Although TLC is the first choice for screening of GABA producers, it also has some limitations such as requirement of many plates and a large volume of solvents (Li et al., 2010). In addition, pre-screening methods are normally developed based on the decarboxylation of glutamate into GABA by glutamate decarboxylase; this reaction eliminates a proton and also produces carbon dioxide. Spectrophotometry assay involves using GABase enzyme. Thus, glutamate decarboxylase-based microtiter plate assay (Tsukatani, Higuchi & Matsumoto, 2005), pH indicator method (Yang et al., 2006) and gas release-based assays are suitable for high-throughput pre-screening of GABA producers with remarkable improvements in testing time and economic practice (Wu & Shah, 2015). Moreover, chromatography-based techniques such as amino acid analyzer (AAA), liquid chromatography (LC), gas chromatography (GC), are also an alternative for determining GABA amounts in many kinds of food samples (Li et al., 2010). The distinctive of AAA is does not require the derivatization process prior to analysis. LC method normally requires pre-column derivatization with dansyl chloride by an addition of chromophore to GABA that could be detected by fluorescence spectroscopy (Wu & Shah, 2015). For GC method, pre-column derivatization of GABA is usually achieved with norvaline to form a volatile compound, which could be separated and detected by GC equipped with a flame ionization detector (FID) or mass spectrometry (MS) (Kagan et al., 2008) and Gas chromatography-mass spectrometry (GC-MS) (Lim et al., 2017). The above methods are time-consuming due to tedious sample preparation processes according to analytical protocols; this requires high work-load for screening GABA-producing microorganisms and is neither economical nor rapid. Hence, a pre-screening method is necessary for identifying GABA producers prior to determining their capability to produce GABA by chromatography-based methods (Kim & Kim, 2012). The advantages and disadvantages of each technique mentioned above are summarized in Table 3.

Table 2. Determination of GABA using TLC and HPLC

TLC technique	HPLC technique		Reference
	Derivative	Condition	
-	Phenylisothiocyanate (PITC)	**MP**, 60% solution A (10.2 g sodium acetate, 0.5 ml triethylamine and 0.7 mL acetic acid in 1 L), 12% solution B (acetonitrile) and 28% solution C (water); **Column**, C18; **Flow rate**, 0.6 ml/min; **Temp**, 27°C; **Detector**, UV at 254 nm	Zareian et al., 2015
SP, silica gel plate; **MP**, n-butyl alcohol: glacial acetic acid: DW(distilled water) (3:1:1 v/v/v); **Visualization**, 0.2% ninhydrin in acetone, heat at 105°C	Phenylisothiocyanate (PITC)	**MP**, 100% Solvent A (140 mM sodium acetate, 0.15% triethylamine, 0.03% EDTA, 6% acetonitrile) and ending at 100%; Solvent B (60% acetonitrile, 0.015%EDTA), during 30 min; **Column**, C18; **Flow rate**, 1.0 ml/min; **Detector**, UV at 254 nm	Kwon et al., 2016
SP, silica gel plate; **MP**, 1-butanol: acetic acid: distilled water (3:2:1 v/v/v); **Visualization**, 0.5% (w/v) ninhydrin and exposing the plates to a heat source	Phenylisothiocyanate (PITC)	**MP**, 80% solution A (1.4 mM sodium acetate, 0.1% triethylamine, and 6% acetonitrile) and 20% solution B (60% acetonitrile); **Column**, Cosmosil 5C18-AR-II; **Flow rate**, 1.0 ml/min; **Temp**, 27°C; **Detector**, UV at 254 nm	Kim &K im, 2012
SP, silica gel plate; **MP**, acetic acid:1-butanol:distilled water (1:4:5 v/v/v); **Visualization**, 0.2% (w/v) ninhydrin	Ortho-phthaladehyde–mercaptoethanol	**MP**, sodium acetate (1.64 g) and triethylamine (200 µl) in 1 L (20%, v/v acetonitrile); **Column**, Hypersil ODS; **Flow rate**, 0.8 ml/min; **Temp**, 30°C; **Detector**, UV at 338 nm	Kangaraju, Murty & Prapulla, 2014
SP, silica gel plate; **MP**, n-butanol:acetic acid:distilled water (5:3:2); **Visualization**, 0.4% (w/v) ninhydrin then heated to visualize the spots	O-phthaldealdehyde 3-mercaptopropionic acid (OPA-3MPA)	**MP**, solution A (40 mM monosodium phosphate) and solution B (acetonitrile: methanol:H$_2$O; 45:45:10 (v/v/v)); **Column**, Hypersil ODS C18 reverse phase; **Flow rate**; 1.0 ml/min; **Temp**, 40°C; **Detector**, Fluorescence detector with excitation at 340 nm, emission at 450 nm	Villegas et al., 2016
-	Phenylisothiocyanate (PITC)	**MP**, solution A (0-100% after 50 min) and solution B (0.1M ammonium acetate: acetonitrile: methanol (44:46:10 v/v/v)) (0-100% after 50 min); **Column**, C18 reversed-phase; **Flow rate**, 1.0 ml/min; **Detector**, UV at 254 nm	Song & Yu, 2018
-	2,4-Dinitrofluorodinitrobe nzene (FDNB)	**MP**, solution A (acetonitrile: H$_2$O (1:1) and solution B (0.02 M phosphate buffer); **Column**, C18 reversed-phase; **Flow rate**, 1.0 ml/min; **Temp**, 35°C; **Detector**, UV at 360 nm	Yingguo Lü et al., 2010
SP, silica gel plate; **MP**, n-butanol: acetic acid: water (5:3:2, v/v/v); **Visualization**, 0.2% (w/v) ninhydrin then develop at 60°C	O-phthaldialdehyde (OPA)	**MP**, solution A (0.05 M sodium acetate) solution B (the mixture of 0.1 M sodium acetate, acetonitrile and methanol (46:44:10 v/v/v); **Column**, XTerra column reversed-phase; **Flow rate**, 1.0 ml/min; **Temp**, 35°C; **Detector**, UV at 358 nm	Kook & Cho, 2013

SP, Stationary phase; MP, Mobile phase; Temp, Temperature.

Table 3. The advantage and disadvantage of GABA determination methods

Determination Methods	Advantage	Disadvantage	Reference
Thin layer chromatography	Expedient, rapid and accurate, not require the derivatization process	Not quantitative analysis, requires a number of plates and a large volume of solvents	Kim & Kim, 2012; Kwon et al., 2016
Gas chromatography	Rapid and accurate quantitative and qualitative measurements	High cost of machine, non-volatile matrices, requires additional derivatives, an experienced technician, and be careful in using gas	Kagan et al., 2008
Gas chromatography-Mass spectrometry	Rapid and accurate, strong analytical tool, good technique for simple mixtures, accurate quantitative and qualitative	High cost of machine and requires an experienced technician	Lim et al., 2017
HPLC	Rapid and accurate, accurate quantitative and qualitative	High cost of machine and derivatives method, time consuming for sample derivatives	Zareian et al., 201; Kwon et al., 2016; Kim & Kim, 2012
Gas release-based assays	Rapid, not require the derivatization process	Expensive, not a completely accurate method to identify high GABA	Wu & Shah, 2015
Determination Methods	*Advantage*	*Disadvantage*	*Reference*
pH indicator method	*Expedient and rapid, not require the derivatization process*	*Using mechanism only, not detect the quantitative analysis*	*Yang et al., 2006*
Enzyme assay	Specific technique, easily measurable reaction	Complicate and expensive reaction	Barla et al., 2016
Amino acid analyzer (AAA)	Expedient and rapid	Using mechanism only, high cost of machine	Lu et al., 2008; Barla et al., 2016; Chung et al., 2009
Determination Methods	Advantage	Disadvantage	Reference
Liquid chromatography (LC)	Accurate and detect in qualitative and quantitative technic	Expensive, not portable, requires an experienced technician, and only moderate throughput	Wu & Shah, 2017; Dhakal, Bajpai & Baek, 2012
Electrospray tandem mass spectroscopy (ESI-MS)	For biological samples analysis, high amount of accuracy, accurate quantitative and qualitative measurements	Cannot analyze mixtures very well, difficult to clean and the multiple charges attached to the molecular ions will make confusing for spectral data	Gangaraju, Murty & Prapulla, 2014

IMPROVEMENT OF γ-AMINOBUTYRIC ACID PRODUCTION

The optimum conditions vary among the fermenting microorganisms due to the different properties of the GAD, characterization of the biochemical properties of the GAD will be required in the interested microorganisms to achieve the highest GABA production. Several fermentation factors affect the rate of GABA production by microorganisms are summarized below, especially on the effects of pH, temperature, cultivation time, glutamate concentration, and media additives of culture.

Effect of pH on GABA Production

LAB produce lactic acid as the end product through a variety of carbohydrate catabolism including Embden-Meyerhof-Parnas (EMP), Leloir pathways and pentose phosphate pathways (Kandler, 1983). This acidic environment causing LAB to be acid resistant and also GABA is an amino acid as the end product of the decarboxylation of glutamic acid, therefore this acid-resistant property has benefit for their high GABA production.

The important role of bacterial amino acids decarboxylase occurred in LAB resulted in the stoichiometric including considered regulate of acidic pH by utilization of proton ions and produced GABA as the end product, also the net effect of this reaction increases the alkalinity of the cytosol and environment (De Biase et al., 1999).

GABA production was depended on GAD and the highest yield of GABA might be obtained when the cultivation conditions were the same as that of enzyme catalysis. However, decarboxylase reaction was not enhanced under lower pH 4 (Komatsuzaki et al., 2005). In another report, the optimum of pH 5.0 which showed the highest GABA production of 1005.81 mM was obtained from *Lb. brevis* NCL912, while the cell growth was inhibited at pH 3.0 and the cell mass was decreased at pH 6.0, resulting in no GABA production. The higher or lower pH may lead to the partial loss of the GAD activity therefore, pH affected significantly on the production of biomass and GABA of *Lb. brevis* NCL912 (Li et al., 2010). Similarly, the optimum pH for GAD activity on extracellular GABA production from *Lb. paracasei* NFRI 7415 was 5.0. The glutamate content 500 mM with the addition of PLP in the culture medium was converted to 302 mM (Komatsuzaki et al., 2005). The GABA production by *Strep. salivarius* subsp. *thermophilus* Y2 was also enhanced by optimizing fermentation at pH 4.5 and by the addition of PLP (Yang et al., 2008). For the GAD of *Lb. rhamnosus* YS9, the optimal pH was determined by the measurement of enzyme activity in cells under various pH conditions. GAD activity dramatically decreased at pH 4.0 than pH more than 4.8. *Lb. rhamnosus* YS9 did not show high GAD activity at low pH, although it was believed that bacterial GAD plays a key role in the acid-resistance mechanism. The optimum pH for GAD activity was 4.4. Accordingly, GABA production was measured by *Lb. rhamnosus* YS9 under pH-regulated conditions at 43°C, 200 μM PLP, and 200 mM MSG), *Lb. rhamnosus* YS9 produced the maximum to 187 mM of GABA by *Lb. rhamnosus* YS9 at pH 4.4, which might be due to the pH condition consistent with the optimal pH for GAD activity (Lin, 2013).

Also, there are many studies that support the effect of pH values for maintaining the activity of LAB. GAD exhibited activity ranged from 4.0 to 5.0 (Komatsuzaki et al., 2008; Huang et al., 2007; Ueno et al., 1997). It was reported that pH affected on the GABA production in LAB (Yang et al., 2008; Komatsuzaki et al., 2005; Li et al., 2010).

Effect of Temperature on GABA Production

The cell growth of *Lb. brevis* NCL912 increased when temperature was increased and the maximum was at 35°C (Li et al., 2010). Likewise, production of GABA was a similar profile to cell growth. This phenomenon could be explained that high cell density was required for effective production of GABA. Nevertheless, GABA concentration at 30°C was almost the same to that at 35°C, but biomass at 30°C was less than that at 35°C. In addition, *Lb. brevis* NCL912 could not produce GABA at 45°C while the strain could grow at this temperature. Accordingly, the production of GABA by *Lb. paracasei* NFRI 7415 was increased rapidly at 37°C but GABA production and growth rate were drastically inhibited at 43°C. These results suggested that 37°C was optimum condition for GABA production by *Lb. paracasei* NFRI 7415 (Komatsuzaki et al., 2005). On the other hand, the optimal cultivation temperature of *Lb. rhamnosus* YS9 for GABA production was 43°C (Lin, 2013), which would dramatically inhibit other LAB's growth rate, such as *Lb. paracasei* NFRI 7415 (Komatsuzaki et al., 2005) and *Lb. brevis* NCL912 (Li et al., 2010). This might be due to the different physiological and enzymatic character (Lin, 2013). The results from the above reports indicated that appropriate temperature is beneficial to produce GABA, and excessively high temperature is unfavorable to the GABA production; and high efficient conversion glutamate to GABA required not only high growth rate but also appropriate temperature.

Effect of Glutamate Concentration on GABA Production

LAB fermented monosaccharide to produce a low molecular organic acid, resulting in acidic conditions to inhibit the growing bacteria. However, some LAB employed glutamic acid decarboxylase (GAD) system for maintaining neutral cytoplasmic pH when the external pH decreased because the decarboxylation of glutamate within the LAB cell exterminates an intracellular proton. Comparison of the adding and not adding glutamic acid appeared that the pH value of the culture without glutamate decreased to 3.4 after 24 h of fermentation by *Lb. brevis* NCL912 while pH values of the cultures supplemented with glutamate were maintained at 5.0. This indicated that GAD system of *Lb. brevis* NCL912 acted under low pH and resulted in an increase of pH in the medium by glutamate, and protected cell survival from the acidic condition. However, the extra high concentration of glutamate was harmful to the cell growth (over 0.5 M). Therefore the concentration of glutamate was optimum in the range of 0.25-0.50 M for the cell growth and GABA production by *Lb. brevis* NCL912 (Li et al., 2010). The optimum of MSG for extracellular and intracellular GABA content produced by *Lb. rhamnosus* YS9 was 100 mM at 37°C (Lin, 2013).

Effect of PLP Addition on Extracellular GABA Production

PLP, a GAD coenzyme is necessary in decarboxylation reaction. Therefore, an addition of PLP to the medium could be increased GAD activity to enhance the GABA production (Tong et al., 2002). The appropriate concentration of PLP added to the culture medium was 200 μM gave the maximum of GABA to 187 mM ((Lin, 2013). The higher concentration of PLP, resulted in higher extracellular released of GABA by *Lb. rhamnosus* YS9.

The result according to the previous research of Komatsuzaki et al. (2005), when 10 μM PLP was added at initial cultivation, the GABA production was enhanced by *Lb. paracasei* NFRI 7415 and even though 100 μM PLP was added in the medium, it did not strongly inhibit the cell growth. However, this was different from that observed with *Strep. thermophilus* Y2 (Yang et al., 2008) and *Lb. brevis* NCL912 (Li et al., 2010). Consequently, it could not be concluded that the PLP will always promote GABA accumulation by LAB because some researchers reported that the addition of PLP did not affected on cell growth and not increased GABA production by *Lb. brevis* NCL912 (Li et al., 2010). They explained that the strain *Lb. brevis* NCL912 could synthesize sufficient PLP for themselves.

Effect of Cultivation Time

GABA is a growth associate product, therefore, the cultivation time has an effect on the production. LAB strains have different growth profiles depending on their biochemical properties. The main isolates from cheese produced GABA under fermentation conditions at 30°C and pH 4.7 for 24 h (Seok et al., 2008). Among the strains, GABA was produced within 24-60 h as described in Table 1.

The production of GABA by the microorganism is depended on several factors, including pH, temperature, glutamate concentration, medium component, and cultivation time. The primary objective of LAB fermentation is the highest yielding of the target product. The factors of the physical and medium component have been optimized by conventional or statistical methods, providing basic information to develop the bioprocess for commercial GABA production (Kumar & Satyanarayana, 2007). The most significant stages in the biological process are modeling and optimization to improve a system and increase the efficiency of the process. The biochemical differences of the species used for the optimal condition for GABA production are different as discussed above. Therefore, the optimum conditions vary among the fermenting microorganisms due to the different properties of the GAD. Currently, a single variable optimization design used as the first step was efficient for identifying which the ranges of fermentation factors, had a significant effect on the GABA production. The response surface methodology was used to optimize the fermentative parameter for the high production of GABA (Tajabadi et al., 2015).

The conventional method is the one-factor-at-a-time (OFAT) method. The limitation of OFAT method varies only one independent factor at a time while fixing all the other interactions between each factor are neglected resulting in just local optimum points being obtained. Eventually, response surface methodology (RSM), as a statistical method, uses an experimental design to obtain the optimum point of response, and experiments can be easily organized and translated using efficient design. In addition, RSM is more rapid, reduces the number of experiments and considering interactions between each factor, which saves the cost of experiment and time (Lim et al., 2017).

The optimizations of GABA production by LAB have focused on the conventional determination of optimal pH, temperature, PLP and glutamate concentration (Komatsuzaki et al., 2005; Li et al., 2010; Lin, 2013). Although the statistical RSM method is more efficient compared with the conventional OFAT method, there have been a few studies on the optimization for GABA production via the RSM method.

The optimum condition for GABA production by *Ped. pentosaceus* HN8 and *Lb. namurensis* NH2 using the Central Composite Design (CCD) has been reported (Ratanaburee et al., 2013) using 0.5% (w/v) MSG, inoculum size of roughly 6 log CFU/g of each two strains. The GABA production gave the maximum of 4051 mg/kg in the 'GABA Nham' product. Using the same experimental design, CCD of RSM was used to define conditions for maximum GABA production by *Lb. plantarum* Taj-Apis362. The maximum GABA of 7.15 mM was obtained under 497.97 mM of MSG, 36°C, at pH 5.31, and 60 h for cultivation time (Tajabadi et al., 2015).

In addition, the Box-Behnken Design (BBD) of RSM was used to optimize GABA production by *Lb. brevis* HYE1. The optimum conditions for maximum GABA production of 18.76 mM were obtained using 2.14% (w/v) maltose, 4.01% (w/v) tryptone, 2.38% (w/v) MSG, and pH 4.74 (Lim et al., 2017). This was similar to the results using BBD of RSM to optimize 3 variables such as level of MSG, PLP, and culture temperature for GABA production by *Lb. plantarum* NDC75017 (Shan et al., 2015). The optimal conditions were 80 mM of MSG, 18 µM of PLP, and temperature of 36°C, gave the maximum GABA to 314.56 mg/100 g. *Lb. plantarum* NDC75017 was a commercial starter culture that produced high GABA production in fermented yogurt. Recently, the adzuki bean milk inoculated with *Lb. rhamnosus* GG increased GABA from 0.05 mg/mL to 1.13 mg/mL after 36 h optimized by RSM, using 1.44% galactose, 2.27% MSG, and 0.20% PLP. The GABA production was increased higher than 22.4 times compared to unfermented adzuki bean milk (Song & Yu, 2018).

The secondary objective is improvement and development of the GABA production process concerning the cost-effectiveness and simple production of bio-products. it is important to select a proper process that allows of the highest yielding of the target product. For batch fermentation systems, the substrate should be put in the tank once only from the start fermentation. The limitation is that the higher initial concentration of fermentation substrate which inhibit the cell growth or residual material resources and the higher yield of product resulted in the product inhibition effect. Consequently, fed-batch fermentation system, one or more components are supplied to the fermenter while cells and products remain in the tank until the end of the operation. The optimum starting substrate concentrations, which solves the problem of substrate inhibition, and limiting fermentation can be added by feeding in the fermentation course. It may be an improvement to obtain a high yield and productivity (Krause et al., 2010). Fed-batch fermentations was used for GABA production by *Lb. brevis* NCL912 at pH, temperature, and initial glutamate concentration was fixed at 32°C, pH 5.0 and 400 mM, respectively and after 280.70 g (1.5 mol) and 224.56 g (1.2 mol) glutamate were supplemented into the bioreactor at 12 h and 24 h, respectively. Under the selected fermentation conditions, GABA was rapidly produced at 36 h other than in stationary phase. The GABA concentration was improved to 1005.81 mM, and the residual glucose and glutamate were 15.28 g/L and 134.45 mM at 48 h (Li et al., 2010). Furthermore, GABA concentration in fed-batch fermentation was improved from 660 mM to 1000 mM by *Lb. brevis* B3-18 in the batch fermentation (Kook & Cho, 2013). These experiments showed that a simple and effective fed-batch fermentation method was developed for GABA production by LAB.

APPLICATION AND BENEFITS OF GABA

The available amount of GABA normally has major inhibitory neurotransmitters in the sympathetic nervous system (Mody et al., 1994) and plays an important role in cardiovascular function (DeFeudis, 1983). Moreover, physiological action of GABA includes induction of hypotensive, diuretic effects, tranquilizer effects, prevention of diabetic conditions in humans, anterograde amnesia and retrograde amnesia and anti-convulsive effects (Wong, Bottiglieri & Snead, 2003). GABA provides beneficial effects for human health, treatments for sleeplessness, depression, autonomic disorders (Okada et al., 2000), inhibits development of cancer cells, helping the recovery from chronic alcohol-related symptoms, controlling stress and lipid levels in serum because it promotes fat loss by the stimulation of human growth hormone production, treatment of anxiety disorders, lowers blood pressure, boosts immune system (Oh & Oh, 2003), antihypertension, anticancer, anti-inflammatory and anti-hepatitis agents (Kang et al., 2011; Ali et al., 2013). In addition, GABA increases a strong secret of insulin from the pancreas and may prevent diabetic conditions (Adeghate & Ponery, 2002) and has preventive effects on Alzheimer's disease such as amnesia and dementia; also it improves memory and the learning abilities (Nakashima et al., 2009).

These multiple benefits of GABA on human health have led to the development of GABA-enriched fermented foods, a bioactive component, a primary element of food ingredients, and application in pharmaceuticals (Kim & Kim, 2012). For GABA in food enrichment, there is an idea of producing GABA-rich wheat germ food in normal daily consumption in a relatively large population that can be beneficial to those suffering pain from chronic hypertension. Several GABA-enriched functional foods are currently manufactured such as tea leaves treated anaerobically, rice germ soaked in water, red mold rice, tempeh-like fermented soybeans, and dairy products such as yogurt, fermented milk products and cheese (Thwe et al., 2011). Therefore, great efforts are focused on mass production of GABA and its utilization as a bioactive food ingredient in the modern food industry (Kim & Kim, 2012). GABA-enriched fermented food can reduce blood pressure after consuming such products (Inoue et al., 2003). It is worth noting that the human body can generally produce its own supply of GABA; however, GABA production is inhibited by an extra amount of salicylic acid or food additives or by the lack of vitamins, zinc or estrogen (Tenpaku, 1997). Therefore, GABA enriched food is required because the GABA content in typical human diets is relatively low.

GABA-enriched fermented foods with LAB included yogurt and cheese (Park & Oh, 2006; Nomura et al., 1998), sourdough (Diana et al., 2014) and kimchi (Seok et al., 2008). The highest content of GABA was found in Pecorino di Filiano (391 mg/kg) among 22 different varieties of Italian cheeses, in which the responsible microorganisms for the GABA production were *Lb. paracasei* PF6, PF8, and PF13, *Lb. plantarum* PF14, *Lactobacillus* sp. PF7 (Siragusa et al., 2007). In shochu distillery-lees, a Japanese distilled alcoholic beverage, most of the free glutamic acid (10.50 mM) was converted to GABA (10.18 mM) by *Lb. brevis* IFO-12005 (Yokoyama, Hiramatsu & Hayakawa, 2002). GABA-enriched grape-free grape preparation must beverage, which has a potential anti-hypertensive effect and dermatological protection was manufactured by fermentation using *Lb. plantarum* DSM19463 (Di Cagno et al., 2010). In addition, (Zareian et al., 2015) reported that biosynthesis activities for the production of glutamic acid and GABA in the fermentation of dosa (contains wheat and flour, a carbohydrate-rich food) by *Lb. plantarum* MNZ provided a GABA-enhanced functional food such as antihypertensive effects on spontaneously hypertensive.

However, the direct addition of chemical GABA to food is considered unnatural and unsafe. So it is necessary to find a natural method to produce an increase GABA in food with safely and effectively. Therefore, great efforts are focused on mass production of

GABA and its utilization as a bioactive food ingredient in the modern food industry (Kim & Kim, 2012).

CONCLUSION

Numerous functions of GABA are summarized to develop foods enrichment, pharmaceutical and to provide the natural products. Therefore, the production of GABA by LAB is interesting; however, many factors have been shown to affect microbial GABA production that is dependent on their biochemical properties and GAD. Also, the optimization of many factors including pH, temperature, medium component, medium concentration and cultivation time is required; thus the first step is to define the appropriate conditions to improve a high yield of GABA by microorganisms. Currently, the conventional method and statistical method are used for optimization condition. Furthermore, the second is the improvement and development of the GABA production process taking into account the cost-effective and simple production of bio-products. This type of fermentation including batch and fed-batch are compared. All these processes will lead to the higher expedient of microbial fermentation for universal applications of GABA.

REFERENCES

Adeghate E., Ponery A. S. (2002). GABA in the endocrine pancreas: cellular localization and function in normal and diabetic rats. *Tissue and Cell*, 34 (1): 1-6.

Ali N. M., Yusof H. M., Long K., Yeap S. K., Ho W. Y., Beh B. K., Koh S. P., Abdullah M. P., Alitheen N. B. (2013). Antioxidant and hepatoprotective effect of aqueous extract of germinated and fermented mung bean on ethanol-mediated liver damage. *Biomed Research International*, 2013: 693613.

Aoki H., Uda I., Tagami K., Furuya Y., Endo Y., Fujimoto K. (2003). The production of a new tempeh-like fermented soybean containing a high level of gamma-aminobutyric acid by anaerobic incubation with *Rhizopus. Bioscience Biotechnology and Biochemistry*, 67 (5): 1018-1023.

Aoshima H., Tenpaku Y. (1997). Modulation of GABA receptors expressed in Xenopus oocytes by 13-L-hydoxylinoleic acid and food additives. *Bioscience, Biotechnology and Biochemistry*, 61 (12): 2051-2057.

Barla F., Koyanagi T., Tokuda N., Matsui H., Katayama T., Kumagai H., Michihata T., Sasaki T., Tsuji A., Enomoto T. (2016). The gamma-aminobutyric acid-producing ability under low pH conditions of lactic acid bacteria isolated from traditional fermented foods of Ishikawa Prefecture, Japan, with a strong ability to produce ACE-inhibitory peptides. *Biotechnology Reports* (Amst), 10: 105-110.

Barrett E. (2014). This article corrects: gamma-Aminobutyric acid production by culturable bacteria from the human intestine. *Journal of Applied Microbiology*, 116 (5): 1384-1386.

Chung H. J., Jang S. H., Cho H. Y., Lim S. T. (2009). Effects of steeping and anaerobic treatment on GABA (gamma-aminobutyric acid) content in germinated waxy hull-less barley. *Lwt-Food Science and Technology*, 42 (10): 1712-1716.

De Biase D., Tramonti A., Bossa F., Visca P. (1999). The response to stationary-phase stress conditions in Escherichia coli: role and regulation of the glutamic acid decarboxylase system. *Molecular Microbiology*, 32 (6): 1198-1211.

DeFeudis F. V. (1983). gamma-Aminobutyric acid and cardiovascular function. *Experientia*, 39 (8): 845-849.

Dhakal R., Bajpai V. K., Baek K. H. (2012). Production of GABA (gamma-aminobutyric acid) by microorganisms: a review. *Brazilian Journal of Microbiology*, 43 (4): 1230-1241.

Di Cagno R., Mazzacane F., Rizzello C. G., De Angelis M., Giuliani G., Meloni M., De Servi B., Gobbetti M. (2010). Synthesis of gamma-aminobutyric acid (GABA) by *Lactobacillus plantarum* DSM 19463: functional grape must beverage and dermatological applications. *Appied Microbiology and Biotechnology*, 86 (2): 731-741.

Diana M., Tres A., Quilez J., Llombart M., Rafecas M. (2014). Spanish cheese screening and selection of lactic acid bacteria with high gamma-aminobutyric acid production. *Lwt-Food Science and Technology*, 56 (2): 351-355.

Divyashri G., Prapulla S. G. (2016). Production and characterization of fermented rice flour containing gamma-aminobutyric acid (GABA). *International Journal of Environmental and Agriculture Research*, 2 (10): 98-106.

Foster A. C., Kemp J. A. (2006). Glutamate- and GABA-based CNS therapeutics. *Current Opinion in Pharmacology*, 6 (1): 7-17.

Franciosi E., Carafa I., Nardin T., Schiavon S., Poznanski E., Cavazza A., Larcher R., Tuohy K. M. (2015). Biodiversity and gamma-aminobutyric acid production by lactic acid bacteria isolated from traditional alpine raw cow's milk cheeses. *BioMed Research International*, 2015, 625740: 1-11.

Gangaraju D., Murty V. R., Prapulla S. G. (2014). Probiotic-mediated biotransformation of monosodium glutamate to gamma-aminobutyric acid: differential production in complex and minimal media and kinetic modelling. *Annals of Microbiology*, 64 (1): 229-237.

Hayakawa K., Kimura M., Kasaha K., Matsumoto K., Sansawa H., Yamori Y. (2004). Effect of a gamma-aminobutyric acid-enriched dairy product on the blood pressure of spontaneously hypertensive and normotensive Wistar-Kyoto rats. *British Journal of Nutrition*, 92 (3): 4 11-417.

Hiraga K., Ueno Y., Sukontasing S., Tanasupawat S., Oda K. (2008). *Lactobacillus senmaizukei* sp. nov., isolated from Japanese pickle. *International Journal of Systematic and Evolutionary Microbiology*, 58 (7): 1625-1629.

Huang J., Mei L. H., Sheng Q., Yao S. J., Lin D. Q. (2007). Purification and characterization of glutamate decarboxylase of *Lactobacillus brevis* CGMCC 1306 isolated from fresh milk. *Chinese Journal of Chemical Engineering*, 15 (2): 157-161.

Inoue K., Shirai T., Ochiai H., Kasao M., Hayakawa K., Kimura M., Sansawa H. (2003). Blood-pressure-lowering effect of a novel fermented milk containing gamma-aminobutyric acid (GABA) in mild hypertensives. *European Journal of Clinical Nutrition*, 57 (3): 490-495.

Jannoey P., Niamsup H., Lumyong S., Suzuki T., Katayama T., Chairote G. (2010). Comparison of gamma-aminobutyric acid production in Thai rice grains. *World Journal of Microbiology and Biotechnology*, 26 (2): 257-263.

Kagan I. A., Coe B. L., Smith L. L., Huo C. J., Dougherty C. T., Strickland J. R. (2008). A validated method for gas chromatographic analysis of gamma-aminobutyric acid in tall fescue herbage. *Journal of Agricultural and Food Chemistry*, 56 (14): 5538-5543.

Kandler O. (1983). Carbohydrate metabolism in lactic acid bacteria. *Antonie Van Leeuwenhoek*, 49 (3): 209-224.

Kang Y. M., Qian Z. J., Lee B. J., Kim Y. M. (2011). Protective Effect of GABA-enriched Fermented Sea Tangle against Ethanol-induced Cytotoxicity in HepG2 Cells. *Biotechnology and Bioprocess Engineering*, 16 (5): 966-970.

Kim J. Y., Lee M. Y., Ji G. E., Lee Y. S., Hwang K. T. (2009). Production of gamma-aminobutyric acid in black raspberry juice during fermentation by *Lactobacillus brevis* GABA100. *International Journal of Food Microbiology*, 130 (1): 12-16.

Kim M. J., Kim K. S. (2012). Isolation and Identification of gamma-Aminobutyric acid (GABA)-producing Lactic Acid Bacteria from Kimchi. *Journal of the Korean Society for Applied Biological Chemistry*, 55 (6): 777-785.

Kinnersley A. M., Turano F. J. (2000). Gamma aminobutyric acid (GABA) and plant responses to stress. *Critical Reviews in Plant Sciences*, 19 (6): 479-509.

Komatsuzaki N., Nakamura T., Kimura T., Shima J. (2008). Characterization of glutamate decarboxylase from a high gamma-aminobutyric acid (GABA)-producer, *Lactobacillus paracasei*. *Bioscience, Biotechnology, and Biochemistry*, 72 (2): 278-285.

Komatsuzaki N., Shima J., Kawamoto S., Momose H., Kimura T. (2005). Production of gamma-aminobutyric acid (GABA) by *Lactobacillus paracasei* isolated from traditional fermented foods. *Food Microbiology*, 22 (6): 497-504.

Kook M. C., Cho S. C. (2013). Production of GABA (gamma amino butyric acid) by Lactic Acid Bacteria. *Korean Journal for Food Science of Animal Resources*, 33 (3): 377-389.

Krause M., Ukkonen K., Haataja T., Ruottinen M., Glumoff T., Neubauer A., Neubauer P., asala A. (2010). A novel fed-batch based cultivation method provides high cell-density and improves yield of soluble recombinant proteins in shaken cultures. *Microbial Cell Factories*, 9: 11.

Kumar P., Satyanarayana T. (2007). Optimization of culture variables for improving glucoamylase production by alginate-entrapped Thermomucor indicae-seudaticae using statistical methods. *Bioresource Technology*, 98 (6): 1252-1259.

Kwon S. Y., Garcia C. V., Song Y. C., Lee S. P. (2016). GABA-enriched water dropwort produced by co-fermentation with *Leuconostoc mesenteroides* SM and *Lactobacillus plantarum* K154. *Lwt-Food Science and Technology*, 73: 233-238.

Lahtinen T. H. E., Tuomi J. O., van Dijken S. (2011). Pattern Transfer and Electric-Field-Induced Magnetic Domain Formation in Multiferroic Heterostructures (vol 23, pg 3187, 2011). *Advanced Materials*, 23 (45): 5340-5340.

Li H., Cao Y. (2010). Lactic acid bacterial cell factories for gamma-aminobutyric acid. *Amino Acids*, 39 (5): 1107-1116.

Li H. X., Qiu T., Huang G. D., Cao Y. S. (2010). Production of gamma-aminobutyric acid by *Lactobacillus brevis* NCL912 using fed-batch fermentation. *Microbial Cell Factories*, 9.

Lim H. S., Cha I. T., Roh S. W., Shin H. H., Seo M. J. (2017). Enhanced Production of Gamma-Aminobutyric Acid by Optimizing Culture Conditions of *Lactobacillus brevis* HYE1 Isolated from Kimchi, a Korean Fermented Food. *Journal of Microbiology and Biotechnology*, 27 (3): 450-459.

Lin Q. (2013). Submerged fermentation of *Lactobacillus rhamnosus* YS9 for gamma-aminobutyric acid (GABA) production. *Brazilian Journal of Microbiology*, 44 (1): 183-187.

Lu X. X., Chen Z. G., Gu Z. X., Han Y. B. (2008). Isolation of gamma-aminobutyric acid-producing bacteria and optimization of fermentative medium. *Biochemical Engineering Journal*, 41 (1): 48-52.

Majumdar D, Guha S. (1988). Conformation, electrostatic potential and pharmacophoric pattern of GABA (γ-aminobutyric acid) and several GABA inhibitors. *Journal of Molecular Structure: Theochem*, 180: 125–140.

Mody I., De Koninck Y., Otis T. S., Soltesz I. (1994). Bridging the Cleft at Gaba Synapses in the Brain. *Trends in Neurosciences*, 17 (12): 517-525.

Nakashima Y., Ohsawa I., Konishi F., Hasegawa T., Kumamoto S., Suzuki Y., Ohta S. (2009). Preventive effects of Chlorella on cognitive decline in age-dependent dementia model mice. *Neuroscience Letters*, 464 (3): 193-198.

Nejati F., Rizzello C. G., Di Cagno R., Sheikh-Zeinoddin M., Diviccaro A., Minervini F., Gobbett M. (2013). Manufacture of a functional fermented milk enriched of Angiotensin-I Converting Enzyme (ACE)-inhibitory peptides and gamma-amino butyric acid (GABA). *Lwt-Food Science and Technology*, 51 (1): 183-189.

Nomura M., Kimoto H., Someya Y., Furukawa S., Suzuki I. (1998). Production of gamma-aminobutyric acid by cheese starters during cheese ripening. *Journal of Dairy Science*, 81 (6): 1486-1491.

Oh C. H., Oh S. H. (2003). Brown rice extracts with enhanced levels of gamma-aminobutyric acid inhibit cancer cell proliferation. *Faseb Journal*, 17 (5): A1157-A1157.

Okada T., Sugishita T., Murakami T., Murai H., Saikusa T., Horino T., Onoda A., Kajimoto O., Takahashi R., Takahashi T. (2000). Effect of the defatted rice germ enriched with GABA for sleeplessness, depression, autonomic disorder by oral administration. *Nippon Shokuhin Kagaku Kogaku Kaishi*, 47 (8): 596-603.

Olsen Richard W, DeLorey Timothy M. (1999). GABA receptors have been identified electrophysiologically and pharmacologically in all regions of the brain. In *Basic Neurochemistry: Molecular, Cellular and Medical Aspects*. G. Siegel; B. W, Agranoff, W. Albers, S. K Fisher and M. D. Uhler (Eds), Lippincott Williams & Wilkins, Philadelphia.

Park K. B., Oh S. H. (2006). Isolation and characterization of *Lactobacillus buchneri* strains with high gamma-aminobutyric acid producing capacity from naturally aged cheese. *Food Science and Biotechnology*, 15 (1): 86-90.

Ratanaburee A., Kantachote D., Charernjiratrakul W., Sukhoom A. (2013). Enhancement of gamma-aminobutyric acid (GABA) in Nham (Thai fermented pork sausage) using starter cultures of *Lactobacillus namurensis* NH2 and *Pediococcus pentosaceus* HN8. *International Journal of Food Microbiology*, 167 (2): 170-176.

Rizzello C. G., Cassone A., Di Cagno R., Gobbetti M. (2008). Synthesis of angiotensin I-converting enzyme (ACE)-inhibitory peptides and gamma-aminobutyric acid (GABA) during sourdough fermentation by selected lactic acid bacteria. *Journal of Agricultural and Food Chemistry*, 56 (16): 6936-6943.

Roth R. J., Cooper J. R., Bloom F. E. 2003. *The Biochemical Basis of Neuropharmacology*. Oxford: Oxford University Press.

Sapse A. M., (2000). *Molecular Orbital Calculations for Amino Acids and Peptides*. Boston: Birkhäuser.

Seok J. H., Park K. B., Kim Y. H., Bae M. O., Lee M. K., Oh S. H. (2008). Production and Characterization of Kimchi with Enhanced Levels of gamma-Aminobutyric Acid. *Food Science and Biotechnology*, 17 (5): 940-946.

Shan Y., Man C. X., Han X., Li L.,Guo Y., Deng Y., Li T., Zhang L. W., Jiang Y. J. (2015). Evaluation of improved gamma-aminobutyric acid production in yogurt using *Lactobacillus plantarum* NDC75017. *Journal of Dairy Science*, 98 (4): 2138-2149.

Siragusa S., De Angelis M., Di Cagno R., Rizzello C. G., Coda R., Gobbetti M. (2007). Synthesis of gamma-aminobutyric acid by lactic acid bacteria isolated from a variety of Italian cheeses. *Applied and Environmental Microbiology*, 73 (22): 7283-7290.

Song H. Y., Yu R. C. (2018). Optimization of culture conditions for gamma-aminobutyric acid production in fermented adzuki bean milk. *Journal of Food and Drug Analysis*, 26 (1): 74-81.

Su Y. C., Wang J. J., Lin T. T., Pan T. M. (2003). Production of the secondary metabolites gamma-aminobutyric acid and monacolin K by *Monascus*. *Journal of Indian Microbiology and Biotechnology*, 30 (1): 41-46.

Tajabadi N., Ebrahimpour A., Baradaran A., Rahim R. A., Mahyudin N. A., Manap M. Y. A., Abu Bakar F., Saari N. (2015). Optimization of gamma-aminobutyric acid production by *Lactobacillus plantarum* Taj-Apis362 from honeybees. *Molecules*, 20 (4): 6654-6669.

Thwe S. M., Kobayashi T., Luan T., Shirai T., Onodera M., Hamada-Sato N.,I mada C. (2011). Isolation, characterization, and utilization of gamma-aminobutyric acid (GABA)-producing lactic acid bacteria from Myanmar fishery products fermented with boiled rice. *Fisheries Science*, 77 (2): 279-288.

Tong J. C., Mackay I. R., Chin J., Law R. H., Fayad K., Rowley M. J. (2002). Enzymatic characterization of a recombinant isoform hybrid of glutamic acid decarboxylase (rGAD67/65) expressed in yeast. *Journal of Biotechnology*, 97 (2): 183-190.

Tsai J. S., Lin Y. S., Pan B. S., Chen T. J. (2006). Antihypertensive peptides and gamma-aminobutyric acid from prozyme 6 facilitated lactic acid bacteria fermentation of soymilk. *Process Biochemistry*, 41 (6): 1282-1288.

Tsukatani T., Higuchi T., Matsumoto K. (2005). Enzyme-based microtiter plate assay for gamma-aminobutyric acid: Application to the screening of gamma-aminobutyric acid-producing lactic acid bacteria. *Analytica Chimica Acta*, 540 (2): 293-297.

Ueno H. (2000). Enzymatic and structural aspects on glutamate decarboxylase. *Journal of Molecular Catalysis B-Enzymatic*, 10 (1-3): 67-79.

Ueno Y., Hayakawa K., Takahashi S., Oda K. (1997). Purification and characterization of glutamate decarboxylase from *Lactobacillus brevis* IFO 12005. *Bioscience Biotechnology and Biochemistry*, 61 (7): 1168-1171.

Villegas J. M., Brown L., de Giori G. S., Hebert E. M. (2016). Optimization of batch culture conditions for GABA production by *Lactobacillus brevis* CRL 1942, isolated from quinoa sourdough. *Lwt-Food Science and Technology*, 67: 22-26.

Wong C. G., Bottiglieri T., Snead O. C., 3rd. (2003). GABA, gamma-hydroxybutyric acid, and neurological disease. *Ann Neurol*, 54 Suppl 6: S3-12.

Wu Q. L., Shah N. P. (2015). Gas release-based prescreening combined with reversed-phase HPLC quantitation for efficient selection of high-gamma-aminobutyric acid (GABA)-producing lactic acid bacteria. *Journal of Dairy Science*, 98 (2): 790-797.

Wu Q., Shah N. P. (2017). High gamma-aminobutyric acid production from lactic acid
    bacteria: Emphasis on *Lactobacillus brevis* as a functional dairy starter. *Critical
    Reviews in Food Science and Nutrition*, 57 (17): 3661-3672.

Yang S. Y., Lu F. X., Lu Z. X., Bie X. M., Jiao Y., Sun L. J., Yu B. (2008). Production of
    gamma-aminobutyric acid by *Streptococcus salivarius* sub sp *thermophilus* Y2 under
    submerged fermentation. *Amino Acids*, 34 (3): 473-478.

Yang S. Y., Lu Z. X., Lu F. X., Bie X. M., Sun L. J., Zeng X. X. (2006). A simple
    method for rapid screening of bacteria with glutamate decarboxylase activities.
    Journal of Rapid Methods and Automation in Microbiology, 14 (3): 291-298.

Yingguo Lü, Hui Zhang, Xiangyong Meng, Li Wang, Guo Xiaona. (2010). A validated
    HPLC method for the determination of GABA by pre-column derivatization with
    2,4dinitrofluorodinitrobenzene and its application to plant gad activity study.
    Analytical Letters, 43: 2663-2671.

Yokoyama S., Hiramatsu J., Hayakawa K. (2002). Production of gamma-aminobutyric
    acid from alcohol distillery lees by *Lactobacillus brevis* IFO-12005. *Journal of
    Biosciences and Bioengineering*, 93 (1): 95-97.

Zareian M., Oskoueian E., Forghani B., Ebrahimi M. (2015). Production of a wheat-
    based fermented rice enriched with gamma-amino butyric acid using *Lactobacillus
    plantarum* MNZ and its antihypertensive effects in spontaneously hypertensive rats.
    Journal of Functional Foods, 16: 194-203.

Zhang Y., Song L., Gao Q., Yu S. M., Li L., Gao N. F. (2012). The two-step
    biotransformation of monosodium glutamate to GABA by *Lactobacillus brevis*
    growing and resting cells. *Applied Microbiology and Biotechnology*, 94 (6): 1619-
    1627.

Zhao A. Q., Hu X. Q., Pan L., Wang X. Y. (2015). Isolation and characterization of a
    gamma-aminobutyric acid producing strain *Lactobacillus buchneri* WPZ001 that
    could efficiently utilize xylose and corncob hydrolysate. *Applied Microbiology and
    Biotechnology*, 99 (7): 3191-3200.

Effect of Lactic Acid Bacteria on Health and Other Applications

In: The Many Benefits of Lactic Acid Bacteria
Editors: J. G. LeBlanc and A. de Moreno

ISBN: 978-1-53615-388-0
© 2019 Nova Science Publishers, Inc.

Chapter 10

LACTOCOCCUS LACTIS AS A DNA VACCINE DELIVERY SYSTEM

Vanessa Bastos Pereira*, Tatiane Melo Preisser*,
Camila Prósperi De Castro, Bianca Mendes Souza,
Meritxell Zurita Turk, Vanessa Pecini da Cunha,
and Anderson Miyoshi†

Laboratório de Tecnologia Genética, Instituto de Ciências Biológicas,
Universidade Federal de Minas Gerais (UFMG), Belo Horizonte-MG, Brazil

ABSTRACT

DNA vaccines, which consist of plasmids encoding antigens of interest, emerge as novel options to protect hosts against new emerging infectious diseases. As most pathogens affect or initiate their infection at mucosal surfaces, this is a strategic route to induce specific immune responses. However, to make it possible, many environmental barriers must be surpassed. In this context, the use of bacteria, like Lactococcus lactis, as vehicles to deliver vaccine plasmids by the mucosal route is a promising strategy. Native Lc. lactis was used for this purpose and, aiming to make the plasmid delivery more efficient, invasive recombinant Lc. lactis strains were constructed and tested, thus showing to be able to deliver DNA vaccines more efficiently than the wild type strain. Therefore, in summary, this chapter presents a general view of DNA vaccines and the use of bacteria, especially Lc. lactis, to deliver them to intestinal eukaryotic cells, representing a new strategy for the tuberculosis control.

---

* These authors contributed equally to this work.
† Corresponding Author's E-mail: miyoshi@icb.ufmg.br

INTRODUCTION

Vaccination stimulates the development of immune responses to avoid infection and protect the individual from establishing a disease. Many diseases like tetanus, diphtheria, polio and hepatitis A and B have decreased significantly with the advent of vaccines (Roush et al., 2007). However, emerging infectious diseases have become challenges to be faced by the vaccinology field and new strategies must be developed to produce more efficient vaccines to protect the population worldwide. Within this scenario, DNA vaccines emerge as good alternatives to invest time and efforts, as they may be a way to overcome these new obstacles. DNA vaccines consist of eukaryotic expression plasmids that have the coding sequence of an antigen of interest inserted into their multiple cloning sites. Although different administration routes exist, the goal of these vaccines is that the plasmid DNA reaches the host cell nucleus so that it can be transcribed, translated and expressed by the eukaryotic cell and, in this way, elicit the expected immune responses (Pereira et al., 2014).

The first findings that reported about the potential of vaccines occurred in 1990, when Wolf and colleagues injected a DNA plasmid containing the coding sequence of β-galactosidase and verified the expression of this protein in quadriceps muscle sections of mice (Wolf et al., 1990). In 1992, after injecting a plasmid encoding the human growth hormone (hGH) into mice´s skin, Tang and colleagues observed the production of specific antibodies against the protein, indicating that the injected DNA was expressed and elicited a specific humoral immune response (Tang et al., 1992). Subsequently, in 1993, two independent research groups vaccinated mice with DNA plasmids encoding different viral proteins and showed success when challenging against lethal influenza (Fynan et al., 1993; Ulmer et al., 1993). These studies were fundamental to understand how DNA could be used for vaccination and thus, protect against a pathogen or an antigen. To date, the scientific community has been striving for the use of DNA vaccines to become a reality around the world.

Since the discovery that DNA vaccines could be valuable tools to generate immune responses, four DNA vaccines have been approved for veterinarian use (Kutzler & Weiner, 2008). In 2005, two prophylactic vaccines against infection with the West Nile Virus in horses and the Hematopoietic Necrosis Virus in salmon were licensed in The United States and Canada, respectively. Following, in 2008, a therapeutic vaccine against fetal loss, which expressed the growth hormone releasing hormone (GHRH), was licensed for use in swine in Australia and in 2010 a therapeutic vaccine to treat oral melanoma in dogs was licensed in the United States. The success of these approved DNA vaccines showed perspectives of an innovative platform to prevent and treat emerging diseases.

STRUCTURE OF DNA VACCINES

DNA vaccines are bacterial-derived plasmids essentially composed of two distinguishable backbones: one for prokaryotic propagation and the other for eukaryotic expression (Liu, 2003). The prokaryotic propagation backbone consists of an origin of replication, which allows the plasmid to multiply during cell division and thus, to be passed on to daughter cells. Additionally, to ensure that all bacterial cells are carrying the plasmid of interest, DNA vaccines have a prokaryotic selectable marker generally consisting of a resistance gene to a specific antibiotic. The eukaryotic expression backbone comprises a transcriptional unit that allows for antigen expression in eukaryotic cells. A constitutive or inducible promoter has to be carefully chosen according to the desired expression, as, for example, to enhance protein expression and immunogenicity, strong viral promoters, such as the cytomegalovirus (CMV) and Simian Virus 40 (SV40) promoters, are good options. For correct recognition of messenger RNA (mRNA) and initiation of the translation process by the eukaryotic machinery, a consensus sequence, called Kozak sequence, is present after the promoter and includes the start codon (ACC<u>AT</u>GG). Following is the coding sequence of the antigen of interest, which is transcribed, translated and, if necessary, post-translationally modified by the eukaryotic host cell. At the end of the gene sequence, there is the stop codon, which follows the correct reading frame to accurately end translation. Finally, the backbone presents the polyadenylation sequence (poly A), which stabilizes the mRNA and ensures that it is exported from the nucleus to the cytoplasm of eukaryotic cells without degradation.

Over the years, strategies aiming at increasing the effectiveness of DNA vaccines have been developed. Because prokaryotes prefer codons that differ from those more frequently used in mammalian cells, in order to achieve optimal expression by the host cells, the coding sequence of the antigen of interest can be adapted for codons with more available corresponding transfer RNA (tRNA) (Wang et al., 2006). Strategies to enhance non-specific and antigen-specific responses and stimulatory cytokines are also encouraged, and include the use of immunostimulatory sequences, co-injection of immunomodulatory plasmids or use of adjuvants (Bode et al., 2011; Marc et al., 2015). Moreover, to increase the number of plasmids that reach the host cell nucleus and, consequently, are expressed, DNA nuclear targeting sequences (DTS) can be added to the eukaryotic backbone of DNA vaccines (Miller & Dean, 2009).

CHARACTERISTICS OF DNA VACCINES

DNA vaccination is a strategy designed to comply with efficient and effective immunization. Clinical trials have shown bottlenecks that need to be improved as well as the advantages of DNA vaccines, highlighting the potential of this innovative vaccination platform (Marc et al., 2015). At first, a recurrent concern was whether plasmids could integrate into the mammalian genome and its consequent unknown implications. However, the probability of plasmid integration into the host genome could be considered negligible, making vaccination with DNA safe (Glenting & Wessels 2005). The time spent to develop and produce DNA vaccines is short, compared to other platforms such as egg-based protein vaccines, and can rapidly be manufactured in large-scale to meet global demands (Moss 2009). In addition, the production of DNA vaccines is inexpensive compared to conventional vaccines, making this vaccination strategy even more attractive. Due to their composition, DNA vaccines are thermally stable, allowing for easy storage and shipping, making it possible to reach locations that do not have the appropriate refrigerated structure storage that some vaccines require (Ghanem et al., 2013).

The ability to elicit humoral and cell-mediated immune responses is one of the main advantages of DNA vaccines. While conventional protein-based vaccines commonly only induce antibody production, this new vaccination strategy can stimulate B and T cells, inducing antibody production and cellular responses, supporting their potential to confer protection against different pathogens (Li et al., 2012). The use of genetic material avoids manipulation of deadly infectious microorganisms and results in more safety during the production of DNA vaccines. Additionally, the formulation of DNA vaccines allows for the insertion of distinct antigen-coding sequences into the same plasmid, resulting in a broader immune response. Finally, DNA vaccines can be administered through different routes, increasing the range of strategies to reach the desired immunization process (Reyes-Sandoval & Ertl, 2001; Laddy & Weiner, 2006).

MUCOSAL DNA VACCINES

The respiratory, gastrointestinal and urogenital tracts are covered by mucosal surfaces, which are large tissues protected by a highly specialized innate and adaptive immune system. As the majority of infections affect or start at mucosal surfaces, the mucosal immune system can prevent infectious agents from attaching, colonizing, penetrating and replicating, blocking microbial toxins in the mucosal epithelium (Holmgren & Czerkinsky, 2005).

The preferred mucosal routes for vaccination are the vaginal (Kanazawa et al., 2008), intranasal (Brave et al., 2008) and oral mucosas (Chatel et al., 2008). However, the oral route is the most commonly used because the gastrointestinal tract has the largest surface area in the body, with more than $200m^2$ of mucous membrane (Andrew et al., 2011). Thus, the gut-associated lymphoid tissue (GALT) is the largest immune compartment in the body, containing the highest number of immune cells (Schroeter & Klaenhammer, 2009; Mowat & Agace, 2014).

The GALT has two prominent non-encapsulated structures, which are the induction sites of immune responses, known as Peyer's patches (PPs), and isolated lymphoid follicles. Additionally, the effector sites include lymphocytes, dendritic cells (DCs) and macrophages that are also dispersed throughout the epithelial layer and lamina propria (Brandtzaeg et al., 2008; Wells, 2011; Mowat & Agace, 2014).

The GALT is characterized by an overlying follicle-associated epithelium (FAE), with low levels of digestive enzymes and less pronounced villi, and is infiltrated by a large number of immune cells. The most notable feature of the FAE is the presence of Microfold cells (M cells), which are specialized for the uptake and transport of particulate antigens from the lumen into an underlying DC-rich subepithelial dome (SED) region, where they can be presented to the adaptive immune cells. PPs, on the other hand, are macroscopic lymphoid follicles located on the antimesenteric side of the small intestine. They are composed of numerous B cell lymphoid follicles, which are flanked by smaller T cell areas. The mesenteric lymph nodes (LM) drain all these sites and connect the mucosal and the systemic immune systems (Mowat, 2003; Wells, 2011; Mowat & Agace, 2014).

The efficacy of oral DNA vaccines depends therefore on whether the administered agents can survive the gastric and intestinal environments, which include pH-induced degradation, enzymes and diffusion across the mucus layers, and whether their residence time in the intestine is sufficient for interaction with and endocytosis by target cells. Therefore, oral administration of vaccines often requires delivery systems that protect against enzymatic degradation and elimination in the gastrointestinal tract, maintaining their high bioavailability (Chadwick et al., 2009). However, results of experimental approaches have proven difficulty to induce strong intestinal immune responses with a single oral administration of naked DNA vaccines and development of effective mucosal delivery systems remains therefore an active area of research.

BACTERIA AS A VEHICLE TOR MUCOSAL DELIVERY OF DNA VACCINES

Initially, DNA vaccines consisted in direct immunization with naked plasmid in tissues, which internalized and expressed the immunogenic antigen (Liu et al., 2011). However, to reach significant levels of cellular transfection, very high doses of plasmids and multiple boosts were required (Van Drunen Littel-van den Hurk et al., 2000; Le et al., 2000; Babiuk et al., 2003), mainly due to high DNA degradation and low efficacy to overcome membrane permeability and cellular uptake, resulting in poor gene expression and low immunity

(Miller & Dean, 2009). This showed that potent transfection methods had to be developed for the success of DNA vaccines.

In this context, the use of bacteria to transfer vaccine plasmids to mammalian host cells showed to be a promising strategy. In 1980, Schaffner was the first to observe the transfer of DNA from bacterial to mammalian cells, when tandem copies of the SV40 genome, carried by Eacherichia (*E.*) coli strains, were transferred to co-cultured mammalian cells (Schaffner, 1980). Since then, many bacterial transfer systems have been developed and improved to transfer plasmid DNA to eukaryotic host cells.

This approach has several advantages, such as (i) protection of the DNA vaccine from physical elimination and enzymatic digestion, (ii) targeting of the inductive sites of the body, allowing stimulation of the immune system and (iii) appropriate stimulation of the innate immune system to generate effective adaptive immunity (Holmgren & Czerkinsky, 2005).

Moreover, bacterial carriers are easy to manufacture and able to maintain the plasmids with high copy numbers. Circumventing the need to co-administrate plasmid selection markers, required for vaccination with naked DNA, stable replication of plasmids can be further improved by the introduction of bacterial genes essential for survival within their host (Pilgrim et al., 2003). Additionally, in contrast to immunization with naked plasmid DNA, no further plasmid purification steps are needed, reducing costs and labor (Schoen et al., 2004; Becker et al., 2008). Furthermore, the World Health Organization (WHO) recommends the use of mucosal vaccines because of economic, logistic and security reasons (Holmgren & Czerkinsky, 2005; Wells, 2011).

IMMUNOLOGICAL ASPECTS OF BACTERIAL DNA DELIVERY TO THE ORAL MUCOSA

After oral administration, bacteria carrying plasmids are recognized by immune cells and intestinal epithelial cells (IECs) coating mucosal surfaces. Some attenuated pathogenic bacteria can invade IECs by the expression of virulence factors (invasins) to deliver DNA vaccines while other bacteria can be sampled by intestinal lamina propria DCs or by specialized epithelial M cells overlying PPs (Weiss & Chakraborty, 2001).

After bacterial delivery, innate immune receptors expressed by IECs and DCs, such as pattern recognition receptors likeToll-like and Nod-like receptors, can respond to some bacterial components, which serve as natural antigens after host cell invasion (Barbosa & Rescigno, 2010). Bacterial recognition by the immune system modulates the innate immune response thus, promoting a robust and lasting adaptive response (Hoebe et al., 2004).

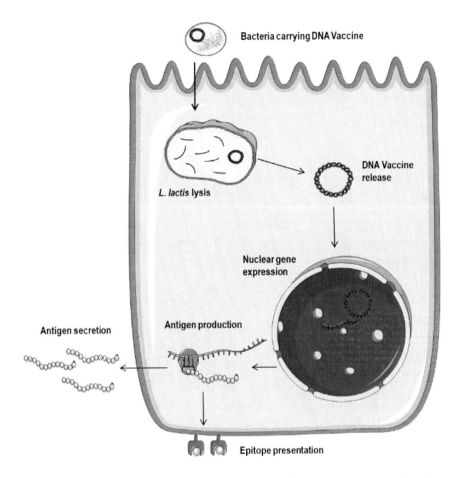

Figure 1. Representation of bacteria-mediated DNA vaccine delivery into intestinal cells. Bacteria enter the host cells and are lysed. The DNA plasmids are released in the cytoplasm and transferred into the nucleus, where expression of the open reading frame (ORF) of interest occurs. After protein synthesis by the host cell machinery, the antigen is processed for immune presentation or for secretion to the extracellular medium, eliciting specific prophylactic immune responses.

Once inside mammalian cells a primary vesicle, the phagosome, envolves the bacteria. This vesicle fuses to a lysosomal compartment where bacterial lysis occurs releasing the DNA plasmids that then reach the host cytosol. The DNA vaccines then migrate to the cell nucleus, where the gene of interest is transcribed for subsequent cytoplasmic protein synthesis by the host cellular machinery (Grillot-Courvalin et al., 1999; Schoen et al., 2004). Exogenous antigens are presented on the surface of epithelial cells or secreted to the extracellular medium, where they can be transported by antigen presenting cells (APCs), especially DCs present in lymphoid follicles, to the nearest mesenteric lymph nodes and be presented to naïve T cells, resulting in the induction of specific cellular and humoral immune responses (Figure 1).

The induction of mucosal immunity, in which antigen-stimulated T cells migrate to the systemic blood and then head to other mucosal surfaces, occurs thus, demonstrating the importance of mucosal immunization to protect against pathogens that infect through this route (Brandtzaeg et al., 2008; Mowat & Agace, 2014). Furthermore, plasma cells primed in the mucosa are also attracted to these sites resulting in secretory IgA (sIgA) in the mucosal tissues around the body (Barbosa & Rescigno, 2010). Consequently, effector T lymphocytes and sIgA-producing plasma cells are able to protect the mucosal surfaces. In addition, effector memory cells are also induced, accumulating preferentially in non-lymphoid tissues, especially in the intestinal mucosa (Sallusto et al., 1999).

LACTOCOCCUS LACTIS AS DNA DELIVERY SYSTEM

The non-pathogenic and food-grade bacteria *Lactococcus* (*Lc.*) *lactis* is an attractive alternative for mucosal delivery of DNA vaccines. *Lc. lactis* is the lactic acid bacteria (LAB) model, is very easy to manipulate and has a large number of genetic tools developed (De Vos & Simons, 1994; Bolotin et al., 2001). Additionally, as these bacteria have been used for food fermentation and preservation for centuries, the Food and Drug Administration (FDA) considers *Lc. lactis* a generally regarded as safe (GRAS) microorganism for human consumption (Van de Guchte et al., 2001).

Therefore, the use of *Lc. lactis* for mucosal delivery of DNA vaccines can be justified by the fact that this bacteria (i) does not contain endotoxins, such as lipopolysaccharides (LPS), on their cell wall, eliminating the risk of anaphylaxis and can be therefore delivered safely at high doses in immunization programs, (ii) can be easily engineered for DNA delivery (Mercenier et al., 2000; Miyoshi et al., 2004; Wells & Mercenier, 2008; Brondyk, 2009), and (iii) its administration induces high production of antigen-specific sIgA (Neutra & Kozlowski, 2006).

USE OF WILD TYPE *LACTOCOCCUS LACTIS* AS A VEHICLE FOR DNA DELIVERY

The use of *Lc. lactis* as a potential DNA vaccine delivery system has been successfully show (Pontes et al., 2011; Chatel et al., 2008). Firstly, in 2006, Guimarães and colleagues used a wild type (wt) *Lc. lactis* strain harboring a DNA plasmid coding for bovine ß-lactoglobulin (BLG) to deliver DNA into mammalian cells. Production and secretion of BLG were observed in the intestinal epithelial cell line Caco-2 after its co-culture with the recombinant *Lc. lactis* strain, demonstrating that this bacterium could efficiently deliver a fully functional plasmid into epithelial cells *in vitro* (Guimarães et al., 2006). Then, Chatel and colleagues (2008) confirmed the ability of wt *Lc. lactis* to transfer DNA to mice IECs *in vivo*. This strain demonstrated its ability to translocate the intestinal membrane, as BLG cDNA was detected in the enterocytes of mice immunized with *Lc. lactis* carrying the same eukaryotic expression plasmid (Chatel et al., 2008).

Although these studies successfully demonstrated that *Lc. lactis* can be used as a DNA delivery system, different strategies have been developed to increase the ratio of gene transfer and therefore boost the immune response. One of these strategies includes chemical treatments to weaken bacterial cell walls: glycine-treated *Lc. lactis* NZ3900 showed a higher transfer frequency of a eukaryotic expression plasmid encoding for the red fluorescent protein (RFP) to Caco-2 cells (Tao et al., 2011).

Another approach was a new plasmid (pExu) constructed for DNA delivery using *Lc. lactis*, with a theta origin of replication and the pCMV expression cassette. An *in vitro* assay showed that 15.8% of CHO cells were able to express the *enhanced green fluorescent protein* (eGFP) protein after pExu:egfp transfection. Additionaly, *Lc. lactis* MG1363 (pExu:egfp) were administered by gavage to Balb/C mice and the eGFP protein was detected in mice enterocytes (Mancha-Agresti et al., 2016). Also using wt *Lc. lactis* for DNA delivery, a different research group showed that this strain efficiently delivered the pPERDBY reporter plasmid encoding the eGFP to Caco-2 cells *in vitro* (Yagnik et al., 2016).

Furthermore, as described in the following section, a very interesting approach based on the use of recombinant invasive *Lc. lactis* strains have already been successfully demonstrated both *in vitro* and *in vivo*.

RECOMBINANT INVASIVE *LACTOCOCCUS LACTIS* AS DNA DELIVERY VEHICLES

Pathogenic bacteria have refined strategies to overcome host defenses and to interact with the host's immune system. To explore this concept and to increase the ability of lactococci to deliver DNA vaccines, some *Lc. lactis* strains expressing invasins have been constructed and shown to improve DNA delivery (Guimarães et al., 2005; Sleator & Hill, 2006). One such strategy engineered *Lc. lactis* to express InlA from *Listeria monocytogenes*. InlA is a cell wall-anchored protein that binds to the extracellular receptor E-cadherin (Gaillard et al., 1991; Mengaud et al., 1996). The recombinant lactococci (*Lc. lactis* InlA+), efficiently displaying the cell wall-anchored form of InlA, showed invasion rates in Caco-2 cells that were approximately 100-fold higher than those of wt lactococci. *In vivo* the invasive InlA+ strain also proved its capacity to enter epithelial Caco-2 cells with subsequent GFP expression after oral administration to guinea pigs and delivery of a functional eukaryotic plasmid (Guimarães et al., 2005).

To optimize the use of *Lc. lactis* as a DNA delivery system, a new plasmid (pValac - Vaccination using lactic acid bacteria) of 3,742 bp was constructed. The pValac plasmid consists of the cytomegalovirus promoter (pCMV) and polyadenylation sequences from the bovine Growth Hormone (BGH), which allows for the transcription of the ORF of interest with a PoliA tail that is essential to stabilize the mRNA transcript in eukaryotic cells. The pValac also contains origins of replication for its propagation in both *E. coli* (OriC) and *Lc. lactis* (OriA), and a chloramphenicol resistance gene for selection of strains harboring the specific plasmid. To evaluate its functionality, the gfp ORF was cloned into the pValac (pValac:gfp) and fluorescence was observed after PK15 cells transfection. Moreover, the *Lc. lactis* InlA+ strain invaded and delivered the pValac:gfp to Caco-2 cells (Guimarães et al., 2009).

However, the use of the *Lc. lactis* InlA+ strain presented a major bottleneck: InlA cannot bind to murine E-cadherin receptors, preventing *in vivo* experiments in these animals and limiting its study only in guinea pigs, which may be difficult to handle and/or expensive (Wollert et al., 2007). In this context and in order to increase the delivery of DNA to mammalian cells (Innocentin et al., 2009), a new recombinant *Lc. lactis* strain, expressing Fibronectin Biding Protein A (FnBPA) of *Staphylococcus aureus*, and that could be used in mice was constructed (Que et al., 2001). FnBPA is a virulence factor, which mediates bacterial adhesion to the host tissue and its entry into host cells (Sinha et al., 1999). FnBPA showed increased interaction and entrance of L. lactis in IECs and thus levels of DNA vaccine delivery. *Lc. lactis* FnBPA+ and *Lc. lactis* InlA+ presented comparable *in vitro* internalization rates by Caco-2 cells. After co-incubated of *Lc. lactis* FnBPA+ with Caco-2 cells, the pValac:gfp plasmid was more efficiently delivered than by non-invasive strains (Innocentin et al., 2009). Additionally, this strain was successfully used to deliver DNA vaccines encoding the BLG allergen and GFP *in vivo*, showing to be more effective than the wt strain (Pontes et al., 2012).

Another interesting work was performed using the mutated form of Internalin A (mInlA), which can bind to murine E-cadherin. mInlA was successfully expressed at the surface of *Lc. lactis* and showed higher invasion rates into Caco-2 cells and consequently higher delivery levels of BLG-coding plasmid when compared with the wt strain. Furthermore, *in vivo* studies demonstrated that *Lc. lactis* mInlA+ tended to increase the number of mice producing BLG (De Azevedo et al., 2012).

All these engineered strains to be used as DNA vaccine delivery systems represent promising tools for mucosal immunization and the strategy to use *Lc. lactis* to deliver prophylactics plasmids has proved to be relevant and possible. Our research group is currently working on the validation of this strategy in the development of DNA vaccines against tuberculosis, a re-emerging infectious disease that deserves great attention in current researches. This approach could therefore allow the use of *Lc. lactis* as efficient DNA vaccine delivery vehicles in a near future.

LACTOCOCCUS LACTIS AS A DNA VACCINE DELIVERY SYSTEM AGAINST TUBERCULOSIS DISEASE

Tuberculosis (TB), which is caused by *Mycobacterium tuberculosis*, is an infectious disease that affects one third of the global population in its latent form. This, in turn, makes TB one of the main social, economic and public health problems worldwide (Lugo & Bewley, 2008).

The only available vaccine for clinical use, the BCG (Bacillus Calmette-Guérin), has a variable efficacy in adolescents and adults, requiring the development of more effective and economically viable vaccines.

In this context, *Lc. lactis* appears as an attractive alternative for a TB DNA vaccine delivery system, because it elicits an immune response as efficient as the other mucosal bacterial DNA delivery systems (Table 1). Moreover, *Lc. lactis* does not present risk in future clinical use when compared to attenuated pathogenic bacteria.

Table 1. Bacterial mucosal DNA vaccines studies against TB

DNA Vaccine	Administration strategy	Immune Response									Protection
		IFN-γ	TNF-α	IL-17	IL-2	IL-4	IL-10	IgG1	Ig2a	sIgA	
Lc. lactis DNA delivery system											
pValac:Ag85A (Mancha-Agresti 2014)	Intranasal *Lc. lactis* FnBPA+	Yes	Yes	No	No	No	Yes	No	No	Yes	-
pValac:ESAT-6 (Pereira et al., 2015)	Intra-gastric *Lc. lactis* FnBPA+	Yes	No	No	-	No	No	No	No	Yes	-
pValac:ESAT6/Ag85A (Prósperi 2016)	Intra-gastric *Lc. lactis* FnBPA+	Yes	Yes	Yes	-	No	No	No	No	Yes	-
BCG + pValac:ESAT-6 (Pereira et al., 2017)	BCG + Intra-gastric *Lc. lactis* FnBPA+	Yes	Yes	Yes	-	No	No	No	No	No	-
Other bacterial DNA delivery system											
pCMVβ::Ag85A (Parida et al., 2005)	Intranasal *Salmonella typhimurium* ΔaroA SL7207	Yes	-	-	Yes	-	-	-	-	-	Yes
pcDNA_htpX or pcDNA_fbpA (Brun et al., 2008)	Intranasal *E. coli* BM2710/pGB2_inv-hly	Yes	-	-	Yes	-	-	-	-	-	Yes
pVax-ESAT6-Ag85B (Wang et al., 2009)	BCG + Intra-gastric *S. typhimurium* ΔaroA SL7207	Yes	-	-	-	-	-	Yes	Yes	Yes	Yes
pORT-mpt64 (Huang et al., 2010)	Intra-gastric RecA+ *S. typhimurium* SLDAPD	Yes	-	-	-	-	-	-	-	-	Yes

Thus, our research group developed new strategies for DNA vaccination against TB using the invasive *Lc. lactis* FnBPA+ strain (Innocentin et al.,, 2009) and the pValac vaccinal plasmid (Guimarães et al., 2009), in which the coding sequences of the immunodominant antigens ESAT-6 and/or Ag85A were inserted. Mancha-Agresti and colleagues constructed a DNA vaccine encoding the Ag85A antigen (pValac:Ag85A) which was intranasally administrated to mice by L. lactis FnBPA+, showing a significant increase in the production of IFN-γ and TNF-α by splenocytes and of sIgA in the bronchoalveolar lavage (Mancha-Agresti et al., 2017).

Then, Pereira and colleagues reported a significant increase in the production of IFN-γ by splenocytes and of specific sIgA in the colons of mice orally immunized with the *Lc. lactis* FnBPA[+] (pValac:ESAT-6) strain (Pereira et al., 2015). Additionally, the administration of this strain enhanced the immune responses after vaccination with BCG, with promising results in relation to IFN-γ, IL-17, IL-6 and TNF-α levels in the spleen of the immunized mice (Pereira et al., 2017).

Finally in 2016, Prósperi developed a *Lc. lactis* strain carrying a multigenic DNA vaccine in which the coding sequences of ESAT-6 and Ag85A were fused [*Lc. lactis* FnBPA[+] (pValac::e6ag85a)]. After oral administration of this strain to mice, INF-γ, TNF-α and IL-17 levels increased significantly and it was also observed activation of specific cytotoxic T lymphocytes. Additionally, this strategy elicited an antigen-specific mucosal immune response shown by a significant increase in sIgA production in the colons of mice (Prósperi, 2016).

CONCLUSION

The applicability and efficacy of DNA vaccines carried by L. lactis strains, administered by mucosal routes, have been successfully established, generating a specific and broad immune response. Therefore, this immunization strategy can be used in research and development of DNA vaccines against other pathogens, with applications in the fields of immunization and immunomodulation.

ACKNOWLEDGMENTS

This work was funded by the Conselho Nacional de Desenvolvimento Científico e Tecnológico (CNPq), Coordenação de Aperfeiçoamento de Pessoal de Nível Superior (CAPES) and Fundação de Amparo à Pesquisa do estado de Minas Gerais (FAPEMIG).

REFERENCES

Andrew H. L., Abul K. A., Shiv P. (2011) Imunidade regional: respostas imunes especializadas em tecidos epiteliais imunoprivilegiados [Regional Immunity: specialized immune responses in immunoprivileged epithelial tissues] In: *Imunologia Celular e Molecular*, Rio de Janeiro, Elsevier, 293-317.

Babiuk L. A., Pontarollo R., Babiuk S., Loehr B., Little V. D., Den Hurk V. S. (2003) Induction of immune responses by DNA vaccines in large animals. *Vaccine*, 21, 649-658.

Barbosa T., Rescigno M. (2010). Host-bacteria interactions in the intestine: homeostasis to chronic inflammation. *Wiley Interdisciplinary Reviews: Systems, Biology and Medicine*, 2, 80-97.

Becker P. D., Noerder M., Guzmán C. A (2008). Genetic immunization: bacteria as DNA vaccine delivery vehicles. *Human Vaccines*, 4, 189-202.

Bode C., Zhao G., Steinhagen F., Kinjo T., Klinman D. M. (2011). CpG DNA as a vaccine adjuvant. *Expert Reviews of Vaccines*, 10, 499-511.

Bolotin A., Wincker P., Mauger S., Jaillon O., Malarme K., Weissenbach J., Ehrlich S. D., Sorokin A. (2001). The complete genome sequence of the lactic acid bacterium *Lactococcus lactis* ssp. lactis IL1403. *Genome Research*, 11, 731–753.

Brandtzaeg P., Kiyono H., Pabst R., Russell M. W. (2008). Terminology: nomenclature of mucosa-associated lymphoid tissue. *Mucosal Immunology*, 1, 31-37.

Brave A., Hallengard D., Schroder U., Blomberg P., Wahren B., Hinkula J.. (2008). Intranasal immunization of young mice with a multigene HIV-1 vaccine in combination with the N3 adjuvant induces mucosal and systemic immune responses. *Vaccine*, 26, 5075-5078.

Brondyk W. H. (2009). Selecting an appropriate method for expressing a recombinant protein. *Methods in Enzymology*, 463, 131-47.

Brun P., Zumbo A., Castagliuolo I., Delogu G., Manfrin F., Sali M., Fadda G., Grillot-Courvalin C., Palù G., Manganelli R. (2008). Intranasal delivery of DNA encoding antigens of Mycobacterium tuberculosis by non-pathogenic invasive Escherichia coli. *Vaccine*, 26, 1934-1941.

Chadwick S., Kriegel C., Amiji, M. (2009). Delivery strategies to enhance mucosal vaccination. *Expert Opinion on Biological Therapy*, 9 427-440.

Chatel J. M., Pothelune L., Ah-Leung S., Corthier G., Wal J. M., Langella P. (2008). *In vivo* transfer of plasmid from food-grade transiting lactococci to murine epithelial cells. *Gene Therapy* 15, 1184-1190.

De Azevedo M., Karczewski J., Lefévre F., Azevedo V., Miyoshi A., Wells J. M., Langella P., Chatel J. M. (2012). *In vitro* and *in vivo* characterization of DNA delivery using recombinant *Lactococcus lactis* expressing a mutated form of L. monocytogenes Internalin A. *BMC Microbiology* 12, 1-9.

De Vos W. M., Simons, G. F. M. (1994). Gene cloning and expression systems in lactococci In: *Genetics and biotechnology of lactic acid bacteria* ed. Gasson, M. J., De Vos, W. M. Netherlands: Springer, 52-105.

Fynan E. F., Webster R. G., Fuller D. H., Hayne J. R. Santoro, J. C. Robinson H. L. (1993). DNA vaccines: protective immunizations by parenteral, mucosal and gene-gun inoculations. *Proceedings of the National Academy of Sciences of US* 90, 11478-11482.

Gaillard J. L., Berche P., Frehel C., Gouin E., Cossart P. (1991). Entry of Listeria monocytogenes into cells is mediated by internalin, a repeat protein reminiscent of surface antigens from Gram-positive cocci. *Cell* 65, 1127–1141.

Ghanem A., Healey R., Adly F. G. (2013). Current trends in separation of plasmid DNA vaccines: A review. *Anal Chim Acta* 760, 1-15.

Glenting J., Wessels S. (2005). Ensuring safety of DNA vaccines. *Microbial Cell Factories* 4, 1-5.

Grillot-Courvalin C., Goussard S., Courvalin P. (1999). Bacteria as gene delivery vectors for mammalian cells. *Current Opinion in Biotechnology* 10, 477–481.

Guimarães V. D., Innocentin S., Lefèvr, F., Azevedo V., Wal J. M., Langella P., Chatel J. M. (2006). Use of Native Lactococci as Vehicles for Delivery of DNA into Mammalian Epithelial Cells. *Applied and Environmenta Microbiology* 72, 7091-7097.

Guimarães V. D., Gabriel J. E., Lefèvre F., Cabanes D., Gruss A., Cossart P., Azevedo V., Langella P. (2005). Internalin-expressing *Lactococcus lactis* is able to invade small intestine of guinea pigs and deliver DNA into mammalian epithelial cells. *Microbes Infectins* 7, 836–844.

Guimarães V., Innocentin S., Chatel J. M., Lefèvre F., Langella P., Azevedo V., Miyosh, A. (2009). A new plasmid vector for DNA delivery using lactococci. *Genetic Vaccines and Therapy* 7, 1-7.

Hoebe K., Janssen E., Beutler B. (2004). The interface between innate and adaptive immunity. *Nature Immunology*5, 971-974.

Holmgren J., Czerkinsky, C. (2005). Mucosal immunity and vaccines. *Nature Medicine* 11, 45-53.

Huang J. M., Sali M., Leckenby M. W. Radford, D. S. Huynh, H. A. Delogu, G. Cranenburgh R. M., Cutting S. M. (2010). Oral delivery of a DNA vaccine against tuberculosis using operator-repressor titration in a *Salmonella enterica* vector. *Vaccine* 3, 7523-7528.

Innocentin S., Guimarães V., Miyoshi A., Azevedo V., Langella P., Chatel J. M., Lefèvre F. (2009). *Lactococcus lactis* expressing either *Staphylococcus aureus* Fibronectin-Binding Protein A or *Listeria monocytogenes* Internalin A can efficiently internalize and deliver DNA in human epithelial cells. *Applied and Environmental Microbiology* 75, 4870-4878.

Kanazawa T., Takashima Y., Hirayama S., Okada H. (2008). Effects of menstrual cycle on gene transfection through mouse vagina for DNA vaccine. *Internationa Journal of Pharmacology* 360, 164–170.

Kutzler M. A., Weiner D. B. (2008). DNA vaccines: ready for prime time? *Nature Reviews in Genetic* 9, 776-788.

Laddy D. J., Weiner D. B. (2006). From Plasmids to Protection: A Review of DNA Vaccines Against Infectious Diseases. *Internationa Review Immunoogyl* 25, 99-123.

Le T. P., Coonan K. M., Hedstrom R. C., Charoenvi, Y., Sedegah M., Epstein J. E., Kumar S., Wang R., Doolan D. L., Maguire J. D., Parker S. E., Hobar, P., Norman J., Hoffman S. L. (2000). Safety, tolerability and humoral immune responses after intramuscular administration of a malaria DNA vaccine to healthy adult volunteers. *Vaccine* 18, 1893–1901.

Li L., Saade F., Petrovsky N. (2012). The future of human DNA vaccines. *Journal of Biotechnology* 162, 171-182.

Liu M. A. (2011). DNA vaccines: an historical perspective and view to the future. *Immunology Reviews* 239, 62–84.

Liu M. A. (2003). DNA vaccines: a review. *J Intern Med* 253, 402-410.

Lugo, M. T. G. & Bewley, C. A (2008) Natural Products, Small Molecules, and Genetics in Tuberculosis Drug Development. *Journal of Medical Chemistry* 51, 2606-2612.

Mancha-Agresti P., Drumond M. M., Carmo F. L., Santos M. M., Santos J. S., Venanzi F., Chatel J. M., Leclercq S. Y., Azevedo V. (2016). A New Broad Range Plasmid for DNA Delivery in Eukaryotic Cells Using Lactic Acid Bacteria: *In Vitro* and *In Vivo* Assays. Molecular therapy. *Methods & Clinical Development*, 4, 83-91.

Mancha-Agresti P., de Castro C. P., Dos Santos J. S. C., Araujo M. A. Pereira, V. B. LeBlanc, J. G. Leclercq, S. Y. Azevedo V. (2017). Recombinant invasive *Lactococcus lactis* carrying a DNA vaccine coding the Ag85A antigen increases INF-γ, IL-6, and TNF-α cytokines after intranasal immunization. *Ogy*, 8, 1-12.

Marć M. A., Domínguez-Álvarez E., Gamazo C. (2015). Nucleic acid vaccination strategies against infectious diseases. *Expert Opinion on Drug Delivery*, 12, 1851-1865.

Mengaud J., Ohayon H., Gounon P. Mege, R. M. Cossart P. (1996). E-cadherin is the receptor for internalin, a surface protein required for entry of *L. monocytogenes* into epithelial cells. *Cell*, 84, 923-932.

Mercenier A., Muller-Alouf H., Grangette C. (2000). Lactic acid bacteria as live vaccines. *Current Issues in Molecular Biology*, 2, 17-25.

Miller A. M., Dean D. A. (2009). Tissue-specific and transcription factor-mediated nuclear entry of DNA. *Advances Drug Delivery Reviews*, 61, 603–613.

Miyoshi A., Jamet E., Commissaire J., Renaul, P., Langella P., Azevedo V. (2004). A xylose-inducible expression system for Lactococcus lactis. *FEMS Microbiology Letters,* 239, 205-212.

Moss R. (2009). Prospects for control of emerging infectious diseases with plasmid DNA vaccines. *Journal of Immune Based Therapy Vaccines*, 7, 1-9.

Mowa, A. M. (2003). Anatomical basis of tolerance and immunity to intestinal antigens. *Nature Reviews Immunology*, 3, 331-341.

Mowa, A. M., Agace W. W. (2014). Regional specialization within the intestinal immune system. *Nature Reviews Immunology*, 14, 667-685.

Neutra, M. R., Kozlowsk, P. A. (2006). Mucosal vaccines: the promise and the chllenge. *Nature Reviews Immunology*, 6, 148-58.

Parida S. K., Huygen K., Ryffel B., Chakraborty T. (2005). Novel Bacterial Delivery system with attenuated *Salmonella typhimurium* carrying plasmid encoding Mtb antigen 85A for mucosal immunization. *Annals of the New York Academy of Sciences*, 1056, 366-378.

Pereira V. B., Zurita-Turk M., Saraiva T. D. L., De Castro C. P., Souz, B. M., Mancha-Agresti P., Lima, F. A., Pfeiffer V. N., Azevedo M. S. P., Rocha C. S., Pontes D. S., Azevedo V., Miyoshi A. (2014). DNA Vaccines Approach: From Concepts to Applications. *World Journal of Vaccines*, 4, 50-71.

Pereira V. B., Saraiva T. D., Souza B. M., Zurita-Turk M., Azevedo M. S., De Castro C. P. Mancha-Agresti, P. Dos Santos, J. S. Santos, A. C. Faria A. M., Leclercq S. Azevedo, V. Miyoshi, A. (2015). Development of a new DNA vaccine based on mycobacterial ESAT-6 antigen delivered by recombinant invasive *Lactococcus lactis* FnBPA+. *Applied Microbiology and Biotechnology*, 99, 1817-1826.

Pereira V. B., Da Cunha V. P., Preisser T. M. Souza, B. M. Zurita-Turk M., De Castro C. P., Azevedo M. S., Miyoshi A. (2017). *Lactococcus lactis* carrying a DNA vaccine coding for the ESAT-6 antigen increases IL-17 cytokine secretion and boosts the BCG vaccine immune response. *Journal of Applied Microbiology*, 122, 1657-1662.

Pilgrim S., Stritzker J., Schoen C. Kolb-Mäurer, A. Geginat, G. Loessner, M. J. Gentschev, I. Goebel, W. (2003). Bactofection of mammalian cells by *Listeria monocytogenes*: improvement and mechanism of DNA delivery. *Gene Therapy*, 10, 2036-2045.

Pontes, D. S., de Azevedo M. S., Chatel J. M., Langella P., Azevedo V., Miyoshi A. (2011). *Lactococcus lactis* as a live vector: Heterologous protein production and DNA delivery systems. *Protein Expression and Purification*, 79 165–75.

Pontes D., Innocentin S., Del Carmen S., Almeida J. F., Leblanc J. G., de Moreno de Leblanc A., Blugeon S., Cherbuy C., Lefèvre F., Azevedo V., Miyoshi A., Langella P., Chatel J. M. (2012). Production of fibronectin binding protein A at the surface of *Lactococcus lactis* increases plasmid transfer *in vitro* and *in vivo*. *PLoS One* 7, 1-6.

Prósperi C. C. (2016). *Utilização de Lactococcus lactis como veículo para a entrega de um plasmídeo vacinal codificando uma proteína de fusão composta pelos antígenos ESAT-6 e Ag85A de Mycobacterium tuberculosis em células mamíferas e avaliação do perfil de resposta imunológica gerado em modelo murinho [Use of Lactococcus lactis as vehicle for the delivery of a vaccine plasmid encoding a fusion protein composed of ESAT-6 and Ag85A antigens of Mycobacterium tuberculosis in mammalian cells and evaluation of the immunological response profile generated in murine model*]. PhD Thesis – Universidade Federal de Minas Gerais, Programa de Pós-Graduação em Genética, Belo Horizonte.

Que Y. A., Francois P., Haefliger J. A., Entenza J. M., Vaudaux P., Moreillon P. (2001). Reassessing the role of Staphylococcus aureus clumping factor and fibronectin-binding protein by expression in *Lactococcus lactis*. *Infection and Immunity*, 69, 6296-6302.

Reyes-Sandoval A., Ertl H. C. (2001). DNA Vaccines. *Current Molecular Medicine*, 1, 217-243.

Roush S. W., Murphy T. V., Vaccine-Preventable Disease Table Working Group. (2007). Historical comparisons of morbidity and mortality for vaccine-preventable diseases in the United States. *JAMA* 298, 2155-2163.

Sallusto F., Lenig D., Forster R., Lipp M., Lanzavecchia A. (1999). Two subsets of memory T lymphocytes with distinct homing potentials and effector functions. *Nature*, 401, 708-712.

Schaffner W. (1980). Direct transfer of cloned genes from bacteria to mammalian cells. *Proceedings of the National Academy of Sciences of US*, 77, 2163–2167.

Schoen C., Stritzker J., Goebel W., Pilgrim S. (2004). Bacteria as DNA vaccine carriers for genetic immunization. *International Journal of Medical Microbiology*, 294 319-335.

Schroeter J., Klaenhammer, T. (2009). Genomics of lactic acid bacteria. *FEMS Microbiology Letters*, 292, 1-6.

Sinha B., François P. P., Nüsse O., Foti M., Hartford O. M., Vaudaux P., Foster T. J., Lew D. P., Herrmann, M., Krause K. H. (1999). Fibronectin-binding protein acts as *Staphylococcus aureus* invasin via fibronectin bridging to integrin alpha5beta1. *Cell Microbiology*, 1, 101-17.

Sleator R. D., Hill C. (2006). Patho-biotechnology: using bad bugs to do good things. *Current Opinion in Biotechnology*, 17, 211-216.

Tang D. C., DeVit M., Johnston S. A. (1992). Genetic immunization is a simple method for eliciting an immune response. *Nature*, 356, 152-154.

Tao L., Pavlova S. I., Ji X., Jin L., Spear G. (2011). A novel plasmid for delivering genes into mammalian cells with noninvasive food and commensal lactic acid bacteria. *Plasmid*, 65, 8-14.

Ulmer J. B. Donnelly J. J., Parker S. E., Rhodes G. H., Felgner P. L., Dwarki V. J., Gromkowski S. H., Deck R. R., DeWitt C. M., Friedman A., Hawe L. A., Leander K. R., Martinez D., Perry H. C., Shiver J. W., Montgomery D. L., Liu M. A. (1993). Heterologous protection against influenza by injection of DNA encoding a viral protein. *Science* 259, 1745-1749.

Van de Guchte M., Ehrlich S. D., Maguin E. (2001). Production of growth-inhibiting factors by *Lactobacillus delbrueckii*. *Journal of Applied Microbiology*, 91, 147-53.

Van Drunen Littel-van den Hurk S., Gerdts V., Loehr B. I., Pontarollo R., Rankin R., Uwiera R., Babiuk L. A. (2000). Recent advances in the use of DNA vaccines for the treatment of diseases of farmed animals. *Advances in Drug Delivery Reviews*, 43, 13-28.

Wang S., Farfan-Arribas D. J., Shen S. (2006). Relative contributions of codon usage, promoter efficiency and leader sequence to the antigen expression and immunogenicity of HIV-1 Env DNA vaccine. *Vaccine*, 24, 4531-4540.

Wang Q. L., Pan Q., Ma Y., Wang K., Sun P., Liu S., Zhang X. L. (2009). An attenuated *Salmonella*-vectored vaccine elicits protective immunity against Mycobacterium tuberculosis. *Vaccine*, 27, 6712-6722.

Weiss S., Chakraborty T. (2001). Transfer of eukaryotic expression plasmids to mammalian host cells by bacterial carriers. *Current Opinion in Biotechnology*, 12, 467-472.

Wells J. (2011). Mucosal Vaccination and Therapy with Genetically Modified Lactic Acid Bacteria. *Annual Review of Food Science and Technology*, 2, 423-445.

Wells J. M., Mercenier A. (2008). Mucosal delivery of therapeutic and prophylactic molecules using lactic acid bacteria. *Nature*. 1038, 1-14.

Wolff J. A., Malone R. W., Williams P., Chong W., Acsadi G., Jani A., Felgner P. L. (1990). Direct gene transfer into mouse muscle *in vivo*. *Science*, 247, 1465-1468.

Wollert T., Pasche B., Rochon M., Deppenmeier S., van den Heuvel J., Gruber A. D., Heinz D. W., Lengeling A., Schubert W. D. (2007). Extending the host range of *Listeria monocytogenes* by rational protein design. *Cell*, 129, 891-902.

Yagnik B., Padh H., Desai P (2016). Construction of a new shuttle vector for DNA delivery into mammalian cells using non-invasive *Lactococcus lactis*. *Microbes Infections*, 18, 237-244.

In: The Many Benefits of Lactic Acid Bacteria
Editors: J. G. LeBlanc and A. de Moreno

ISBN: 978-1-53615-388-0
© 2019 Nova Science Publishers, Inc.

Chapter 11

BILE SALT HYDROLASE ACTIVITY OF LACTIC ACID BACTERIA: A POTENTIAL FOR HEALTH

Engkarat Kingkaew and Somboon Tanasupawat*
Department of Biochemistry and Microbiology,
Faculty of Pharmaceutical Sciences, Chulalongkorn University,
Bangkok, Thailand

ABSTRACT

Many species of lactic acid bacteria (LAB) were usually found and isolated from various sources such as fermented foods (fermented meats, fermented dairy products, fermented fish, fermented vegetables, fermented fruits, fermented cereals and fermented legumes), dairy products, gastrointestinal tract of humans and animals, plant materials, feces, and sewage. LAB are a major group of probiotics, live bacteria that are good for host health. Several LAB express bile salt hydrolase (BSH) activity. Frequently, the bile salt hydrolase (BSH) activity of LAB have been involved with cholesterol-lowering effects and are exhibited by numerous genera, especially *Enterococcus (Ec)*, *Bifidobacterium (Bf)*, *Lactobacillus (Lb)*, *Lactococcus (Lc)*, *Leuconostoc (Leuc)*, *Pediococcus (Ped)*, *Clostridium (Cl)*, and *Bacteroides (B)*. They have shown the ability to lower serum cholesterol. Cholesterol is an essential building block for biological synthesis in the human body (hormone, vitamin D and bile acid) and is the important hypocholesterolemic role of bile salt hydrolase (BSH). One of the reasons that LAB have a hypocholesterolemic effect is bile salt hydrolase production or deconjugation of bile acid. Bile acids are made from cholesterols and amino acids (Taurine or Glycine) and metabolized *de novo* in the human liver via conjugation. Conjugated bile acid then flows into the intestine, where the amide bond of conjugated bile acid react with the bile salt hydrolase (BSH) that is produced by LAB. The amino acids were disintegrated from steroid core, and then transformed into a final form of bile acid is called deconjugated or

---

* Corresponding Author's Email: Somboon.T@chula.ac.th.

unconjugated bile acid, which is less effectively reabsorbed than its conjugated form. Deconjugated bile salt results in the excretion of greater amount of free bile acids in feces, that is less solubility and absorption in intestine; therefore, enterohepatic recirculation, the homeostasis of bile acid, and could be consequently dragged into a reduction in serum cholesterol by augmentation the demand of cholesterol for *de novo* synthesis of bile acid to substitute those are eliminated in feces or by decreasing cholesterol solubility and thereby absorption of cholesterol through the intestinal lumen.

INTRODUCTION

Cholesterol is an essential building block for maintaining good health; however, hypercholesterolemia has been recognized as a major risk factor in atherosclerosis and coronary heart diseases (Brown & Goldstein, 1984; Khedkar, 1993; Smet et al., 1994). Bile salt hydrolase (BSH) activity of LAB may reduce the serum cholesterol level as suggested by (De Smet et al., 1995; Ahn et al., 2003). BSH enzyme is synthesized by several LAB and hydrolyzes either glycine- or taurine-link bile acids deconjugation reaction. The deconjugated bile acids are less soluble and less reabsorbed than conjugated bile acids, which results in excretion of larger amounts of free bile acids in feces; consequently, the cholesterol is taken up for *de novo* bile acid synthesis to maintain bile acid levels. Therefore, the deconjugation of bile acids by LAB bacteria could lead to a reduction in serum cholesterol (Pereira, McCartney & Gibson, 2003). Bile salt hydrolase (BSH) activity has been detected in many LAB living in the various sources (Jarocki et al., 2014; Ding et al., 2017; Xu et al., 2016; Saravanan & Shetty, 2016; Hou et al., 2015; Du Toit et al., 1998; Tanaka et al., 2000; Grill, Manginot-Durr et al., 1995; Öner, Aslim & Aydaş, 2014; Shekh, Dave & Vyas, 2016; Abushelaibi et al., 2017; Sirilun et al., 2010; Liu et al., 2017; Argyri et al., 2013; Franz et al., 2001; da Silva Ferrari et al., 2016; Elkins, Moser & Savage, 2001; Shehata et al., 2016; Corzo & Gilliland, 1999; Pato et al., 2004; Patel et al., 2010). Additionally, other researchers suggested the importance roles of BSH enzyme so that the bacterial cells can be detoxificated of conjugated bile acids for microbial bile tolerance and survival in the intestine, liberating of amino acids from bile acids for bacterial growth, and alteration of bacterial membrane (Tanaka et al., 2000; Tannock, Dashkevicz & Feighner, 1989; De Smet et al., 1995; Dambekodi & Gilliland, 1998; Taranto et al., 1998; Taranto et al., 2003).

Furthermore, some studies suggested undesirable effects of bacterial BSH activity on human health. It is possible that the deconjugated bile acids may impair digestive system and colonic mucosal function, play a role in gallstone formation and also cause DNA damage that promotes colon cancer (Jarocki et al., 2014). Thus, the bacterial BSH activity has been included among the criteria for the selection of probiotic strains. However, the bacterial BSH activity has been argued to be potentially harmful to human, and therefore it is as yet not clear whether BSH activity is, in fact, a desirable trait in a probiotic bacterium. This chapter gathered, described and summarized the available literature in the characteristic of bile, bile salt hydrolase (BSH) enzyme, mechanisms of cholesterol-lowering by LAB, bile salt hydrolase producing LAB, the effects of bacterial bile salt hydrolase on hosts and the beneficial effects of BSH activity to bacteria.

CHARACTERISTIC OF BILE AND FUNCTION

Bile is a dark green to a yellowish brown aqueous solution whose major constituents include cholesterol, bile acids, phospholipid, and the pigment biliverdin (Begley, Hill & Gahan, 2006). Bile is produced in the pericentral hepatocytes of the liver, stored and concentrated in the gallbladder interdigestively, and secreted into the first part of the small intestine which is named duodenum. Bile plays an essential role in fat digestion, which acts as a biological detergent, helping to emulsify and solubilize fats in food. Because of its biological detergent and membranolytic activity, bile is also an antimicrobial agent, a feature considered important for inhibiting the bacterial colonization in the proximal small intestine (Tremblay et al., 2017).

Bile acids are hydroxylated steroids. The two primary bile acids are cholic acid and chenodeoxycholic acid that are synthesized *de novo* from cholesterol in the liver. After that, the primary bile acid is also metabolized via conjugation (N-acyl amidation) to taurine (tauroconjugated) or glycine (glycoconjugate). The conjugation in the liver increases the solubility of the hydrophobic steroid nucleus; consequently, these molecules are amphipathic and can solubilize lipid to form mix micelles. Bile acids are conserved by a process named enterohepatic recirculation. Both conjugated and deconjugated bile acids are absorbed via active transport in the distal part of ileum and via passive transport (Kumar et al., 2012).

Reabsorbed bile acids flow into the portal bloodstream and are taken up by hepatocytes. Later, reabsorbed bile acids are reconjugated and resecreted into bile. Secondary bile acids are produced by intestinal bacteria. According to Bortolini, Medici & Poli (1997), "approximately 5% of the total bile acid pool (0.3 g to 0.6 g) per day eludes epithelial absorption and may be extensively modified by the indigenous intestinal bacteria." One of the main bile acids conversations is deconjugation (Batta et al., 1990). Deconjugation is the reaction that hydrolyzes the amide bond and liberates the glycine or taurine moiety from the steroid core. Finally, the bile acids are termed deconjugated or unconjugated.

BILE SALT HYDROLASE ENZYME

Bile salt hydrolase enzyme (BSH) (Cholylglycine hydrolase, E.C.3.5.1.24) has been found and characterized from many bacteria such as LAB, and mammalian gut microbiota. BSH enzyme is responsible for bile acid metabolism and plays a vital role in their colonization and survival in the mammalian intestine or harsh environments. BSHs are characterized as intracellular enzymes which belong to choloylglycine hydrolase family (Patel et al., 2010; Bi et al., 2013). The characteristics of BSHs are oxygen insensitivity, and the optimal pH normally between pH 5 and 6. In addition, one of the important factors is biomass production, since BSHs activity depends on biomass production (Begley, Gahan &Hill, 2005). Furthermore, bile deconjugation also leads to decreasing of serum cholesterol and alterations in energy homeostasis. The presence of an active BSH enzyme has long been considered as a criterion for the selection of potential probiotics (Food & Group, 2002).

MECHANISMS OF CHOLESTEROL-LOWERING BY BILE SALT HYDROLASE OF LAB

Numerous mechanisms to lower cholesterol have been proposed in previous researches, such as enzymatic deconjugation of bile salt by LAB. LAB reducing cholesterol activities are described here. The mechanism of cholesterol-lowering by LAB is to deconjugated bile acids by bile salt hydrolase enzyme. Bile acids are hydroxylated steroids, synthesized in the pericentral hepatocytes of the liver from cholesterol. The primary bile acids, such as cholic and chenodeoxycholic acids, are *de novo* synthesized in the liver and conjugated to either glycine (glycoconjugated) or taurine (tauroconjugated) via an amide linkage at the C24 carboxyl. They are excreted across the canaliculi to the biliary system. Over than 95% of bile acids secreted in bile are reabsorbed in the distal part of ileum and return to the liver because bile acids are conserved under normal condition by a process which is called enterohepatic circulation. In addition, bile acids play an important role in lipid-soluble nutrient and dietary lipid absorption (Korpela, Adlercreutz & Turunen, 1988; Kurdi et al., 2003). Unabsorbed bile acids flow into the colon and then they are catalyzed by bacterial metabolism such as deconjugation with bile salt hydrolase enzyme (BSH) which is generated by LAB. The conjugated bile acids are transformed to the deconjugated form. Most deconjugated bile acids are excreted and found in human feces because of their lower solubility and absorption. Consequently, it causes a reduction in serum cholesterol by increasing cholesterol demand for *de novo* bile acid synthesis to compensate the eliminated part. In a homeostatic response, new bile acids are synthesized from cholesterol for maintaining bile acid level, resulting in decreasing of serum cholesterol (Moser & Savage, 2001; Ahn et al., 2003; Ooi & Liong, 2010; Kumar, Ghosh 6 Ganguli, 2012; Bi et al., 2013). In conclusion, the BSH enzyme performance has a cholesterol-lowering effect.

BILE SALT HYDROLASE PRODUCING LAB

Bile salt hydrolase activity has been detected in LAB such as *Lactobacillus* (Tannock, Dashkevicz & Feighner, 1989; Lundeen & Savage, 1990; Christiaens et al., 1992; Lundeen & Savage, 1992; Bateup et al., 1995; De Smet et al., 1995; Taranto et al., 1996; Du Toit et al., 1998; Elkins & Savage, 1998; Corzo & Gilliland, 1999; De Boever & Verstraete, 1999; Tanaka et al., 1999; Grill et al., 2000; Elkins, Moser & Savage, 2001;

Table 1. Bile salt hydrolase positive strains of LAB isolated from various sources

Species	Strain	Source	Reference
Lb. acidophilus	ATCC 4356	Human	Elkins, Moser & Savage, 2001
	ATCC 53544	Human infant, rectal swab	
	JCM 1034	Human intestine	
	O16	Human intestine	Corzo & Gilliland, 1999
	BFE 1059	Pig faeces	Du Toit et al., 1998
Lb. brevis	BCCM 7944	Human feces	Elkins, Moser & Savage, 2001
	BCCM 11998	Starter from dairy	
	BCCM 18022	Zabady (yogurt)	
	UNIVASF CAP 16, UNIVASF CAP 279	Goat milk	da Silva Ferrari et al., 2016
Lb. rhamnosus	BO3	Boza	Shehata et al., 2016
Lb. reuter	Iso66	Camel milk	Abushelaibi et al., 2017
Lc. lactis subsp. *lactis*	BO37	Boza	Shehata et al., 2016
	RM39	Rayeb milk	
	IS-10285	Dadih	Pato et al., 2004
Lc. lactis	Iso15, Iso76	Camel milk	Abushelaibi et al., 2017
Lc. garvieae	Iso47	Camel milk	Abushelaibi et al., 2017
Lb. paracasei	BO51, BO52	Boza	Shehata et al., 2016
	UNIVASF CAP 45, UNIVASF CAP 84	Goat milk	da Silva Ferrari et al., 2016
Lb. delbrueckii subsp. *bulgaricus*	BO34	Boza	Shehata et al., 2016)
	D11, D14	Dongbei kimchi	Xu et al., 2016
Lb. fermentum	ATCC 11976	Infant intestine	Elkins, Moser & Savage, 2001
Lb. gasseri	BCCM 9203	Human	Elkins, Moser & Savage, 2001
	JCM 1025	Human intestine	Elkins, Moser & Savage, 2001
	RM28	Rayeb milk	Shehata et al., 2016
Lb. johnsonii	JCM 1022	Human intestine	Elkins, Moser & Savage, 2001
	JCM 8791	Human feces	Elkins, Moser & Savage, 2001
	BFE 1061	Pig feces	Du Toit et al., 1998

Table 1. (Continued)

Species	Strain	Source	Reference
Lb. plantarum	BCCM 18021	Milk	Elkins, Moser & Savage, 2001
	BCCM 18027	Laban rayeb	Elkins, Moser & Savage, 2001
	D24, D25	Dongbei kimchi	Xu et al., 2016
	B282, E45, E10, E73	Naturally fermented olives	Argyri et al., 2013
	RC	Raw cheese	Shekh, Dave & Vyas, 2016
	GV	Guava	
	SG	Sugarcane	
	GP	Grapes	
Lb. plantarum	OP	Prickly pear	Shekh, Dave & Vyas, 2016
	Iso34, Iso70	Camel milk	Abushelaibi et al., 2017
	TGCM 15, TGCM 33	Thai fermented food	Sirilun et al., 2010
	LA3, GD2	Breast-fed infants' feces	Öner, Aslim &Aydaş, 2014
	Lp3	Fermented yak milk	Ding et al., 2017
	LP96	Fermented food	Liu et al., 2017
Lb. pentosus	B279, B283, E43, E100, E128	Naturally fermented olives	Argyri et al., 2013
Ec. faecium	B20, B21	Stinky soybean	Xu et al., 2016
	C10	Rubing	
	P1, P7,P8	Guizhou kimchi	
	V24	Sichuan kimchi	
	FAIR-E 154	Food	Franz et al., 2001
Ec. durans	C12, C5, C3	Rubing	Xu et al., 2016
	V18	Sichuan kimchi	
	FAIR-E 231	Food	Franz et al., 2001
Ped. ethanolidurans	D13	Dongbei kimchi	Xu et al., 2016
	K2	Kimchi	
Leuc. mesenteroides	V12, V21	Sichuan kimchi	Xu et al., 2016
	D10	Dongbei kimchi	
Leuc. lactis	KC117496	*Idli* batter	Saravanan & Shetty, 2016
Lb. reuteri	NCIMB 30242	Pig	Hou et al., 2015
	BFE 1058	Pig feces	Du Toit et al., 1998
Bf. longum	SBT2928	Human feces	Tanaka et al., 2000
	BB536	Human	Grill et al., 1995
	BASO9, BASO15	Breast-fed infants' feces	Öner, Aslim &Aydaş, 2014
Bf. longum subsp. *longum*	NRRL B-41409	Adult intestine	Jarocki et al., 2014
Bf. adolescentis	DSM 20087	Bovine serum	Jarocki et al., 2014
Bf. animalis subsp. *animalis*	NRRL B-41406	Rat feces	Jarocki et al., 2014
Bf. animalis subsp. *lactis*	NRRL B-41405	Yoghurt	Jarocki et al., 2014
Bf. longum subsp. *infantis*	ATCC 15697	Infant intestine	Jarocki et al., 2014
Bf. longum subsp. *suis*	NRRL B-41407	Pig feces	Jarocki et al., 2014
Bf. bifidum	DSM 20456	Breast-fed infant stool	Jarocki et al., 2014
Bf. breve	DSM 20091, NRRL B-41408	Infant intestine	Jarocki et al., 2014
	A19, A26	Breast-fed infants' feces	Öner, Aslim &Aydaş, 2014
Bf. catenulatum	DSM 20224	Sewage	Jarocki et al., 2014
Bf. pseudocatenulatum	DSM 20439	Infant intestine	Jarocki et al., 2014
Bf. longum subsp. *suis*	NRRL B-41407	Pig feces	Jarocki et al., 2014
Bf. pseudolongum subsp. *pseudolongum*	DSM 20095	Chicken feces	Jarocki et al., 2014

Bf, Bifidobacterium; Ec, Enterococcus; Lb, Lactobacillus; Lc, Lactococcus; Leuc, Leuconostoc.

Pereira, McCartney & Gibson, 2003; McAuliffe, Cano & Klaenhammer, 2005; Argyri et al., 2013; Hou et al., 2015; da Silva Ferrari et al., 2016; Shehata et al., 2016; Shekh, Dave & Vyas, 2016; Abushelaibi et al., 2017), *Lactococcus* (Pato et al., 2004; Shehata et al., 2016; Abushelaibi et al., 2017), *Bifidobacterium* (Grill, Manginot-Durr et al., 1995; Grill et al., 1995; Grill et al., 2000; Tanaka et al., 2000; Kim et al., 2004; Kim, Yi & Lee, 2004; Kim, BrochetLee, 2005; Jarocki et al., 2014; Öner, Aslim & Aydaş, 2014), *Enterococcus* (Franz et al., 2001; Knarreborg et al., 2002; Veysey et al., 2001; Xu et al., 2016), *Pediococcus* (Xu et al., 2016), *Leuconostoc* (Saravanan & Shetty, 2016; Xu et al., 2016), and also in *Clostridium* (Coleman & Hudson, 1995; Pereira, McCartney & Gibson, 2003), and *Bacteroides* strains (Kawamoto, Horibe & Uchida, 1989; Stellwag & Hylemon, 1976) which have been isolated from traditional fermented foods, dairy, plant, sewage, feces, animal, and the human gastrointestinal tract. The presence of BSH producing LAB is mostly found in the bile acid-rich environment (Tanaka et al., 1999). Generally, bifidobacteria and lactobacilli are used as probiotic, whereas *Enterococcus*, *Clostridium*, and *Bacteroides* strains are mostly found in the gastrointestinal tract. Numerous studies have reported BSH-positive strains as shown in Table 1.

EFFECTS OF BACTERIAL BILE SALT HYDROLASE ON HOSTS

Cholesterol-lowering Effect of LAB: Hypocholesterolemic Benefits

Hypercholesterolemia is the major risk factors of cardiovascular disease (CVDs) that is the main cause of the mortality (Labarthe & Dunbar, 2012; Roth et al., 2011). The coronary arteries are affected by high serum cholesterol. Atherosclerosis, the hardening of the arteries, is a condition where the arteries become narrowed and hardened due to a buildup of plaque around the artery wall. Atherosclerosis begins when the endothelium becomes damaged, allowing the harmful type of cholesterol to make up part of the artery wall. Thus, dietary modification and functional foods are becoming an interesting issue to reduce the risk of heart diseases (Aronow, 2013; Lichtenstein & Goldin, 1993; St-Onge, Farnworth & Jones, 2000). Functional food is an interesting alternative choice to manage serum cholesterol due to the effect of LAB or other probiotics. LAB consist as a major group of probiotic bacteria: live bacteria that are good for host health. Several LAB express cholesterol-lowering effects (Jones, 2002).

Even though there have been several medicines for the treatment of hypercholesterolemia, these medicines are expensive and have negative side effects. In these senses, it is possible that the using of BSH enzyme to reduce serum cholesterol levels in the patient with hypercholesterolemia and also prevent hypercholesterolemia in normal people has arisen as an interesting alternative to drugs available today (Chae et al., 2013).

Impaired Digestive Functions

In lipid digestion and absorption, the BSH activity may compromise the mechanisms of lipid emulsification and micelle formation due to lack of conjugated bile acids which are more efficient than deconjugated bile acids (Nagpal et al., 2012). The growth defects in chicken have been found with microbial BSH activity correlation (Walker &Gilliland, 1993; Du Toit et al., 1998). However, there is no association between the growth defects and the BSH activity in mice (Kumar et al., 2012).

Disruption of Intestinal Conditions and/or Gallstones

Several studies have reported that the deconjugated bile acids may disturb colonic mucosal function which would be the causes of diarrhea or inflammation. In addition, it is assumed that colon cancer as a result of DNA damage can be caused by deconjugated bile acids. (Kandell & Bernstein, 1991; Bernstein et al., 2005; Marteau et al., 1995; Nagengast, Grubbenvan & Munster, 1995; Pazzi et al., 1997).

Additionally, the alterations of the bile acids concentration may lead to bile being supersaturated with cholesterol. This cholesterol may precipitate together with calcium salts and bile pigments to form solid build-ups of crystallized bile termed gallstones, which may build up and impede the biliary ducts (Low-Beer & Nutter, 1978; Marcus & Heaton, 1986; Veysey et al., 1999). Interestingly, the increasing deconjugated bile acid formation has been observed in gallstone sufferers (Berr et al., 1996; Mamianetti et al., 1999).

BENEFICIAL EFFECTS OF BSH ACTIVITY TO BACTERIA

Firstly, studies strongly support the hypothesis that bacterial BSH depress the antimicrobial properties of bile acid; consequently, the bacterial BSH producing strains can better survive and persist in the human gastrointestinal tract. Since the deconjugated bile acids have decreased solubility and diminished detergent activity, these can be less toxic to bacteria in the intestine. Secondly, other studies suggested that some bacterial species may be given carbon, nitrogen and sulfur sources by deconjugation with BSH enzyme (Tanaka et al., 2000; Van Eldere et al., 1996; Jarocki et al., 2014). Finally, there is a possibility that the BSH facilitates incorporation of cholesterol moiety or bile into bacterial membranes. This incorporation may probably modify and improve fluidity, permeability and the net charge of the bacterial membranes; consequently, this alteration improves the colonization ability and survival of these microbes in the gut. Because the bacterial cell modification, which is the result from BSH activity, could protect the bacteria from the immune system (Dambekodi & Gilliland, 1998; Taranto et al., 1997; Taranto et al., 2003). However, this incorporation may be the important resistance mechanisms that are the cause of persistent pathogen infection (Begley, Hill 6 Gahan, 2006).

In conclusion, BSH activity is desirable for enhancing the overall probiotic beneficial effects correlated with each strain. For the human supplementation, the safety of administering BSH producing bacteria as probiotic must be considered in order to prevent undesirable side effects (Ahn et al., 2003; Gilliland & Speck, 1977; Takahashi & Morotomi, 1994).

CONCLUSION

Increasingly, it is becoming apparent that microbial BSH activity may confer a selective benefit on probiotic strains in the highly competitive environment of the human intestinal tract. Moreover, the manipulation of BSH activity may ultimately lead to the development of more robust probiotics with improved competitiveness and performance. Finally, targeting the bacterial BSH to modify cholesterol and bile acid metabolisms might be interestingly alternative choices to prevent or treat various diseases such as hypercholesterolemia and cardiovascular diseases.

REFERENCES

Abushelaibi A., Al-Mahadin S., El-Tarabily K., Shah N. P., Ayyash M.(2017). Characterization of potential probiotic lactic acid bacteria isolated from camel milk. *LWT-Food Science and Technology*, 79, 316-325.

Ahn Y. T., Kim G. B., Lim K. S., Baek Y. J., Kim H. U. (2003). Deconjugation of bile salts by *Lactobacillus acidophilus* isolates. *International Dairy Journal*, 13 (4), 303-311.

Argyri A. A., Zoumpopoulou G., Karatzas K. A., Tsakalidou E., Nychas G. J., Panagou E. Z., Tassou C. C. (2013). Selection of potential probiotic lactic acid bacteria from fermented olives by *in vitro* tests. *Food Microbiology*, 33 (2), 282-291.

Aronow W. A. (2013). Treatment of hypercholesterolemia. *Journal of Clinical and Experimental Cardiology*, 1, 1-8.

Bateup J. M., McConnell M. A., Jenkinson H. F., Tannock G. W. (1995). Comparison of Lactobacillus strains with respect to bile salt hydrolase activity, colonization of the gastrointestinal tract, and growth rate of the murine host. *Applied and Environmental Microbiology*, 61 (3), 1147-1149.

Batta A. K., Salen G., Arora Renu, Shefer S., Batta M., Person A. (1990). Side chain conjugation prevents bacterial 7-dehydroxylation of bile acids. *Journal of Biological Chemistry,* 265 (19), 10925-10928.

Begley M., Gahan C. G. M., Hill C. (2005). The interaction between bacteria and bile. *FEMS Microbiology Reviews*, 29 (4), 625-651.

Begley M., Hill C., Gahan C. G. M. (2006). Bile salt hydrolase activity in probiotics. *Applied and Environmental Microbiology*, 72 (3), 1729-1738.

Bernstein H., Bernstein C., Payne C. M., Dvorakova K., Garewal H. (2005). Bile acids as carcinogens in human gastrointestinal cancers. *Mutation Research/Reviews in Mutation Research*, 589 (1), 47-65.

Berr F., Kullak-Ublick G. A., Paumgartner G., Munzing W., Hylemon P. B. (1996). 7 alpha-dehydroxylating bacteria enhance deoxycholic acid input and cholesterol saturation of bile in patients with gallstones. *Gastroenterology,* 111 (6), 1611-1620.

Bi J., Fang F., Lu., Du G., Chen J. (2013). New insight into the catalytic properties of bile salt hydrolase. *Journal of Molecular Catalysis B: Enzymatic*, 96, 46-51.

Bortolini O., Medici A., Poli S. (1997). Biotransformations on steroid nucleus of bile acids. *Steroids,* 62 (8-9), 564-577.

Brown M. S., Goldstein J. L. (1984). How LDL receptors influence cholesterol and atherosclerosis. *Scientific American,* 251 (5), 58-69.

Chae J. P., Valeriano V. D., Kim G. B., Kang D. K. (2013). Molecular cloning, characterization and comparison of bile salt hydrolases from *Lactobacillus johnsonii* PF 01. *Journal of Applied Microbiology,* 114 (1), 121-133.

Christiaens H., Leer R. J., Pouwels P. H., Verstraete W. (1992). Cloning and expression of a conjugated bile acid hydrolase gene from *Lactobacillus plantarum* by using a direct plate assay. *Applied and Environmental Microbiology*, 58 (12), 3792-3798.

Coleman J. P., Hudson L. L. (1995). Cloning and characterization of a conjugated bile acid hydrolase gene from *Clostridium perfringens. Applied and Environmental Microbiology*, 61 (7), 2514-2520.

Corzo G., Gilliland S. E. (1999). Bile salt hydrolase activity of three strains of *Lactobacillus acidophilus. Journal of Dairy Science*, 82 (3), 472-480.

da Silva Ferrari I., de Souza J. V., Ramos C. L., da Costa M. M., Schwan R. F., Dias F. S.. (2016). Selection of autochthonous lactic acid bacteria from goat dairies and their addition to evaluate the inhibition of *Salmonella typhi* in artisanal cheese. *Food Microbiology,* 60, 29-38.

Dambekodi P. C., Gilliland S. E. (1998). Incorporation of Cholesterol into the Cellular Membrane of *Bifidobacterium longum. Journal of Dairy Science*, 81 (7), 1818-1824.

De Boever P., Verstraete W. (1999). Bile salt deconjugation by *Lactobacillus plantarum* 80 and its implication for bacterial toxicity. *Journal of Applied Microbiology*, 87 (3), 345-352.

De Smet I., Van Hoorde L., Vande Woestyne M., hristiaens H., Verstraete W. (1995). Significance of bile salt hydrolytic activities of lactobacilli. *Journal of Applied Bacteriology*, 79 (3), 292-301.

Ding W., Shi C., Chen M., Zhou J., Long R., Guo X. (2017). Screening for lactic acid bacteria in traditional fermented Tibetan yak milk and evaluating their probiotic and cholesterol-lowering potentials in rats fed a high-cholesterol diet. *Journal of Functional Foods*, 32, 324-332.

Du Toit M., Franz C. M. A. P., Dicks L. M. T., Schillinger U., Haberer P., Warlies B., Ahrens F., Holzapfel W. H. (1998). Characterisation and selection of probiotic lactobacilli for a preliminary minipig feeding trial and their effect on serum cholesterol levels, faeces pH and faeces moisture content. *International Journal of Food Microbiology*, 40 (1-2), 93-104.

Elkins C. A., Moser S. A., Savage D. C. (2001). Genes encoding bile salt hydrolases and conjugated bile salt transporters in *Lactobacillus johnsonii* 100-100 and other *Lactobacillus* species. *Microbiology*, 147 (Pt 12), 3403-3412.

Elkins C. A., Savage D. C. (1998). Identification of genes encoding conjugated bile salt hydrolase and transport in *Lactobacillus johnsonii* 100-100. *Journal of Bacteriology*, 180 (17), 4344-4349.

Food Joint, *Group Agriculture Organization/World Health Organization Working. 2002. Guidelines for the evaluation of probiotics in food*. In Joint FAO/WHO Working Group Report on Drafting Guidelines for the evaluation of probiotics in food: Joint Food and Agriculture Organization and World Health Organization.

Franz C. M., Specht I., Haberer P., Holzapfel W. H. (2001). Bile salt hydrolase activity of Enterococci isolated from food: screening and quantitative determination. *Journal of Food Protection*, 64 (5), 725-729.

Gilliland S. E., Speck M. L. (1977). Deconjugation of bile acids by intestinal lactobacilli. *Applied and Environmental Microbiology*, 33 (1), 15-18.

Grill J. P., Manginot-Durr C., Schneider F., Ballongue J. (1995). Bifidobacteria and probiotic effects: action of *Bifidobacterium* species on conjugated bile salts. *Current Microbiology*, 31 (1), 23-27.

Grill J., Schneider F., Crociani J., Ballongue J. (1995). Purification and Characterization of Conjugated Bile Salt Hydrolase from *Bifidobacterium longum* BB536. *Applied and Environmental Microbiology*, 61 (7), 2577-2582.

Grill J. P., Cayuela C., Antoine J. M., Schneider F. (2000). Isolation and characterization of a *Lactobacillus amylovorus* mutant depleted in conjugated bile salt hydrolase activity: relation between activity and bile salt resistance. *Journal of Applied Microbiology*, 89 (4), 553-563.

Hou C., Zeng X., Yang F., Liu H., Qiao S. (2015). Study and use of the probiotic Lactobacillus reuteri in pigs: a review. *Journal of Animal Science and Biotechnology*, 6 (1), 14.

Jarocki P., Podleśny M., Glibowski P., Targoński Z. (2014). A new insight into the physiological role of bile salt hydrolase among intestinal bacteria from the genus *Bifidobacterium*. *PloS One*, 9 (12), e114379.

Jones P. J. (2002). Clinical nutrition: 7. Functional foods—more than just nutrition. *Canadian Medical Association Journal,* 166 (12), 1555-1563.

Kandell R. L., Bernstein C. (1991). Bile salt/acid induction of DNA damage in bacterial and mammalian cells: implications for colon cancer. *Nutrition and Cancer*, 16 (3-4), 227-238.

Kawamoto K., Horibe I., Uchida K. (1989). Purification and characterization of a new hydrolase for conjugated bile acids, chenodeoxycholyltaurine hydrolase, from Bacteroides vulgatus. *Journal of Biochemistry*, 106 (6), 1049-1053.

Khedkar CD. (1993). Hypocholesterolemic effect of fermented milks: A review. *Cultured Dairy Products Journal,* 28, 14-18.

Kim G. B., Brochet M., Lee B. H. (2005). Cloning and characterization of a bile salt hydrolase (bsh) from *Bifidobacterium adolescentis*. *Biotechnology Letters*, 27 (12), 817-822.

Kim G. B., Miyamoto C. M., Meighen E. A., Lee B. H. (2004). Cloning and characterization of the bile salt hydrolase genes (bsh) from *Bifidobacterium bifidum* strains. *Applied and Environmental Microbiology*, 70 (9), 5603-5612.

Kim G. B., Yi S. H., Lee B. H. (2004). Purification and characterization of three different types of bile salt hydrolases from *Bifidobacterium* strains. *Journal of Dairy Science*, 87 (2), 258-266.

Knarreborg A., Engberg R. M., J ensen S. K., Jensen B. B. (2002). Quantitative determination of bile salt hydrolase activity in bacteria isolated from the small intestine of chickens. *Applied and Environmental Microbiology*, 68 (12), 6425-8.

Korpela J. T., Adlercreutz H, Turunen M. J. (1988). Fecal free and conjugated bile acids and neutral sterols in vegetarians, omnivores, and patients with colorectal cancer. *Scandinavian Journal of Gastroenterology,* 23 (3),277-283.

Kumar M., Nagpal R., Kumar R., Hemalatha R, Verma V., Kumar A., Chakraborty Ch., Singh B., Marotta F., Jain S. (2012). Cholesterol-lowering probiotics as potential biotherapeutics for metabolic diseases. *Experimental Diabetes Research*, 2012, 902917.

Kumar M., Ghosh M., Ganguli A. (2012). Mitogenic response and probiotic characteristics of lactic acid bacteria isolated from indigenously pickled vegetables and fermented beverages. *World Journal of Microbiology and Biotechnology*, 28 (2), 703-711.

Kurdi P., Tanaka H., van Veen H. W, Asano K., Tomita F., Yokota A. (2003). Cholic acid accumulation and its diminution by short-chain fatty acids in bifidobacteria. *Microbiology,* 149 (8), 2031-2037.

Labarthe D. R, Dunbar S. B. (2012). Global cardiovascular health promotion and disease prevention: 2011 and beyond. *Circulation,* 125 (21), 2 667-2676.

Lichtenstein A. H, Goldin B. R. (1993). Lactic acid bacteria and intestinal drug and cholesterol metabolism. *Lactic Acid Bacteria*, 227-235.

Liu Y., Zhao F., Liu J., Wang H., Han X., Zhang Y., Yang Z. (2017). Selection of cholesterol-lowering lactic acid bacteria and its effects on rats fed with high-cholesterol diet. *Current Microbiology*, 74 (5), 623-631.

Low-Beer T. S., Nutter S. (1978). Colonic bacterial activity, biliary cholesterol saturation, and pathogenesis of gallstones. *The Lancet*, 312 (8099), 1063-1065.

Lundeen S. G., Savage D. C. (1992). Multiple forms of bile salt hydrolase from *Lactobacillus* sp. strain 100-100. *Journal of Bacteriology*, 174 (22), 7217-7220.

Lundeen S. G, Savage D. C. (1990). Characterization and purification of bile salt hydrolase from *Lactobacillus* sp. strain 100-100. *Journal of Bacteriology*, 172 (8), 4171-4177.

Mamianetti A., Garrido D., Carducci Clyde N. C., Vescina M. (1999). Fecal bile acid excretion profile in gallstone patients. *Medicina,* 59, 269-273.

Marcus S. N., Heaton K. W. (1986). Intestinal transit, deoxycholic acid and the cholesterol saturation of bile--three inter-related factors. *Gut,* 27 (5), 550-8.

Marteau P., Gerhardt M. F., Myara A., Bouvier E., Trivin F., Rambaud J. C. (1995). Metabolism of bile salts by alimentary bacteria during transit in the human small intestine. *Microbial Ecology in Health and Disease,* 8 (4), 151-157.

McAuliffe O., Cano R. J., Klaenhammer T. R. (2005). Genetic analysis of two bile salt hydrolase activities in *Lactobacillus acidophilus* NCFM. *Applied and Environmental Microbiology*, 71 (8), 4925-4929.

Moser S. A., Savage D. C. (2001). Bile salt hydrolase activity and resistance to toxicity of conjugated bile salts are unrelated properties in lactobacilli. *Applied and Environmental Microbiology,* 67 (8), 3476-3480.

Nagengast F. M., Grubben M. J., van Munster I. P. (1995). Role of bile acids in colorectal carcinogenesis. *European Journal of Cancer*, 31A (7-8), 1067-1070.

Nagpal R., Behare P. V., Kumar M., Mohania D., Yadav M., Jain S., Menon S., Parkash O., Marotta F., Minelli E., Henry C. J., Yadav H. (2012). Milk, milk products, and disease free health: an updated overview. *Critical Reviews in Food Sciences and Nutrition,* 52 (4), 321-333.

Öner Ö., Aslim B., Aydaş S. B. (2014). Mechanisms of cholesterol-lowering effects of lactobacilli and bifidobacteria strains as potential probiotics with their bsh gene analysis. *Journal of Molecular Microbiology and Biotechnology*, 24 (1), 12-18.

Ooi L. G., Liong M. T. (2010). Cholesterol-lowering effects of probiotics and prebiotics: a review of *in vivo* and *in vitro* findings. *International Journal of Molecular Sciences*, 11 (6), 2499-2522.

Patel A. K., Singhania R. R., Pandey A., Chincholkar S. B. (2010). Probiotic bile salt hydrolase: current developments and perspectives. *Applied Biochemistry and Biotechnology,* 162 (1), 166-180.

Pato U., Surono I. S., Koesnandar K., Hosono A. (2004). Hypocholesterolemic effect of indigenous dadih lactic acid bacteria by deconjugation of bile salts. Asian-australasian *Journal of Animal Sciences*, 17 (12), 1741-1745.

Pazzi P., Puviani A. C, Dalla M. L., Guerra G., Ricci D., Gullini S., Ottolenghi C. (1997). Bile salt-induced cytotoxicity and ursodeoxycholate cytoprotection: in-vitro study in perifused rat hepatocytes. *European Journal of Gastroenterology and Hepatology*, 9 (7), 703-709.

Pereira D. I., McCartney A. L., Gibson G. R. (2003). An in vitro study of the probiotic potential of a bile-salt-hydrolyzing *Lactobacillus fermentum* strain, and determination of its cholesterol-lowering properties. *Applied and Environmental Microbiology*, 69 (8), 4743-4752.

Roth G. A., Fihn S. D., Mokdad A. H., Aekplakorn W., Hasegawa T., Lim S. S. (2011). High total serum cholesterol, medication coverage and therapeutic control: an analysis of national health examination survey data from eight countries. *Bulletin of the World Health Organization*, 89, 92-101.

Saravanan C., Shetty P. Kumar H. (2016). Isolation and characterization of exopolysaccharide from *Leuconostoc lactis* KC117496 isolated from idli batter. *International Journal of Biological Macromolecules*, 90, 100-106.

Shehata M. G., El Sohaimy S. A., El-Sahn Malak A., Youssef M. M. (2016). Screening of isolated potential probiotic lactic acid bacteria for cholesterol lowering property and bile salt hydrolase activity. *Annals of Agricultural Sciences*, 61 (1), 65-75.

Shekh S. L., Dave J. M, Vyas B. R. M. (2016). Characterization of *Lactobacillus plantarum* strains for functionality, safety and γ-amino butyric acid production. *LWT-Food Science and Technology,* 74, 234-241.

Sirilun S., Chaiyasut C., Kantachote D., Luxananil P. (2010). Characterisation of non-human origin probiotic *Lactobacillus plantarum* with cholesterol-lowering property. *African Journal of Microbiology Research*, 4 (10), 994-1000.

Smet I. De, Hoorde L. Van, Saeyer N. De, Woestyne M. Vande, Verstraete W. (1994). In vitro study of bile salt hydrolase (BSH) activity of BSH isogenic *Lactobacillus plantarum* 80 strains and estimation of cholesterol lowering through enhanced BSH activity. *Microbial Ecology in Health and Disease*, 7 (6), 315-329.

St-Onge M. P., Farnworth E. R., Jones P. J. (2000). Consumption of fermented and nonfermented dairy products: effects on cholesterol concentrations and metabolism. *American Journal of Clinical Nutrition*, 71 (3), 674-681.

Stellwag E. J., Hylemon P. B. (1976). Purification and characterization of bile salt hydrolase from *Bacteroides fragilis* subsp. *fragilis. Biochimica et Biophysica Acta*, 452 (1), 165-176.

Takahashi T., Morotomi M. (1994). Absence of cholic acid 7α-dehydroxylase activity in the strains of *Lactobacillus* and *Bifidobacterium*. *Journal of Dairy Science*, 77 (11), 3275-3286.

Tanaka H., Doesburg K., Iwasaki T., Mierau I. (1999). Screening of lactic acid bacteria for bile salt hydrolase activity. *Journal of Dairy Science*, 82 (12), 2530-2535.

Tanaka H., Hashiba H., Kok J., Mierau I. (2000). Bile salt hydrolase of *Bifidobacterium longum*-biochemical and genetic characterization. *Applied and Environmental Microbiology*, 66 (6), 2502-2512.

Tannock G. W., Dashkevicz M. P., Feighner S. D. (1989). Lactobacilli and bile salt hydrolase in the murine intestinal tract. *Applied and Environmental Microbiology*, 55 (7), 1848-1851.

Taranto M. P., Medici M., Perdigon G., Ruiz Holgado A. P., Valdez G. F. (1998). Evidence for hypocholesterolemic effect of *Lactobacillus reuteri* in hypercholesterolemic mice. *Journal of Dairy Science*, 81 (9), 2336-2340.

Taranto M. P., De-Llano D., González,Rodriguez A., De-Ruiz Holgado A. P., Font de Valdez G. (1996). Bile tolerance and cholesterol reduction by *Enterococcus faecium*, a candidate microorganism for the use as a dietary adjunct in milk products. *Milchwissenschaft*, 51 (7), 383-385.

Taranto M. P., Fernandez Murga M. L., Lorca G., Font de Valdez G. (2003). Bile salts and cholesterol induce changes in the lipid cell membrane of *Lactobacillus reuteri*. *Journal of Applied Microbiology*, 95 (1), 86-91.

Taranto M. P., Sesma F., de Ruiz Holgado A. P., Font de Valdez G. (1997). Bile salts hydrolase plays a key role on cholesterol removal by *Lactobacillus reuteri*. *Biotechnology Letters*, 19 (9), 845-847.

Tremblay S., Romain G., Roux M., Chen X. L., Brown K., Gibson D. L., Ramanathan S., Menendez A. (2017). Bile Acid Administration Elicits an Intestinal Antimicrobial Program and Reduces the Bacterial Burden in Two Mouse Models of Enteric Infection. *Infection and Immunity*, 85 (6), e00942-16.

Van Eldere J., Celis P., De Pauw G., Lesaffre E., Eyssen H. (1996). Tauroconjugation of cholic acid stimulates 7 alpha-dehydroxylation by fecal bacteria. *Applied and Environmental Microbiology*, 62 (2), 656-61.

Veysey M. J., Thomas L. A., Mallet A. I., Jenkins P. J., Besser G. M., Murphy G. M., Dowling R. H. (2001). Colonic transit influences deoxycholic acid kinetics. *Gastroenterology*, 121 (4), 812-822.

Veysey M. J., Thomas L. A., Mallet A. I., Jenkins P. J., Besser G. M., Wass J. A., Murphy G. M., Dowling R. H. (1999). Prolonged large bowel transit increases serum deoxycholic acid: a risk factor for octreotide induced gallstones. *Gut*, 44 (5), 6756-81.

Walker D. K, Gilliland S. E. (1993). Relationships among bile tolerance, bile salt deconjugation, and assimilation of cholesterol by Lactobacillus acidophilus1. *Journal of Dairy Science*, 76 (4), 956-961.

Xu S., Liu T., Radji C., Akorede I., Yang J., Chen L. (2016). Isolation, identification, and evaluation of new lactic acid bacteria strains with both cellular antioxidant and bile salt hydrolase activities in vitro. *Journal of Food Protection*, 79 (11), 1919-1928.

In: The Many Benefits of Lactic Acid Bacteria
Editors: J. G. LeBlanc and A. de Moreno

ISBN: 978-1-53615-388-0
© 2019 Nova Science Publishers, Inc.

Chapter 12

EXOPOLYSACCHARIDES FROM *LACTOBACILLUS*: BIOSYNTHESIS, CHARACTERIZATION, FUNCTIONAL ASPECTS AND APPLICATIONS IN FOOD INDUSTRY

Elisa C. Ale, Jorge A. Reinheimer and Ana G. Binetti[*]
Instituto de Lactología Industrial (INLAIN; UNL-CONICET),
Santa Fe, Argentina

ABSTRACT

Exopolysaccharides (EPS) are high molecular weight biopolymers produced by a wide range of microorganisms, including lactic acid bacteria (LAB). These polymers can be secreted to the external environment (slime EPS) or adhered to the cell bacterial surface (CPS). Once they are released to the medium, they are able to change its rheology, property that was frequently used by the food industry. Likewise, EPS have been associated with beneficial health effects, making them attractive molecules to be applied as food ingredients.

Lactobacillus is a wide genus of LAB including more than 200 species which has been of great interest because strains with technological, as well as functional (probiotic) properties were described. The present chapter reviews the current knowledge about the role of EPS production in lactobacilli strains, as well the aspects regarding biosynthesis, chemical structure, screening and quantification methods, focusing on the impact of their chemical structure and composition on the possible health benefits and their technological application in novel food products.

[*] Corresponding Author's Email: anabinetti@fiq.unl.edu.ar.

INTRODUCTION

Some lactic acid bacteria (LAB) are able to produce exocellular polymers or exopolysaccharides (EPS) which can be secreted to the medium (slime EPS) or remained attached to the cell wall (capsular polysaccharides or CPS). These molecules are frequently used by the food industry, for example by the application of EPS-producing strains in the manufacture of fermented milk and different types of cheese, improving their textural and organoleptic properties. Furthermore, some EPS are able to reduce syneresis in yogurts due to their water-retention properties and, besides, they can be used as fat replacers in low-fat dairy products, allowing the design of healthy products with acceptable sensory characteristics. Furthermore, EPS were associated with positive health effects, among these are their bifidogenic/prebiotic role, stomach ulcer protection, cholesterol-lowering and immunomodulatory properties. It has also been described that EPS provide protection against different pathogens (virus and bacteria) and present antioxidant and antitumor effects.

Particularly for LAB, the ecological or physiological role of EPS for the producing bacterium remains unclear, and has been associated with a protective role in various niches (biofilm formation) and with carbon catabolism (Schwab et al., 2007; Zannini et al., 2016). In general, it is believed that EPS are not used as an energy source by the producer strain, but there might be some exceptions (Baruah et al., 2018). In some cases, the presence of CPS may protect the producing cell by preventing phagocytosis (Tahoun et al., 2017; Meijerink et al., 2012). A possible hypothesis indicates that some protective roles of these molecules can be associated to the adaptation of the producing bacteria to the stressful conditions of their natural ecosystems such as the gastrointestinal tract and fermented foods (Ruas-Madiedo et al., 2008). EPS have been associated with both rheological and functional effects on the matrix where they are added or synthetized, but the final response strongly depends on their chemical structure and the way of interaction with the medium components. Among LAB, *Lactobacillus* is a wide genus including more than 200 species (Salvetti et al., 2018) which has been of great interest because strains with technological, as well as functional (probiotic) properties were described, frequently associated to the synthesis of EPS.

The aim of this chapter is to review the current state of the art about the chemical structure and genetic organization of EPS in the *Lactobacillus* genus, the methods for EPS isolation and optimization, the potential benefits on human health and the technological applications for the design of novel techno-functional foods, focusing on the structure-function correlation.

CLASSIFICATION, BIOSYNTHESIS AND GENETIC ORGANIZATION

Bacterial EPS are carbohydrate polymers which generally present glucose, fructose, galactose, rhamnose and fucose in their structure (Ruas-Madiedo et al., 2002). In some cases, the presence of different isomers, linkage types and organic and inorganic monosaccharide substituents could be found as well. The oligosaccharide repeating units can present different degrees of polymerisation and branching patterns and are responsible for the great diversity among lactobacilli EPS.

These relatively long-chain and high-molecular-mass polymers can be classified into two main groups: they can be composed of only one type of monosaccharide (homopolysaccharides, HoPS), or can present more than one type of monosaccharide in their structure (heteropolysaccharides, HePS); the mechanisms of biosynthesis are different between them (Cerning, 1990;, 1995). HoPS are formed by the action of enzymes called glucansucrases (or glucosyltransferases, GTF) and fructansucrases (or fructosyltrasnferases, FTF), depending on the composition of the polysaccharide in formation (glucans and fructans, composed exclusively by glucose or fructose, respectively). Both enzymes are glucoside hydrolases that catalyze transglycosylation reactions in the extracellular environment, using sucrose as substrate and obtaining the necessary energy from the molecule breakdown (Figure 1A). The structures of these enzymes are very different, glucansucrases have numerous domains that belong exclusively to LAB, and synthesize α-glucans that differ in their type of glycosidic bounds, degree and type of branches, molecular mass and solubility. On the other hand, fructansucrases are present in Gram positive and Gram negative bacteria, and synthesize β-fructans with β(2,6) (inulin) or β(2,1) linkages (levan) (Hijum et al., 2006). HoPS can be classified in four subgroups according to the monomers and linkages involved: α-D-glucans (dextrans), β-D-glucans, β-fructans (such as levan and inulin), and others like polygalactans. They are produced by several LAB genera, including *Lactobacillus*, *Streptococcus*, *Leuconostoc*, *Oenococcus*, and *Weissella* (Dimopoulou et al., 2017; Hijum et al., 2006). Thus, due to the great diversity of species that form part of the *Lactobacillus* genus, the EPS synthesized by them seem to be very attractive molecules for numerous novel applications. In the case of β-glucans, only one gene is required for their biosynthesis, which encodes a β-glycosyltransferase, a membrane protein whose topological prediction indicates that β-glucan, or at least its repeating unit precursors, are synthesized in the cytosol (Werning et al., 2006; Werning et al., 2012). Besides, *in vitro* recent experiments indicate that the β-glucan is synthesized directly from activated UDP-glucose (Fraunhofer et al., 2018; Werning et al., 2008), similarly to HePS. This could explain the production of a β-glucan with only glucose as the C source for *Lactobacillus* (*Lb.*) *fermentum* Lf2 (a widely studied strain from our group), and not sucrose, as it would be expected for HoPS (Ale 2018; Vitlic et al., 2018). Werning et al., (2012) suggested that the association of several GTF monomers could form a pore for extrusion of the EPS. However, other alternative mechanisms might allow the translocation or secretion of β-glucans across LAB membranes, such as those mediated by ABC-like transporters, related to the export of various bacterial polysaccharides (Silver et al., 2001). In the presence of certain sugar acceptors (like maltose), glycansucrases can form low-molecular-mass oligosaccharides from sucrose (Korakli & Vogel, 2006). Their synthesis occurs via successive transfer of glucosyl or fructosyl units to the acceptor rather than the glucan or fructan polymer, with the consequent formation of glucooligosaccharides (GOS) or fructooligosaccharides (FOS), respectively (Lynch et al., 2018).

On the other hand, HePS are molecules composed by repetitive units which present mainly D-glucose, D-galactose and L-rhamnose, and are formed by 2 to 8 monomers. Other monosaccharides, such as fructose, fucose, mannose, N-acetylglucosides, and glucuronic acid, as well as noncarbohydrate moieties (phosphate or acetyl groups) can be present as well (De Vuyst & Degeest, 1999a; Werning et al., 2012), conferring them particular characteristics. HePS are produced by the polymerization of sugar-nucleotide precursors (e.g., UDP-glucose and UDP-galactose) provided by the cell, and the biosynthesis is controlled by several glycosyltransferase genes that are involved in the regulation, chain-length determination, polymerization and export of the repeating units. These genes are generally located together in an operon and constitute an EPS biosynthesis cluster (Werning et al., 2012). All the details of the mechanism are not yet fully understood, but a flippase enzyme may be involved in the transfer of the lipid-bound polymer to the external side of the membrane, where the polymerization of the repeating units takes place (Lebeer et al., 2009). Figure 1B shows the main steps for the synthesis of HePS in LAB. Briefly, activated sugar nucleotides bind to a phosphorylated lipid carrier (such as C55- polyprenyl phosphate) located on the cytoplasmic side of the cell wall by the enzymatic activity of a priming glycosyltransferase associated to the membrane. After this, a series of additional glycosyltransferases catalyze the synthesis of the repetitive unit of the EPS, which is then transported to the outer surface of the bacterial membrane by a flippase (Flippase-like Transmembrane Lipid Transporter, FTLT, also identified as Wzx/Wzy-dependent pathway, according to the nomenclature described for *Escherichi* (*E.*) *coli*; Schmid 2018). Once it is outside, the repetitive units are polymerized and the chain length seems to be regulated by a phosphatase that hydrolyzes the lipid diphosphate anchor of the membrane. These genetic clusters are located mainly in the bacterial chromosome, as it was described for *Streptococcus (Strep.) thermophilus*, *Lb. plantarum* or *Lb. fermentum* (De Vuyst & Degeest, 1999b; Siezen et al., 2010; Stingele et al., 1996; Dan et al., 2009). As it was mentioned above, the genes involved in the HePS synthesis are usually located in operons, with all of them oriented in the same direction, allowing the transcription of a single mRNA molecule (Jolly & Stingele, 2001; Lebeer, Vanderleyden & Keersmaecker, 2008) and are highly conserved among lactobacilli (Lebeer, Vanderleyden & Keersmaecker, 2008). Until now, the genes involved in the regulation of HePS synthesis were detected in lactobacilli but not in bifidobacterial *eps* clusters (Jiang and Yang 2018). A conserved "consensus" *eps* cluster structure for *Lactobacillus* (among other LAB) was proposed starting (from 5' to 3') with genes responsible for the transcriptional regulation and chain-length determination, followed by a core of GTF and ending with genes for polymerisation/export, all flanked by transposases (and derivatives) or insertion sequences (Figure 2) (Castro Bravo et al., 2018).

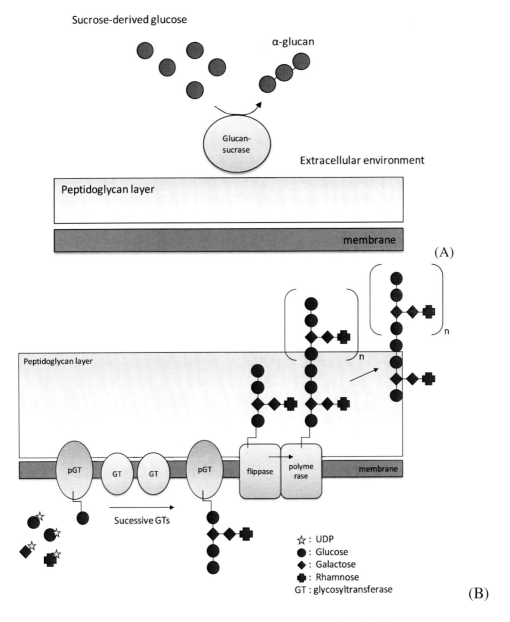

Figure 1. EPS-biosynthesis pathways proposed for homopolysaccharides (HoPS, A) and heteropolysaccharides (HePS, B).

Figure 2. Organization described for the HePS gene clusters.

A typical characteristic found in *eps* clusters is the presence of mobile genetic elements in the DNA region involved in HePS or HoPS synthesis (Péant et al., 2005; Tieking et al., 2005; Hijum 2004). These elements allow horizontal transfer between different genera and their presence could justify the instability of the HePS- producer phenotype of some strains (Ruas-Madiedo et al., 2009). The *gtf* genes (coding for β-glycosyltransferases) of LAB are usually flanked by genes which could be involved in functions of conjugation and recombination. Thus, horizontal transfer mediated by plasmids, or transposition events, might explain the wide distribution and high degree of *gtf* gene preservation in β- glucan-producing strains along different genera (Werning et al., 2006).

METHODS FOR ISOLATION AND PURIFICATION OF EPS AND CPS

The total yield of EPS produced by the LAB strongly depends on the composition of the medium (C and N sources) and the conditions in which the strains develop, such as temperature, pH, oxygen tension, and incubation time (Degeest, Vaningelgem & De Vuyst; 2001). It has been described that the adjustment of the fermentation conditions can lead to higher EPS levels (Ale et al., 2016b; Zisu &Shah, 2003; Petry et al., 2000). For example, temperature modifications presented varying effects on EPS production (De Vuyst & Degeest, 1999b). It has been shown that lower temperatures are better for HoPS synthesis, as glycansucrases have a lower optimum temperature range (Kang, Oh & Kim, 2009). In a recent study by our group, the statistical optimization of the EPS production by *Lb. fermentum* Lf2 was achieved, doubling the yield obtained under non-optimized conditions (1 to 2 g/L, approximately). In this case, sucrose was more effective than glucose as the C source, and the proportions of the three nitrogen sources of the semidefined medium applied were critical (SDM, Kimmel & Roberts, 1998), as well as the pH and time of fermentation (Ale, 2018).

When quantification of the EPS yield is approached, it is important to highlight that the presence of glucomannans (originating from yeast extract and peptone, both included in general media for lactobacilli, like MRS) can co-precipitate with EPS. For this reason, the use of semi-defined media for EPS quantification and purification is recommended (Ruas-Madiedo & de los Reyes-Gavilán, 2005).

In general, the molecular mass of HoPS is greater than 10^3 kDa (Ruas-Madiedo et al., 2009) and are produced in amounts higher than HePS, in the range of g/L, reaching values of up to 10 g/L (Lynch et al., 2018). HePS are synthesized in low amounts, in the range of mg/L, and their molecular mass ranges from 40 to 9×10^3 kDa (Ruas-Madiedo et al., 2009).

Several screening strategies have been applied to detect EPS-producing bacteria. For example, Vedamuthu & Neville (1986) used 1.0 mL pipettes to test the ropiness of acidified reconstituted skimmed milk, assessing mucoidness by the stringiness of the free flowing gel. This method was also described by Mozzi, Torino & de Valdez (2004), defining ropy as strings longer than 6 mm. A variant to the ropy-screen method employs an inoculation wire loop to assess string formation upon touching a colony. In this case, when the string was longer than 5 mm, the strain was considered ropy (Dierksen, Sandine & Trempy, 1997). Although these techniques are easy to perform, they are difficult to standardize since the results obtained are qualitative rather than quantitative. A screening approach using microhaematocrit capillaries to measure efflux, a parameter that correlates with viscosity measurements was also described (Ricciardi, Parente & Clementi, 1997). In general, visualization of slimy or mucoid colonies on solidified media is probably the most frequently used screening method for EPS production, with the disadvantage of only detecting strains with an evident ropy phenotype.

Staining methods are also used for detecting EPS-producing cells, for example by the use of ruthenium red, neutral red, calcofluor white, congo red or Indian ink. Ruthenium red was used to detect *Strep. thermophilus* mutants with loss of ropy phenotype, which appeared as red colonies on milk agar media, while ropy wild-type strains appeared white (Stingele et al., 1996). Although staining assays are relatively easy to perform, the visual interpretation can be difficult when small differences between positive and negative signals occur (Zeidan et al., 2017).

All the screening approaches mentioned above rely on the presence of a physical change in the environment that can be observed by visual inspection. A more direct approach involves gravimetric quantification of the EPS produced. Depending on the medium, certain pre-treatment steps are required in order to obtain the EPS extract with the minor interferences as possible. For example, heating steps are used to inactivate endogenous enzymes that could degrade the polysaccharides (Shao et al., 2014). In general, for broths like MRS, centrifugation is typically carried out to separate the cells, and then the EPS are precipitated from the supernatant with chilled absolute ethanol (Juvonen et al., 2015; Dilna et al., 2015; Hermann, Petermeier & Vogel, 2015; Zarour et al., 2017). Briefly, the protocol applied could be summarized as follows: (i) cell removal by centrifugation or filtration; (ii) EPS precipitation from the supernatant by the addition of chilled absolute ethanol or acetone (2-3 volumes); (iii) dialysis to remove low molecular mass contaminants (12-14 kDa cut off); and (iv) freeze-drying of the precipitated polysaccharide. In some cases, a second precipitation and dialysis step is carried out (Torino, de Valdez & Mozzi, 2015; Ruas-Madiedo & de los Reyes-Gavilán, 2005). Polak-Berecka et al., (2013) found that a heat treatment at first step is critical to inactivate EPS or CPS-degrading enzymes. Besides, they do not recommend too high centrifugal rates (> 10,000 g) to discard the cell pellet, since part of the EPS could be eliminated as well.

Further purification could include more sophisticated steps such as membrane-filtration, anion-exchange and/or gel permeation chromatography (Leivers et al., 2011; Sanz & Martínez-Castro, 2007; Shao et al., 2015).

The quantification of EPS yield may be gravimetric (expressing the yield as mg/L) or colorimetric, by the phenol-sulphuric acid method (Dubois et al., 1956). In this last case, simple sugars, oligosaccharides, polysaccharides and their derivatives (including the methyl ethers) give an orange-yellow colour when they are treated with phenol and sulphuric acid.

When the EPS is isolated from high protein content media, such as dairy products, proteins are usually precipitated with trichloroacetic (TCA) acid, or removed by hydrolysis with proteases or by a combination of both (Torino, de Valdez & Mozzi, 2015), followed by the alcohol precipitation, dialysis and freeze-drying steps.

Regarding CPS, they can be visualized by microscopy after negative staining with Indian ink (Mozzi et al., 1995), a technique that evidences a bright white layer surrounding the cells when CPS are present. For the same purpose, Malang et al., (2015) used crystal violet to stain the cells blue to facilitate the differentiation of CPS in the dark background. To isolate the CPS, the cell pellet can be washed with PBS buffer and, after resuspending the cells in NaCl (1 M), a sonication step is applied to detach the CPS from the cells. Then samples are centrifuged to eliminate insoluble material and the supernatant is precipitated with chilled ethanol as described previously for EPS (Fraunhofer et al., 2018; Tallon, Bressollier & Urdaci, 2003; Toba, Kotani & Adachi, 1991). An alternative method with ethylenediamine tetraacetic acid (EDTA) has also been described (Tallon, Bressollier & Urdaci, 2003); in this case, the cell pellets are resuspended in 0.05 M EDTA. Then, the mixture is incubated under gentle agitation for 4 h at 4 °C and centrifuged and a precipitation step with ethanol is done as described above. The efficiency of this last technique will depend on the nature of the bonds between CPS and the cell wall. In another work (Mende et al., 2013), CPS were detached from the harvested cells suspended in water by applying a heating step (90 °C, 15 min), cells were then removed by centrifugation and the detached CPS was isolated from the supernatant.

CHEMICAL AND STRUCTURAL CHARACTERIZATION OF EPS

In relation to the chemical characterization of EPS, the polymer molecular weight (MW) can be measured by HPSEC-RI (High Performance Size Exclusion

Chromatography- Refractive Index) and an increased use of SEC-MALLS (Size Exclusion Chomatography - Multi-Angle Laser Light Scattering) was observed lately (Picton, Bataille & Muller, 2000; Leivers et al., 2011; Patten et al., 2014). There are other complex techniques that can also be used to determine the average MW, such as Field Flow Fractionation (FFF) and Hydrodynamic Chromatography (HDC) (Isenberg et al., 2010; Galle et al., 2011).

The EPS monomer composition can be determined by total acid hydrolysis (sulphuric or trifluoroacetic acid at 120 °C) followed by detection using high performance anion exchange chromatography (Ale et al., 2016b; Fraunhofer et al., 2018). Additional information about the structural characteristics can be studied by ^{1}H and ^{13}C NMR spectroscopy (Leivers et al., 2011; Balzaretti et al., 2017). This method provides information of the type of monosaccharides, ring size, anomeric configuration and positions of the glycosidic linkages (Torino, de Valdez and Mozzi 2015). In fact, more than fifty different NMR-EPS repeating unit structures have been described for lactobacilli and bifidobacteria (Castro-Bravo et al., 2018), representing in some cases unique structures (Ruas-Madiedo, 2005). In Table 1, some characteristics of EPS from lactobacilli were summarized, together with their potential food applications. For example, in a recent study of our group (Ale, 2018; Vitlic et al., 2018), the structure of two EPS produced by *L. fermentum* Lf2 was elucidated by NMR. The two EPS were separated by preparative size exclusion chromatography to give a high molecular mass β-glucan (1.23 x 10³ kDa) and a medium molecular mass heteroglycan (88 kDa), the last composed of glucose and galactose. Their structures were determined using a combination of NMR spectroscopy, monomer and linkage analysis.

Recently, a high-throughput screening method was described for the determination of EPS concentration and monosaccharide composition (Rühmann, Schmid, and Sieber 2016). The method consists of the following steps: cultivation, cell removal by centrifugation and filtration, sugar monomer removal by gel filtration, EPS hydrolysis with trifluoroacetic acid and derivatization with 1-phenyl-3-methyl-5- pyrazolone (PMP) and monomer detection by UHPLC-UV-ESI-MS (Ultra High Performance Liquid Chromatography with Ultra Violet and Electrospray Ionization Ion Trap Mass Spectrometer).

Molecular screening approaches have also been applied to the search for new EPS-producing isolates. The strategy comprises the design of degenerate primers based on LAB gene sequences followed by PCR reaction and sequencing of PCR products. This approach has been mainly applied to identify glucan and fructan-producing strains (Krajl et al., 2003; Malik et al., 2009). With genome sequencing, it is currently easier to study EPS-related gene sequences and the structure of the clusters involved in EPS synthesis. However, this technique is not suitable for arising to conclusions regarding yield and type of polysaccharides, or even rheological properties when applied in different food matrices.

Table 1. Lactobacilli EPS-producing strains, chemical characteristics and potential applications

Strain	EPS	Application	Reference
Lb. sanfranciscensis LTH2590 (= TMW 1.392)	β(2,6) levan	Improvement of bread texture and dough rheology	Brandt et al., 2003
Lb. buchneri FUA3154	Galactose, glucose, one unidentified; MW 10-10^2 kDa	Wheat and sorghum sourdough rheology	Galle et al., 2011
Lb. plantarum BR2	Glucomannan; MW 2.38×10^3 kDa	Antioxidant and lowering cholesterol activities, antidiabetic potential	Sasikumar et al., 2017
Lb. kefiranofaciens 1P3 *Lb. satsumensis* 10P and 10P2	Glucans, most of the residues linked by α-(1,6) glycosidic bonds (~90%); estimated MW 8×10^2 kDa	Increase in IgA^+ B cells, suitable for food ingredient	Paiva et al., 2016
Lb. paracasei CIDCA 83123 *Lb. paracasei* CIDCA 8339 *Lb. paracasei* CIDCA 83120 *Lb. paracasei* CIDCA 83121 *Lb. paracasei* CIDCA 83124	Main component: a fraction of MW 10- 10^2 kDa, but a high MW fraction was also observed: EPS produced by *Lb. paracasei* CIDCA 8339 and *Lb. paracasei* CIDCA 83124 had a MW fraction of 10^2 Da–10^3 kDa. EPS produced by *Lb. paracasei* CIDCA 83120, *Lb. paracasei* CIDCA 83121 and *Lb. paracasei* CIDCA 83123 had a fraction with MW > 10^3 kDa	Interesting rheological properties in fermented milk, which would be useful for food application	Hamet, Piermaria & Abraham, 2015
Lb. mucosae DPC 6426	Mannosyl residues, with mannose, glucose and galactose found to be the major sugar residues present in an approximate ratio of 3:2:2	Decrease in syneresis in low-fat yogurt and viscosity increased for the EPS-rich yogurts; improved sensory properties of Swiss-type cheese	London et al., 2014; 2015; Ryan et al., 2015
Lb. fermentum Lf2	Glucose and galactose 2:1, composed of a high MW fraction (>1×10^3 kDa) and a low MW fraction (90 kDa)	Increase of the hardness and consistency index of low-fat yogurts, with no sensory defects	Ale et al., 2016b; Vitlic et al., 2018
Lb. suebicus CUPV225 *Lb. suebicus* CUPV226	Glucose, galactose, N-acetylglucosamine and phosphate (in different ratios), mixture of polysaccharides; two fractions: 2×10^4 and 1.5-7.8×10^6 g/mol, depending on the C source	Potential immunological and technological properties	Ibarburu et al., 2015
Lb. paracasei 20408	EPS with unknown structure	EPS^+ mutant showed a positive impact on the textural, melt, proteolysis, and sensory characteristics of half-fat Cheddar cheese, useful as fat replacer	Zhang et al., 2015

POTENTIAL APPLICATIONS OF EPS FROM THE GENUS *LACTOBACILLUS* IN THE FOOD INDUSTRY

EPS are attractive molecules with great potential to be used as ingredients in food matrices since they are natural and safe components when they are produced by GRAS (General Recognized as Safe) microorganisms, such as lactobacilli. Although they have low yields, their production *in situ* is suggested as an option for the food industry, for example by the application of EPS producing adjunct or starter cultures (Hassan, Awad & Muthukumarappan, 2005; Ryan et al., 2015; Han et al., 2016; Amatayakul, Sherkat & Shah, 2006b; Hassan et al., 2003; Yilmaz et al., 2015). EPS, mainly HePS, have been widely used in the dairy industry, since they are able to improve certain attributes due to their thickening properties. Besides, they have been related to syneresis reduction and firmness increase (Ale, et al., 2016b; Dabour et al., 2006; Zisu & Shah, 2005).

For the design of novel foods with improved technological and functional characteristics, it is crucial to find new EPS-producing strains able to impart the expected characteristics to a specific food matrix. In this sense, an important number of EPS-producing *Lactobacillus* was isolated from diverse food matrices, mainly those in which a spontaneous fermentation was implicated as part of processes (Table 1). For example, it was described that certain polymers produced by lactobacilli could beneficially affect the water absorption of dough, dough rheology and machinability, its stability during frozen storage, loaf volume and bread staling (Tieking & Gänzle, 2005). Besides, the EPS produced from sucrose by *Lb. sanfranciscensis* LTH2590 presented the potential to affect the rheological properties of the dough, as well as the texture, volume, and keepability of bread (Korakli et al., 2001). In general, EPS isolated from sourdough include mainly HoPS; however, Meulen et al., (2007) described the production of HePS from a sourdough isolate, a strain of *Lb. curvatus*, whose EPS is composed of galactosamine, galactose, and glucose (in relation 2:3:1, respectively). Galle et al., (2011) studied the HePS-producing strains *Lb. casei* FUA3185, *Lb. casei* FUA3186, and *Lb. buchneri* FUA3154, the last was a ropy strain isolated from bread drink (a German fermented grain beverage), regarding their impact on wheat and sorghum sourdough rheology. *Lb. buchneri* FUA3154 showed the highest viscosity at low shear rate when it was grown in MRS. Despite the fact that glycosyltransferase genes responsible for HePS synthesis from the three strains were expressed in sorghum and wheat sourdough, only HePS from b*L. buchneri* impacted on the rheological properties of sorghum sourdoughs, decreasing resistance to deformation and elasticity.

Regarding the culture conditions that favor the application of EPS-producing strains, Bengoa et al., (2018) analyzed the effect of growth temperature on EPS production of different *L. paracasei* strains isolated from kefir grains and showed that suboptimal temperatures enhanced the EPS yield and polimerization, but their probiotic properties (adhesion to Caco-2 cells and modulation of the epithelial innate immune response) were not affected by this parameter.

Paiva et al., (2016) characterized the EPS produced by novel *Lactobacillus* strains from Brazilian milk or sugar water kefir when they are grown with sucrose as C source. *Lb. kefiranofaciens* 1P3 and *Lb. satsumensis* 10P and 10P2 synthesized EPS composed of glucose, and most of the glucose residues were linked by α-(1,6) glycosidic bonds (~90%) and the estimated molecular weight was 800 kDa. Since the intestinal IgA⁺ B cells were significantly higher in mice that received the EPS, the authors suggested that these EPS might be useful as food additives.

In this direction, Hamet, Piermaria& Abraham (2015) studied the effect of 28 lactobacilli on the texture of acid milk gels and observed that those gels obtained from the fermentation with all the *Lb. paracasei* strains (5 in total) were the ones that presented the highest viscosities, probably due to the presence of a high MW fraction. Nevertheless, the viscoelastic characteristics of the resulting acid gels were different; one of the strains produced a viscous behavior whereas the others presented gel structures.

Another species studied regarding its EPS production was *Lb. fermentum* (Zhang et al., 2011; Leo et al., 2007; Shi et al., 2014). In a recent study of our group (Ale et al., 2016b), the EPS extract from *Lb. fermentum* Lf2 (a strain isolated from an Argentinian cheese with blowing defects) and its impact on the technological properties when applied to yogurts as a natural ingredient, were analyzed. Yogurts with EPS showed higher hardness values than the controls and this difference remained stable during time. The consistency index was also higher for the treated samples at both times evaluated (3 and 25 days), being significantly different for samples with 300 mg/L of EPS extract, while the flow behavior index was lower for EPS-added yogurts. The thixotropic index was lower for samples with the highest EPS extract concentration at the end of the storage time. Regarding the sensory analysis, no defects were detectable, making it suitable for food applications. Syneresis was also decreased for samples with 600 mg/L of EPS extract at 25 days of storage. These results showed the potential of this extract to be used as a technofunctional natural ingredient, also considering its positive impact on health according to previous studies (Ale et al., 2016a). From another report, sixteen EPS-producing *Lb. fermentum* strains isolated from West African fermented millet dough presented desirable technological and probiotic characteristics *in vitro* (Owusu-Kwarteng et al., 2015). The authors suggested that these strains could function as potential probiotic cultures.

London et al., (2015) applied an EPS-producing strain of b*L. mucosae* as adjunct culture in low-fat yoghurt manufacture. The treated yogurts presented a significant decrease in syneresis and increased viscosity as a consequence of the EPS produced *in situ* during fermentation.

Diverse works described the specific interactions among these polymers and the food matrix in which they are employed. In yogurts, it has been shown by confocal scanning laser microscopy that caseins form a gel and polysaccharides are located in pores containing whey. It is probable that the chemical nature of this EPS was responsible for its incompatibility with proteins aggregates (Hassan et al., 2003). In another study, Folkenberg et al., (2005) described two different sensory profiles which depended on the protein-EPS interactions. They observed two types of microstructures: in one of them EPS were associated with the proteins surrounding the pores, while in the other, incompatibility between the EPS and the protein network existed, causing EPS depletion. In this case, the polymers were found inside the pores. Yogurts with the first type of microstructure presented high ropiness, low serum separation and were more resistant to stirring, and yogurts with the second type of microstructure were less ropy, had high serum separation and showed increased mouth thickness as a result of stirring.

In cheeses, HePS produced by lactobacilli strains such as *Lb. delbrueckii* subsp. *bulgaricus*, *L. helveticus* and *L. casei* promoted water retention and positively modified the texture, avoiding alteration of the structure (Zisu & Shah, 2007). This modification of the textural characteristics often permits the reduction of calories in the final product, for example in low-fat cheeses (Oberg et al., 2015; Zisu & Shah, 2005; Zhang et al., 2016). Accordingly, EPS-producing lactobacilli or EPS (as ingredients) were described as interesting fat replacers to improve the rheological parameters of the final product (Broadbent et al., 2001; Perry, McMahon & Oberg, 1998; Ryan et al., 2015). In this direction, Zhang et al., (2015) studied the yield, textural, proteolysis, melting, and sensory properties of half-fat Cheddar cheese including an EPS-producing *Lb. paracasei* strain. The results revealed that cheeses made with the high yield EPS-producing strain, in combination with the commercial Cheddar culture, presented increased moisture content and yield compared with the half-fat controls. The melting properties were significantly improved as well, and the sensory characteristics were similar to the full-fat controls.

From non-dairy industries, the EPS production by two b*L. suebicus* strains (isolated as responsible of an alteration from ropy ciders) was examined in a semidefined medium (Ibarburu et al., 2015). Both strains were able to use glucose, ribose and xylose as C sources, and synthesize a mixture of complex HePS.

In the light of this information, it is important to remark the relevance of the isolation and characterization of new EPS- producing strains, mainly from artisanal or spontaneously fermented products (even when they were present in a food matrix as non-desirable organisms), representing the principal source of new cultures and ingredients (such as EPS), which are essential to diversify the market of techno-functional food.

Functional Aspects of EPS From the Genus Lactobacillus

The presence of *Lactobacillus* within the intestinal microbiota has been linked to a healthy status of the host and, in general, low levels have been associated with specific health problems as IBS (Irritable Bowel Syndrome; Liu et al., 2017). Although lactobacilli represent just 0.01% of total faecal bacteria (Walter 2008), it seems that this genus plays an important role in maintaining the normal intestinal homeostasis, knowing that its addition as a probiotic supplement could help balance an aberrant intestinal microbiota (Staudacher et al., 2017). Probiotics are "live microorganisms which, when administered in adequate amounts, confer a health benefit on the host" (Hill et al., 2014). Some strains with scientifically demonstrated probiotic properties were included in diverse functional foods and, although the mechanisms by which they exert their action are uncertain, it has been suggested that the health-promoting effects of EPS-producing strains are related to the biological activities of these biopolymers (Figure 3). In fact, diverse scientific reports indicate that EPS from diverse *Lactobacillus* species can contribute to human health by means of numerous demonstrated effects: prebiotic role, modulation of the immune system, antioxidant, antitumor, antiulcer or cholesterol-lowering activities, etc. (Ruas-Madiedo, Hugenholtz & Zoon, 2002). However, since not all EPS can improve the technological properties of fermented foods (Hassan, 2008), not all are able to promote health benefits (Amrouche et al., 2006), being their chemical structure and MW relevant characters to define their health-promoting function (Hidalgo Cantabrana et al., 2012). The topics mostly supported by an adequate amount of scientific evidence refer to the ability of EPS to act as immunomodulators and prebiotic compounds and will be discussed in independent sections of this chapter.

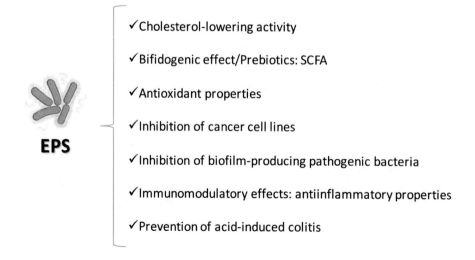

Figure 3. Health-promoting effects related to EPS produced by lactobacilli.Immunomodulatory properties of EPS.

Besides the physiological role of EPS for the producing bacteria and their relevance for industrial applications, in the last decade there has been increasing evidence about the impact on the human health bias the interaction of the EPS-producing lactobacilli and the host by modulating the immune response.

Notably, carbohydrates were considered as traditional T-cell-independent antigens that do not elicit cell-mediated responses and fail to induce immunologic memory and immunoglobulin class-switching. However, in the last decades, certain microbial polysaccharides have been described as potent immunomodulators with specific activity against both T-cells and antigen-presenting cells such as monocytes and macrophages (Cobb & Kasper, 2005).

Several works in the literature reported the ability of EPS from lactobacilli to elicit immune responses and some authors linked the intrinsic physicochemical characteristics of these molecules with their functional properties (López, 2012), indicating two different types of behaviors or patterns. The first pattern associates functional properties with the presence of acidic substituents as phosphate groups (conferring negative charges) in the molecule, which seem to induce the immune response compared with neutral polymers. For example, the EPS from b*L. delbrueckii* subsp. *bulgaricus* OLL-1073-R1 is composed of two fractions of glucose and galactose: the acidic one (containing phosphate) showed a strong immune profile, inducing the proliferation and activity of macrophages; while the neutral fraction evidenced the opposite behavior, being unable to elicit stimulation (Kitazawa et al., 1998; Nishimura-Uemura et al., 2003; Uemura et al., 1998). The second proposed pattern links the immune response with the MW of the polymer. It has been postulated that EPS having high MW (HMW, > 10^3 kDa) could act as strong suppressors of the immune response (Hidalgo Cantabrana et al., 2012), as occurred with the EPS from *Lb. casei* Shirota, a HMW polysaccharide able to induce (by cell culture) the production of diverse cytokines (TNF-α, IL-12, IL-10 and IL-6) in lower extent that EPS$^-$ knockout-mutants (Yasuda, Serata & Tomoyuki, 2008). Bleau et al., (2010) showed that murine peritoneal macrophages stimulated with the HMW-EPS-producing *Lb. rhamnosus* RW-9595M, an isogenic mucous variant of the strain ATCC9595, induced lower levels of TNFα and IL-6 than the parental strain, which also evidenced a decrease in IL-10 production. Similarly, Nicolik et al., (2012) reported a suppressive immune response when the HMW-EPS-producing *Lb. paraplantarum* BGCG11 was used for inducing human peripheral blood mononuclear cells (PBMC), in comparison with its novobiocin EPS$^-$ mutants. In a study from our group, the EPS from *Lb. fermentum* Lf2 showed a mild suppressive profile when administered to mice, increasing s-IgA levels and reducing the proinflammatory cytokine IL-6 in the small intestine when added in a dairy matrix (Ale et al., 2016a). In this case, based on preliminary evidence indicating a relatively low MW (< 100 kDa in average) and the absence of charged substituents, the observed mild stimulation of immune cells when administered to mice could obey the patterns previously proposed by Bleau et al., (2010) and López et al., (2012). In these cases, a

HMW-EPS could be purified from the crude extract, consisting of a β-glucan (1.23×10^3 kDa) which was able to inhibit the rises of TNF-α levels in peripheral blood mononuclear cells pre-treated with lipopolysaccharide (LPS) or phytohemagglutinin (PHA). The results indicate a role of this polysaccharide in modulating the immune response after exposure to agonists such as bacterial LPS.

The cytokine profiles produced by EPS are not always clear. For example, EPS from *Lb. rhamnosus* KL37 and *Lb. reuteri* 115 were able to stimulate the simultaneous secretion of both anti-inflammatory (IL-10) and proinflammatory (TNF-α and IL-6) cytokines from peritoneal mouse macrophages (Ciszek-Lenda et al., 2011). Vinderola et al., (2006) demonstrated that the oral administration of a high dose of kefiran, an EPS produced by *Lb. kefiranofasciens*, showed an immunomodulation (mediated by IgA, Il-2, IL-6, IL-10, TNF-α) at the level of large intestine and blood, suggesting a potential role in intestinal homeostasis. Additionally, another study showed that the systemic administration of *Lb. rhamnosus* KL37C EPS stimulated a significant decrease in the serum concentration of type II collagen-specific IgG antibodies in a murine model of mild rheumatoid arthritis, consequently improved the condition (Nowak et al., 2012).

The role of polysaccharides in the balance of the immune status associated with inflammatory disorders caused by infections has been also reported. One of the traditional phenotypic features tested during *in vitro* probiotic characterization is the ability to coaggregate with pathogens, since it is believed that co-aggregation decreases the accessibility of the harmful microorganisms to the intestinal epithelium (Castro-Bravo et al., 2018). In this sense, the activities against *Candida* of *Lb. rhamnosus* GG, its EPS deficient mutant, and the purified EPS were analyzed, showing that the EPS layer might be involved in the reduction of hyphal formation and in the competition with *Candida* during adhesion to vaginal epithelial cells (Allonsius et al., 2017). Another study with a *Lb. paracasei* subsp. *paracasei* BGSJ2-8 EPS-deficient mutant, producing an EPS with a different composition, indicated that the wild-type EPS is essential to reduce *Escherichia* (*E.*). *coli* association with Caco-2 cells, while the mutant strain was not able to counteract this interaction (Živković et al., 2016). This highlights the importance of EPS composition and not only EPS production in the potential interaction with pathogens. Also, EPS isolated from *Lb. delbrueckii* subsp. *delbrueckii* were able to significantly down-regulate the expression of enterotoxic *E. coli*- induced proinflammatory cytokines (IL-6, IL-8 and MCP-1) in porcine intestinal epithelial cells (Wachi et al., 2014). Regarding *in vivo* experiments, different animal models have been used to study how EPS-producing strains interfere with pathogen activity. For instance, *Lb. johnsonii* FI9785, a poultry derived isolate producing two different EPS, was effective to suppress the colonization and persistence of *Clostridium* (*Cl.*) *perfringens* in poultry and to reduce colonization by *E. coli* in the small intestine (Dertli et al., 2013). In case of the crude EPS from *Lb. fermentum* Lf2, it showed a mild protection against *Salmonella* infection in a survival murine model when suspended in milk, in a dose feasible to be uses as a food ingredient. In this same study, the survival proportion for the treated group (31%) resulted significantly higher than that of the control group (9%), thus representing another mode of action of this EPS to elicit positive health benefits (Ale et al., 2016a). This property could be related with the increased s-IgA levels in the intestinal fluid of mice also observed for this extract when used as an ingredient in yogurts. Similar survival rates were observed by Nagai et al., (2011) when evaluating the effect of an EPS from a strain of *Lb. delbrueckii* subsp. *bulgaricus* on the influenza virus infection after 21 days of administration of yogurts. Although they evaluated the capacity of EPS against a virus, the survival of mice ranged from 0% for the control groups to 38.9% for diverse EPS-fractions.

The molecular mechanisms by which lactobacilli interact with immune cells have not been entirely elucidated. The highly multifunctional Toll-like receptor (TLR) family, which are able to recognize a vast range of extracellular bacterial products (Patten & Collett, 2013; Patten et al., 2014), could mediate these interactions. Particularly, TLR-2 is known to interact with a diverse range of microbial components (Akira, Uematsu & Takeuchi, 2006; Takeda, Kiasho & Akira, 2003). The LPS receptor, TLR-4 (Hoshino et al., 1999), is also a promising candidate, and a response in which a combination of TLR-2 and TLR-4 takes place was also suggested (Wachi et al., 2014).

It has been previously proposed that bacterial EPS could exert their immunomodulatory action via the mannose receptor (Chabot et al., 2001). However, the feasibility of this hypothesis is questionable due to the structural heterogeneity of microbial EPS. Taking into account the immune parameters and the diverse responses according to their physicochemical characteristics (high or low MW, presence or absence of charged substituents and even, ramifications), EPS from lactobacilli with an evident positive health effect could be included in the diet of patients suffering diseases associated with an increased inflammatory status, such as allergy and inflammatory bowel disease, or other auto-immune disorders, as well as in functional foods formulated for the elder population.

EPS as Prebiotics

Prebiotics are non-viable substrates that function as nutrients for beneficial microorganisms harbored by the host (probiotic strains and indigenous microorganisms). Thus, a prebiotic should favor a metabolism biased towards health-promoting microorganisms within the indigenous ecosystem. According to the consensus statement by ISSAP (International Scientific Association for Probiotics and Prebiotics), the definition of a prebiotic has been recently modified to "a substrate that is selectively utilized by host microorganisms providing a health benefit" (Gibson et al., 2017). Health effects of these substances include benefits not only to the gastrointestinal tract (inhibition of pathogens and immune stimulation), but also to the cardiometabolism (reduction in blood lipid levels and effects upon insulin resistance), mental health (metabolites that influence brain function, energy and cognition) and bone (mineral bioavailability), among others (Gibson et al., 2017; Collins & Reid, 2016). In addition, the term synbiotics refers to a combination of both, probiotics and prebiotics, and a synergic interaction among them (Cencic & Chingwaru, 2010).

The catabolism of prebiotic carbohydrates by the metabolic activity of intestinal bacteria produces mainly three SCFA (Short Chain Fatty Acids): acetate, propionate and butyrate. Thus, the prebiotic substrates can induce increased values of the levels of these acids in healthy individuals as end products of human colonic bacteria metabolism (Lecerf et al., 2012). SCFA contribute to the normal large bowel function and prevent pathology through their actions in the lumen and on the colonic musculature and vasculature, and through their metabolism by colonocytes. The most abundant SCFA in the colon is acetate and represents, in general, more than half of SCFA detected in faeces (Louis et al., 2007). It has been shown that bifidobacteria can protect the host against enteropathogenic infections through the production of this acid (Fukuda et al., 2011). In addition, this acid and propionate are produced mainly by species belonging to the *Bacteroidetes* phylum, while butyrate is produced predominantly by the metabolism of some species of the *Firmicutes* phylum (Walker et al., 2005). Acetate has also been related to appetite reduction through its interaction with the central nervous system (Frost et al., 2014) and the three SCFA have been associated with intestinal anti-inflammatory properties (Tedelind et al., 2007). Particularly, butyrate is thought to play a role in maintaining a normal colonocyte population and participates in the motility of the colon, reducing inflammation, increasing visceral irrigation, inducing apoptosis and inhibiting the progression of tumor cells, properties that contribute with the prevention of colorectal cancer (Canani, 2011; Zhang et al., 2010). It was also described that, as a result of a lower luminal pH, all SCFA can inhibit pathogens (Gibson & Roberfroid, 1995).

The most studied prebiotic compounds include inulin (composed by fructose), fructooligosaccharides (FOS), galactooligosaccharides (GOS) and glucooligosaccharides such as β-glucans. In this scenario and taking into account their analogous chemical composition, EPS from lactobacilli can be considered *per se* as prebiotic substances. In some cases, and depending on their chemical nature, they can be fermented by the gut microbiota and, in consequence, associated with a SCFA increase. For example, Zhang et al., (2017) found that the EPS produced by a strain of *Lb. plantarum* was able to improve the oxidative and health conditions of the host intestinal tract in an aging mice model induced with D-galactose. They reported high contents of nitrogen oxides together with gut microbiota regulating activities, evidenced by increased levels of SCFA and higher counts of some members considered beneficial (*Blautia* and *Butyricicoccus*), together with a simultaneous reduction of microbial groups recurrent in humans with inflammatory diseases. Bello et al., (2001) and Korakli, Gänzle & Vogel (2002) revealed that EPS from strains of *Lb. sanfranciscensis* have bifidogenic effects on cultures of fecal bacteria and monocultures of *Bifidobacterium* (*Bf.*) *breve*, *Bf. bifidum*, *Bf. infantis* and *Bf. adolescentis*. More recent studies also demonstrated the bifidogenic properties of EPS isolated from strains of *Lb. rhamnosus* and *Lb. plantarum* (Das, Baruah & Goyal, 2014; Polak-Berecka et al., 2013). Hamet et al., (2016) evaluated the activity of kefiran on intestinal bacterial populations in BALB/c mice and confirmed by DGGE its bifidogenic effect. Furthermore, a study by Tsuda and Miyamoto (2010), for which an EPS isolated from an EPS-overproducing mutant strain of *L. plantarum* was used, described an increase in the proliferation of a number of other known beneficial bacteria, including *Pediococcus* (*Ped.*) *acidilactici* ID-7 and 5885, b*L. casei* L-49, *Lb. plantarum* IFO 3070 and *Lb. delbrueckii* subsp. *lactis* 306701. Besides, the EPS from *Lb. plantarum* mutant can be used by the probiotic parental strain *Lb. plantarum* 301102.

In a recent study (Ale 2018), we demonstrated using a mice model that when added as a food ingredient in yogurts, the EPS from *Lb. fermentum* Lf2 exerted a possible prebiotic role reflected in an increase of SCFA levels (mainly acetate and butyrate) in faeces, accompanied by an increase in the levels (determined by qPCR) of the bacterial cluster XIVa of *Clostridium* (also known as *Cl. coccoides* group) that are known to produce these beneficial acids for health. When this EPS was combined with the probiotic strain *B. animalis* subsp. *lactis* INL1, a bifidogenic effect throughout the treatment was observed, improving the effect conferred by both individual ingredients, a fact that could be associated with a possible synergism between them suggesting a possible symbiotic role.

In all the cases described, it would be necessary to demonstrate the effective health benefit associated with the recent definition of prebiotic by means of human clinical trials or, at least, animal *in vivo* assays. Taking into account the current difficulty of some countries in carrying out clinical studies, as well as their high costs, it would seem reasonable that these results will be provided by research groups from highly developed countries.

Other Health Effects Associated with EPS

A recent perspective for EPS is related to their antitumor and antioxidant (property related to the anticancer activity) bioactivities, and their ability to inhibit the biofilm formation in pathogenic bacteria as well, as revised by Jiang and Yang (2018). In this review, it has been summarized that eight strains of *Lb. plantarum* had antitumor and antioxidant activities, being effective (at least *in vitro*) on colonic, liver, gastric, colon and pancreatic cancer, as well as on breast adenocarcinoma and oxidative stress. The authors suggested the use of these EPS as antitumor adjuvants. However, this claim is too hasty considering that anti-cancer properties are attributed to these molecules based exclusively on *in vitro* tests and, *in vivo* or human clinical studies result mandatory to effectively demonstrate these results, as mentioned above. In another recent work, Riaz et al., (2018) described the functional properties of the EPS from strains of *Lb. rhamnosus* isolated from breast milk. They found the EPS exhibited potent antioxidant activity, antitumor activity against Caco-2 cells and strong antibacterial activities against *E. coli* and *Salmonella* Typhimurium, corresponding to an excellent biofilm inhibition activity against *S.* Typhimurium *in vitro*. Furthermore, cell bound polysaccharides (CPS) were studied regarding their functional properties. Li et al., (2015) studied the CPS of a strain of *Lb. helveticus*, a HePS with an average MW of 1.83×10^2 kDa composed of glucose, mannose, galactose, rhamnose and arabinose. Preliminary *in vitro* tests revealed that this polymer significantly inhibited the proliferation of cancer cell lines. In another previous work, Li et al., (2014) showed that the same EPS presented potent antioxidant effects, chelating ability on ferrous ion and antibiofilm activities by inhibition of three biofilm forming foodborne pathogenic bacteria.

Antimutagenic activity (binding ability to different mutagens such as heterocyclic amines, 4-Nitroquinoline-N-oxide and 2-Nitrofluorine) from EPS-producing lactobacilli (*Lb. plantarum* and *Lb. rhamnosus*) was also reported (Tsuda, Hara & Miyamoto, 2008; Thapa & Zhang, 2009). In one of these works, the authors proposed a binding mechanism of ion-exchange and hydrophobic bonds by which the EPS would bind mutagens in the intestine (Tsuda et al., 2008).

In addition, there is some evidence that certain EPS have a protective role against microbial toxins, as reported for the EPS of *Lb. rhamnosus* ATCC9595, with demonstrated cholera toxin-binding activity (Kim et al., 2006). Similarly, kefiran counteracted the cytotoxic effect of extracellular toxins produced by *Bacillus cereus* in cell cultures (Ruas-Madiedo et al., 2010).

EPS have also been related to alleviate acetic acid-induced experimental colitis model in rats (Şengül et al., 2006). In this work, the effect of the EPS-producing *Lb. delbrueckii* subsp. *bulgaricus* B3 strain was compared with another one with low ability to synthesize EPS (*Lb. delbrueckii* subsp. *bulgaricus* A13), analyzing the role of the EPS dose on colitis. Authors observed that those rats treated with both EPS-producing strains presented lower histologic scores of colonic damage than the control group, demonstrating an improvement by *Lb. bulgaricus* B3. Consequently, a probiotic role associated with the high production of EPS was suggested for this particular strain, showing its therapeutic potential.

An important aspect of the EPS functionality obeys to their general resistance to gastrointestinal conditions (Khalil et al., 2018) resulting from the action of the gastrointestinal tract of the host. Sönmez, Önal Darilmaz, and Beyatli (2018) evaluated the protective effects of EPS production on resistance to gastrointestinal digestive conditions and to oxalate by probiotic strains of the species *Lb. rhamnosus*, *Lb. fermentum* and *Lb. brevis*. The high-EPS-producing *Lb. fermentum* IP5 and *Lb. brevis*

YG7 strains showed high oxalate-degrading activity, whereas the low-EPS-producing strain *Lb. fermentum* BP5 demonstrated low ability to degrade oxalate. These results suggest that dietary supplementation with *Lb. fermentum* IP5 could prevent oxalate stone disease, assuming that the EPS produced by this strain is the responsible of this beneficial response.

EPS from lactobacilli have also been related to the adsorption of heavy metals. For example, Feng et al., (2012) demonstrated the biosorption of Pb(II) ions by an EPS isolated from *Lb. plantarum* 70810. EPS also demonstrated antioxidant and metal ion chelating activities (Li et al., 2014; Kodali & Sen, 2008). EPS from *Lb. rhamnosus* E/N (Polak-Berecka et al., 2013), *Lb. plantarum* C88 (Zhang et al., 2013), *Lb. plantarum* LP6 (Li et al., 2013) and *Lb. helveticus* MB2-1 (Li et al., 2014) have all exhibited moderate free radical scavenging and metal ion chelating activities in *in vitro* systems.

Some other studies associate the administration of EPS with modifications on the cardiometabolism. For example, spontaneously hypertensive stroke-prone rats fed with kefiran exhibited significantly lower blood pressure and lower levels of cholesterol and triglycerides in serum and liver, when compared to untreated rats (Maeda et al., 2004). Additionally, EPS fractions from *Lb. casei* LC2W exhibited a moderate reduction in blood pressure (Ai et al., 2008), without significantly affecting the heart rate. In this direction, Tok and Aslim (2010) suggested that an EPS isolated from *Lb. delbrueckii* subsp. *bulgaricus* was able to sequester cholesterol from conditioned media *in vitro*. Sasikumar et al., (2017), who characterized the EPS extract produced by the strain *Lb. plantarum* BR2, described that this polymer consisted of a HMW glucomannan with antioxidant and lowering cholesterol activities, together with antidiabetic potential.

CHEMICAL STRUCTURE AND FUNCTIONAL/ TECHNOLOGICAL PROPERTIES RELATIONSHIP

Many structural factors were reported to affect EPS properties, such as monosaccharide composition, molecular size, glucosidic bond, charge, presence of side chains and rigidity of the molecules (Salazar et al., 2014). From their functional aspects, EPS having negative charges (phosphate or sulfate), glucose and/or galactose in their composition, and/or low molecular weights were shown to act as mild stimulators of immune cells, whereas neutral polymers with higher molecular weights exhibited a suppressive profile (Caggianiello, Kleerebezem & Spano, 2016; Yasuda, Serata & Sako, 2008). In general, small negative EPS are thought to be able to stimulate professional immune cells and larger neutral EPS believed to possess more immune-suppressive qualities (López et al., 2012; Hidalgo-Cantabrana et al., 2012). Besides, an interrelation between the types of bonds of the HePS and their rigidity has also been found: $\beta(1,4)$

bonds result in more rigid chains than α(1,4) or β(1,3) bonds. In consequence, β bonds are able to give more chain stiffness compared to α bonds (Jolly et al., 2002). Side chains can also affect stiffness and, therefore, increase their efficiency as thickeners.

It was also described that EPS surrounding the bacterial surface, mainly those having a high MW, might reduce adherence to intestinal cells, and even to abiotic surfaces, due to shielding of surface macromolecules acting as adhesins. For example, deletion of the *cps2*-like gene cluster on *Lb. plantarum* Lp90 improved its adhesion to Caco2 cells in comparison with its parental strain (Lee et al., 2016). In accordance to this result, Dertli, Mayer, and Narbad (2015) suggested that EPS may hinder molecular interactions with host epithelial membranes. HMW-EPS reduce the adhesion capacity of the EPS-producing strain *in vitro*, which may also have a negative impact on intestinal colonization, as demonstrated for some *Bifidobacterium* strains (Castro-Bravo et al., 2017). In fact, the same conclusion was previously reported for *Lb. rhamnosus* GG through deletion of the priming-GTF, which abolished the synthesis of a long, galactose-rich EPS (Lebeer et al., 2009). However, while the non EPS-producing mutant presented good adhesion *in vitro*, an *in vivo* study showed that it was more sensitive toward host innate defenses; thus the EPS would favor its persistence in the murine gut (Lebeer et al., 2011). Branches may also be related to EPS functionality, Huazano-García and López (2013) suggested that there is a relationship between the structure of fructans (chain length and branches) with the production of SCFA. They observed that branched fructans induced higher levels of SCFA than linear polysaccharide.

Apart from the influence of the chemical characteristics of these molecules on their functionality, their impact on the technological properties was studied as well. Some structure/function relationships have been proposed, for example, polysaccharides that present β(1,4) linkages were associated with stiffer chains able to provide high consistency to the food matrices. Those containing α(1,6) linkages in the main chain, such as dextran, presented both good thickening and stabilizing effects (Zannini et al., 2016). Furthermore, polysaccharides with high molecular masses and a stiff conformations are important in order to obtain an increase in viscosity (De Vuyst & De Vin, 2007).

Other important physicochemical characteristic is the degree of branching, which may contribute to the stiffness of the polymer, while their primary structure (MW, monomer composition, type of linkages, presence of side-groups) also affect the rheology of EPS solutions (Tuinier et al., 2001).

Regarding the influence of the charge on the technological properties, it has been described that neutral EPS contributed to viscosity but not to elasticity. The contrary behavior was observed for negatively charged EPS which were able to interact with positive charges of caseins (Duboc & Mollet, 2001). The amount of EPS is not always the most important parameter for viscosity, in contrary to its structure which is the determinant for the interactions with casein. For this reason, although LAB synthesize

EPS in small quantities, these could be enough to deliver rheological changes on the texture of dairy products.

When neutral or positively low charged EPS interact with caseins, a depletion effect can occur, leading to the formation of protein and EPS aggregates in the serum phase (Buldo et al., 2016; Costa et al., 2012). This phenomenon has been related to decreasing viscoelastic properties due to the interference of the EPS with the protein network formation, or increasing viscoelastic properties, as a consequence from the formation of a dense casein network (Amatayakul et al., 2006a; Hassan et al., 2003; Buldo et al., 2016). The presence of EPS in the serum phase also leads to higher water-binding capacity, resulting in lower syneresis and high viscosity (Gentès, St-Gelais & Turgeon, 2013). Future studies are required to establish clearer structure-functional/technological properties relationship.

CONCLUSION

An increasing number of EPS produced by *Lactobacillus* strains have been isolated and characterized in the last few years, evidencing a high diversity among them. Despite the unknown physiological role of EPS in the producing bacteria, it seems to be related to the recognition of the external environment, thus exerting a protective effect.

The integral characterization (structure, MW, presence of charged residues, type of branches, etc.) of these macromolecules is essential when EPS (as an ingredient) or the EPS-producing strain is included in a food matrix, since recent studies strongly associated these characteristics with their rheological and health-promoting properties. In this way, the characterization of the EPS molecule to be added in a food matrix would allow to infer the techno-functional properties of the final product.

Although their low production compared with EPS produced by non-LAB bacteria, EPS from lactobacilli are, from their technological point of view, able to improve the rheological properties of dairy products. According to their structural and chemical properties, in some cases, even present in low concentrations in the final product, they are adequate to positively change the rheology, as occurs in some low-fat cheeses. Furthermore, some EPS produced by lactobacilli have the ability to exert a wide range of beneficial effects on human health: immunomodulation, prebiotic role, antioxidant, anticarcinogenic, among others. As these properties were extensively demonstrated *in vitro* and at high EPS concentrations, an interesting challenge for the future will be the elucidation of the physiological mechanisms of actions and the demonstration, by *in vivo* studies, of the real effect on the intestinal environment when the EPS are included in applicable concentrations in a real food matrix.

The use of HoPS or HePS from *Lactobacillus* or the EPS-producing bacteria could have a promising future for their application as potential techno-functional ingredients for the design of novel foods. An interesting alternative is the optimization of the environment conditions to increase the yield of EPS when they are to be used as food ingredients. A further possibility is to use engineered food-grade lactobacilli strains with the aim of specifically modifying certain genes associated with the polymer synthesis, for the design of new fermented foods.

REFERENCES

Ai L., Zhang H. Z., Guo B., Chen W., C. Wu, Z. W., Wu Y. (2008). Preparation, partial characterization and bioactivity of exopolysaccharides from *Lactobacillus casei* LC2W. *Carbohydrate Polymers,* 74(3), 353–357.

Akira S., Uematsu S., Takeuchi O. (2006). Pathogen recognition and innate immunity. *Cell,* 124(4), 783–801.

Ale E. C, Perezlindo M.J., Burns P., Tabacman E., Reinheimer J. A., Binetti A. G. (2016a). Exopolysaccharide from *Lactobacillus fermentum* Lf2 and its functional characterization as a yogurt additive. *Journal of Dairy Research,* 83(04), 487–492.

Ale E. C. (2018). *Functional (in vitro and in vivo) and technological potential of exopolysaccharides (EPS) produced by lactic acid bacteria* (Doctoral Thesis). Universidad Nacional del Litoral, Santa Fe, Argentina.

Ale E. C., Perezlindo M. J., Pavón Y., Peralta G.H., Costa S., Sabbag N., Bergamini C., Reinheimer J. A., Binetti A. G. (2016b). Technological, rheological and sensory characterizations of a yogurt containing an exopolysaccharide extract from *Lactobacillus fermentum* Lf2, a new food additive. *Food Research International,* 90, 259–267.

Allonsius C. N., van den Broek M. F. L., De Boeck I., Kiekens S., Oerlemans E. F. M., Kiekens F., Foubert K., Vandenheuvel D., Cos P., Delputte P., Lebeer S. (2017). Interplay between *Lactobacillus rhamnosus* GG and Candida and the involvement of exopolysaccharides. *Microbial Biotechnology,* 10(6), 1753–1763.

Amatayakul T., Halmos A. L., Sherkat F., Shah N. P. (2006a). Physical characteristics of yoghurts made using exopolysaccharide-producing starter cultures and varying casein to whey protein ratios. *International Dairy Journal,* 16(1), 40–51.

Amatayakul T., Sherkat F., Shah N. P. (2006b). Syneresis in set yogurt as affected by EPS starter cultures and levels of solids. *International Journal of Dairy Technology,* 59(3), 216–221.

Amrouche T., Boutin Y., Prioult G., Fliss I. (2006). Effects of bifidobacterial cytoplasm, cell wall and expolysaccharide on mouse lymphocyte proliferation and cytokine production. *International Dairy Journal,* 16, 70–80.

Balzaretti S., Taverniti V., Guglielmetti S. Fiore, W. Minuzzo M., Ngo H. N. Ngere, J. B. Sadiq S., Humphreys P. N., Laws A. P. (2017). A novel rhamnose-rich heteroexopolysaccharide isolated from *Lactobacillus paracasei* DG activates THP-1 human monocytic cells. *Applied and Environmental Microbiology, 83(3).*

Baruah R., Deka B., Kashyap N., Goyal A. (2018). Dextran utilization during its synthesis by *Weissella cibaria* RBA12 can be overcome by fed-batch fermentation in a bioreactor. *Applied Biochemistry and Biotechnology,* 184(1), 1–11.

Bello F. D., Walter J., Hertel C., Hammes W. P. (2001). In vitro study of prebiotic properties of levan-type exopolysaccharides from lactobacilli and non-digestible carbohydrates using denaturing gradient gel electrophoresis. *Systematic and Applied Microbiology,* 24(2), 232–237.

Bengoa A. A., Llamas M. G., Iraporda C., Dueñas M. T., Abraham A. G., Garrote G. L. (2018). Impact of growth temperature on exopolysaccharide production and probiotic properties of *Lactobacillus paracasei* strains isolated from kefir grains. *Food Microbiology,* 69, 212–218.

Bleau C., Monges A., Rashidan K., Laverdure J-P., Lacroix M., Van Calsteren M-R., Millette M., Savard R., Lamontagne L. (2010). Intermediate chains of exopolysaccharides from *Lactobacillus rhamnosus* RW-9595M increase IL-10 production by macrophages. *Journal of Applied Microbiology,* 108, 666–675.

Brandt M. J., Roth K., Hammes W. P. (2003). Effect of an exopolysaccharide produced by *Lactobacillus sanfranciscensis LTH1729* on dough and bread quality. In L. de Vuyst (Ed.), *Sourdough from fundamentals to applications.* Brussels, Vrije Universiteit Brussel (VUB), IMDO, 80.

Broadbent J. R., McMahon D. J., Oberg C. J., Welker D. L. (2001). Use of exopolysaccharide-producing cultures to improve the functionality of low fat cheese. *International Dairy Journal,* 11(4–7), 433–439.

Buldo P., Benfeldt C., Folkenberg D. M., Jensen H. B., Amigo J. M., Sieuwerts S., Thygesen K., van den Berg F., Ipsen R. (2016). The role of exopolysaccharide-producing cultures and whey protein ingredients in yoghurt. *LWT - Food Science and Technology,* 72, 189–198.

Caggianiello G., Kleerebezem M., Spano G. (2016). Exopolysaccharides produced by lactic acid bacteria, from health-promoting benefits to stress tolerance mechanisms. *Applied Microbiology and Biotechnology,* 100(9), 3877–3886.

Canani R. B., Di Costanzo M., Leone L., Pedata M., Meli R., Calignano A. (2011). Potential beneficial effects of butyrate in intestinal and extraintestinal diseases. *World Journal of Gastroenterology,* 17(12), 1519-1528.

Castro-Bravo N., Hidalgo-Cantabrana C., Rodriguez-Carvajal M. A., Ruas-Madiedo P, Margolles A. (2017). Gene replacement and fluorescent labeling to study the functional role of exopolysaccharides in *Bifidobacterium animalis* subsp. *lactis. Frontiers in Microbiology,* 8.

Castro-Bravo N., Wells J. M., Margolles A., Ruas-Madiedo P. (2018). Interactions of surface exopolysaccharides from Bifidobacterium and Lactobacillus within the intestinal environment. *Frontiers in Microbiology, 9.*

Cencic A., Chingwaru W. (2010). The role of functional foods, nutraceuticals, and food supplements in intestinal health. *Nutrients, 2*(6), 611–625.

Cerning J. (1990). Exocelular polysaccharides produced by lactic acid bacteria. *FEMS Microbiology Reviews, 87*, 113–130.

Cerning J. (1995). Production of exopolysaccharides by lactic acid bacteria and dairy propionibacteria. *Lait, 75*(1), 463–472.

Chabot S., Han-Ling Y., De Léséleuc L., Cloutier D., Van Calsteren M. R. Lessard, M. Roy D., Lacroix M., Oth D. (2001). Exopolysaccharides from Lactobacillus rhamnosus RW-9595M stimulate TNF, IL-6 and IL-12 in human and mouse cultured immunocompetent cells, and IFN-γ in mouse splenocytes. *Le Lait, 81*(6), 683–697.

Ciszek-Lenda M., Nowak B., Śróttek M., Gamian A., Marcinkiewicz J. (2011). Immunoregulatory potential of exopolysaccharide from Lactobacillus rhamnosus KL37. Effects on the production of inflammatory mediators by mouse macrophages. *International Journal of Experimental Pathology, 92*(6), 382–391.

Cobb B. A., Kasper D. L. (2005). Coming of age, Carbohydrates and immunity. *European Journal of Immunology, 35*(2), 352–356.

Collins S., Reid G. (2016). Distant site effects of ingested prebiotics. *Nutrients, 8*(9), 523.

Costa N. E., O'Callaghan D. J., Mateo M. J., Chaurin V., Castillo M., Hannon J. A., McSweeney P. L. H., Beresford T. P. (2012). Influence of an exopolysaccharide produced by a starter on milk coagulation and curd syneresis. *International Dairy Journal, 22*(1), 48–57.

Dabour N., Kheadr E., Benhamou N., Fliss I., LaPointe G. (2006). Improvement of texture and structure of reduced-fat Cheddar cheese by exopolysaccharide-producing lactococci." *Journal of Dairy Science, 89*(1), 95–110.

Dan T., Fukuda K., Sugai-Bannai M., Takakuwa N., Motoshima H., Urashima T. (2009). Characterization and expression analysis of the exopolysaccharide gene cluster in *Lactobacillus fermentum* TDS030603. *Bioscience, Biotechnology and Biochemistry, 73*(12), 2656–2664.

Das D., Baruah R., Goyal A. (2014). A food additive with prebiotic properties of an α-D-Glucan from *Lactobacillus plantarum* DM5. *International Journal of Biological Macromolecules, 69*, 20–26.

De Vuyst L., De Vin F. (2007). Exopolysaccharides from lactic acid bacteria. In *Comprehensive Glycoscience* (pp. 477–519). Elsevier.

De Vuyst L. De, Degeest B. (1999a). Heteropolysaccharides from lactic acid bacteria. *FEMS Microbiology Reviews, 23*(2), 153–177.

De Vuyst L. De, Degeest B. (1999b). Expolysaccharides from lactic acid bacteria: technological bottlenecks and practical solutions. *Macromolecular Symposia,* 140(1), 31–41.

Degeest B., Vaningelgem F., De Vuyst L. (2001). Microbial physiology, fermentation kinetics, and process engineering of heteropolysaccharide production by lactic acid bacteria. *International Dairy Journal,* 11(9), 747–757.

Dertli E., Colquhoun I. J., Gunning A. P., Bongaerts R. J., Le Gall G., Bonev B. B., Mayer M. J., Narbad A. (2013). Structure and biosynthesis of two exopolysaccharides produced by *Lactobacillus johnsonii* FI9785. *Journal of Biological Chemistry,* 288(44), 31938–31951.

Dertli E., Mayer M. J., Narbad A. (2015). Impact of the exopolysaccharide layer on biofilms, adhesion and resistance to stress in *Lactobacillus johnsonii* FI9785. *BMC Microbiology,* 15(1), 8.

Dierksen K. P., Sandine W. S., Trempy J. E. (1997). Expression of ropy and mucoid phenotypes in *Lactococcus lactis. Journal of Dairy Science,* 80(8), 1528–1536.

Dilna S. V., Surya H., Aswathy R. G. Varsha, K. K. Sakthikumar, D. N. Pandey A., Nampoothiri K. M. (2015). Characterization of an exopolysaccharide with potential health-benefit properties from a probiotic *Lactobacillus plantarum* RJF4. *LWT - Food Science and Technology,* 64(2), 1179–1186.

Dimopoulou M., Claisse O., Dutilh L., Miot-Sertier, C., Ballestra P., Lucas P. M., Dols-Lafargue M. (2017). Molecular cloning, expression and characterization of *Oenococcus oeni* priming glycosyltransferases. *Molecular Biotechnology,* 59(8), 323–333.

Duboc P., Mollet B. (2001). Applications of exopolysaccharides in the dairy industry. *International Dairy Journal,* 11(9), 759–768.

Dubois M., Gilles K. A., Hamilton J. K. Rebers, P. T., Smith F. (1956). Colorimetric method for determination of sugars and related substances. *Analytical Chemistry,* 28(3), 350-356.

Fanning S., Hall L. J., Cronin M., Zomer A., MacSharry J., Goulding D., O'Connell-Motherway M., Shanahan F., Nally K., Dougan G., van Sinderen D. (2012). Bifidobacterial surface-expolysaccharide facilitates commensal-host interaction through immune modulation and pathogen protection. *Proceedings of the National Academy of Sciences USA,* 109, 2108–2113.

Feng M., Chen X., Li C., Nurgul R., Dong M. (2012). Isolation and identification of an exopolysaccharide-producing lactic acid bacterium strain from chinese paocai and biosorption of Pb(ii) by its exopolysaccharide. *Journal of Food Science* 77(6), 111–117.

Folkenberg D. M., Dejmek P., Skriver A., Ipsen R. (2005). Relation between sensory texture properties and exopolysaccharide distribution in set and in stirred yoghurts produced with different starter cultures. *Journal of Texture Studies,* 36(2), 174–189.

Fraunhofer M. E., Geissler A. J., Wefers D., Bunzel M., Jakob F., Vogel R. F. (2018). Characterization of β-glucan formation by *Lactobacillus brevis* TMW 1.2112 isolated from slimy spoiled beer. *International Journal of Biological Macromolecules,* 107, 874-881.

Frost G., Sleeth M. L., Sahuri-Arisoylu M. Lizarbe, B. Cerdan, S. Brody L., Anastasovska J., Ghourab S., Hankir M., Zhang S., Carling D., Swann J. R., Gibson G., Viardot A., Morrison D., Thomas E. L., Bell J. D. (2014). The short-chain fatty acid acetate reduces appetite via a central homeostatic mechanism. *Nature Communications,* 5, 3611.

Fukuda S., Toh H., Hase K., Oshima K., Nakanishi Y., Yoshimura K., Tobe T., Clarke J. M., Topping T. L., Suzuki T., Taylor T. D., Itoh K., Kikuchi J., Morita H. Hattori M., Ohno H. (2011). Bifidobacteria can protect from enteropathogenic infection through production of acetate. *Nature,* 469(7331), 543–547.

Galle S., Schwab C., Arendt E. K., Gänzle M. G. (2011). Structural and rheological characterisation of heteropolysaccharides produced by lactic acid bacteria in wheat and sorghum sourdough. *Food Microbiology,* 28(3), 547–553.

Gentès M-C., St-Gelais D.. Turgeon S. L. (2013). Exopolysaccharide–milk protein interactions in a dairy model system simulating yoghurt conditions. *Dairy Science & Technology,* 93(3), 255–271.

Gibson G. R., Roberfroid M. B. (1995). Dietary modulation of the human colonic microbiota: introducing the concept of prebiotics. *The Journal of Nutrition,* 125(6), 1401–1412.

Gibson G. R., Hutkins R., Sanders M. E., Prescott S. L., Reimer R. A., Salminen S. J., Scott K., Stanton C., Swanson K. S., Cani P. D., Verbeke K., Reid G. (2017). Expert Consensus Document, The International Scientific Association for Probiotics and Prebiotics (ISAPP) Consensus Statement on the definition and scope of prebiotics. *Nature Reviews Gastroenterology & Hepatology,* 14(8), 491-502.

Hamet M. F., Medrano M., Pérez P. F., Abraham A. G. (2016). Oral administration of kefiran exerts a bifidogenic effect on BALB/c mice intestinal microbiota. *Beneficial Microbes,* 7(2), 237–246.

Hamet M. F., Piermaria J. A., Abraham A. G. (2015). Selection of EPS-producing *Lactobacillus* strains isolated from kefir grains and rheological characterization of the fermented milks. *LWT - Food Science and Technology,* 63(1), 129–135.

Han X., Yang Z., Jing X., Yu P., Zhang Y., Yi H., Zhang L. (2016). Improvement of the texture of yogurt by use of exopolysaccharide producing lactic acid bacteria." *BioMed Research International,* 2016, 7945675.

Hassan A. N, Awad S., Muthukumarappan K. (2005). Effects of exopolysaccharide-producing cultures on the viscoelastic properties of reduced-fat cheddar cheese. *Journal of Dairy Science,* 88(12), 4221–4227.

Hassan A. N. (2008) Possibilities and challenges of exopolysaccharide-producing lactic cultures in dairy foods. *Journal of Dairy Science,* 91, 1282–1298.

Hassan A. N., Ipsen R., Janzen T., Qvist K. B. (2003). Microstructure and rheology of yogurt made with cultures differing only in their ability to produce exopolysaccharides. *Journal of Dairy Science,* 86(5), 1632–1638.

Hermann M., Petermeier H., Vogel R. F. (2015). Development of novel sourdoughs with in situ formed exopolysaccharides from acetic acid bacteria. *European Food Research and Technology,* 241(2), 185–197.

Hidalgo-Cantabrana C., López P., Gueimonde M., de los Reyes-Gavilán C. G., Suárez A., Margolles A., Ruas-Madiedo P. (2012). Immune modulation capability of exopolysaccharides synthesised by lactic acid bacteria and bifidobacteria. *Probiotics and Antimicrobial Proteins,* 4(4), 227–237.

Hijum S. A. F. T. van, Kralj S., Ozimek L. K., Dijkhuizen L., Geel-Schutten I. G. H. van. (2006). Structure-function relationships of glucansucrase and fructansucrase enzymes from lactic acid bacteria. *Microbiology and Molecular Biology Reviews,* 70(1), 157–176.

Hijum S. A. F. T. van, Szalowska E., van der Maarel M. J. E. C., Dijkhuizen L. (2004). Biochemical and molecular characterization of a levansucrase from *Lactobacillus reuteri. Microbiology,* 150(3), 621–630.

Hill C., Guarner F., Reid G., Gibson G. R., Merenstein D. J., Pot B., Morelli L., Canani R. B., Flint H. J., Salminen S., Calder P. C., Sanders M. E. (2014). Expert consensus document: The International Scientific Association for Probiotics and Prebiotics Consensus Statement on the Scope and Appropriate Use of the Term Probiotic. *Nature Reviews Gastroenterology and Hepatology,* 11(8), 506–514.

Hoshino K., Takeuchi O., Kawai T., Sanjo H., Ogawa T., Takeda Y., Takeda K., Akira S. (1999). Cutting edge, Toll-like receptor 4 (TLR4)-deficient mice are hyporesponsive to lipopolysaccharide, evidence for TLR4 as the LPS gene product. *Journal of Immunology,* 162(7), 3749–3752.

Huazano-García A., López M. G. (2013). Metabolism of short chain fatty acids in the colon and faeces of mice after a supplementation of diets with agave fructans. *Lipid Metabolism,* 163–182.

Ibarburu I., Puertas A. I., Berregi I., Rodríguez-Carvajal M. A., Prieto A., Dueñas M. T. (2015). Production and partial characterization of exopolysaccharides produced by two *Lactobacillus suebicus* strains isolated from cider. *International Journal of Food Microbiology,* 214, 54–62.

Isenberg S. L., Brewer A. K., Côté G. L., Striegel A. M. (2010). Hydrodynamic versus size exclusion chromatography characterization of alternan and comparison to off-line MALS. *Biomacromolecules* 11(9) 2505–2511.

Jiang Y., and Z. (2018). A functional and genetic overview of exopolysaccharides produced by *Lactobacillus plantarum. Journal of Functional Foods,* 47, 229–240.

Jolly L., Stingele F. (2001). Molecular organization and functionality of exopolysaccharide gene clusters in lactic acid bacteria. *International Dairy Journal,* 11(9), 733–745.

Jolly L., Vincent S. J. F., Duboc P., Neeser J-R. (2002). Exploiting exopolysaccharides from lactic acid bacteria. *Antonie van Leeuwenhoek,* 82, 367–374.

Juvonen R., Honkapää K., Maina N. H., Shi Q., Viljanen K., Maaheimo H., Virkki L., Tenkanen M., Lantto R. (2015). The impact of fermentation with exopolysaccharide producing lactic acid bacteria on rheological, chemical and sensory properties of pureed carrots (Daucus carota L.). *International Journal of Food Microbiology,* 207, 109–18.

Kang H-K, Oh J-S., Kim D. (2009). Molecular characterization and expression analysis of the glucansucrase DSRWC from *Weissella cibaria* synthesizing a $\alpha(1\rightarrow6)$ glucan. *FEMS Microbiology Letters,* 292(1), 33–41.

Khalil E., Abd Manap M., Mustafa S., Alhelli A., Shokryazdan P. (2018). Pobiotic properties of exopolysaccharide-producing *Lactobacillus* strains isolated from tempoyak. *Molecules,* 23(2), 398.

Kim J. U., Kim Y. H., Han K. S., Oh S. J., Whang K. Y., Kim J. N.,and Kim S. H. (2006). Function of cell-bound and released exopolysaccharides produced *Lactobacillus rhamnosus* ATCC 9595. *Journal of Microbiology and Biotechnology,* 16(6), 939–945.

Kimmel S. A., Roberts R. F. (1998). Development of a growth medium suitable for exopolysaccharide production by *Lactobacillus delbrueckii* ssp. *bulgaricus* RR. *International Journal of Food Microbiology,* 40(1–2), 87–92.

Kitazawa H., Harata T., Uemura J., Saito T., Kaneko T., Itoh T. (1998) Phosphate group requirement for mitogenic activation of lymphocytes by an extracellular phosphopolysaccharide from *Lactobacillus delbrueckii* ssp. *bulgaricus*. *International Journal of Food Microbiology,* 40, 169–175.

Kodali V. P., Sen R. (2008). Antioxidant and free radical scavenging activities of an exopolysaccharide from a probiotic bacterium. *Biotechnology Journal,* 3(2), 245–251.

Korakli M., Vogel R. F. (2006). Structure/function relationship of homopolysaccharide producing glycansucrases and therapeutic potential of their synthesised glycans. *Applied Microbiology and Biotechnology,* 71(6), 790–803.

Korakli M., Gänzle M. G., Vogel R. F. (2002). "Metabolism by bifidobacteria and lactic acid bacteria of polysaccharides from wheat and rye, and exopolysaccharides produced by *Lactobacillus sanfranciscensis*. *Journal of Applied Microbiology,* 92(5), 958–65.

Korakli M., Rossmann A., Gänzle M. G., Vogel R. F. (2001). Sucrose metabolism and exopolysaccharide production in wheat and rye sourdoughs by *Lactobacillus sanfranciscensis*. *Journal of Agricultural and Food Chemistry,* 49(11), 5194–5200.

Krajl S., Geel-Schutten G. H. van, Maarel M. J. E. C. Van Der, Dijkhuizen L. (2003). Efficient screening methods for glucosyltransferase genes in *Lactobacillus* strains. *Biocatalysis and Biotransformation,* 21(4–5), 181–187.

Lebeer S., Claes I. J. J., Verhoeven T. L. A., Vanderleyden J., Keersmaecker S. C. J. De. (2011). Exopolysaccharides of *Lactobacillus rhamnosus* GG form a protective shield against innate immune factors in the intestine. *Microbial Biotechnology,* 4(3), 368–74.

Lebeer S., Vanderleyden J., Keersmaecker S. C. J. De (2008). Genes and molecules of lactobacilli supporting probiotic action. *Microbiology and Molecular Biology Reviews,* 72(4), 728–764.

Lebeer S., Verhoeven T. L. A., Francius G., Schoofs G., Lambrichts I., Dufrêne Y., Vanderleyden J., Keersmaecker S. C. J. De. (2009). Identification of a gene cluster for the biosynthesis of a long, galactose-rich exopolysaccharide in *Lactobacillus rhamnosus* GG and functional analysis of the priming glycosyltransferase. *Applied and Environmental Microbiology,* 75 (11), 3554–3563.

Lecerf J-M, Dépeint F., Clerc E., Dugenet Y., Niamba C. N., Rhazi L., Cayzeele A., Abdelnour G., Jaruga A., Younes H., Jacobs H., Lambrey G., Abdelnour A. M., Pouillart P. R. (2012). Xylo-Oligosaccharide (XOS) in combination with inulin modulates both the intestinal environment and immune status in healthy subjects, while XOS alone only shows prebiotic properties. *British Journal of Nutrition,* 108(10), 1847–1858.

Lee I-Chiao, Caggianiello G., Swam I. I. van, Taverne N., Meijerink M., Bron P. A., Spano G., Kleerebezem M. (2016). Strain-specific features of extracellular polysaccharides and their impact on *Lactobacillus plantarum*-host interactions. *Applied and Environmental Microbiology* 82(13) 3959–3970.

Leivers S., Hidalgo-Cantabrana C., Robinson G., Margolles A., Ruas-Madiedo P., Laws A. P. (2011). Structure of the high molecular weight exopolysaccharide produced by *Bifidobacterium animalis* subsp. *lactis* IPLA-R1 and sequence analysis of its putative eps cluster. *Carbohydrate Research,* 346(17), 2710–2717.

Leo F., Hashida S., Kumagai D., Uchida K., Motoshima H., Arai I., Asakuma S., Fukuda K., Urashima T. (2007). Studies on a neutral exopolysaccharide of *Lactobacillus fermentum* TDS030603. *Journal of Applied Glycoscience,* 54(4), 223–229.

Li J. Y., Jin M. M., Meng J., Gao S. M., Lu R. R. (2013). Exopolysaccharide from *Lactobacillus plantarum* LP6, antioxidation and the effect on oxidative stress. *Carbohydrate Polymers,* 98(1), 1147–1152.

Li W., Ji J., Chen X., Jiang M., Rui X., Dong M. (2014). Structural elucidation and antioxidant activities of exopolysaccharides from *Lactobacillus helveticus* MB2-1. *Carbohydrate Polymers,* 102(1), 351–359.

Li W., Ji J., Rui X., Yu J., Tang W., Chen X., Jiang M., Dong M. (2014). Production of exopolysaccharides by *Lactobacillus helveticus* MB2-1 and its functional characteristics in vitro. *LWT - Food Science and Technology*, 59(2), 732–739.

Li W., Xia X., Tang W., Ji J., Rui X., Chen X., Jiang M., Zhou J., Zhang Q., Dong M. (2015). Structural characterization and anticancer activity of cell-bound exopolysaccharide from *Lactobacillus helveticus* MB2-1. *Journal of Agricultural and Food Chemistry*, 63(13), 3454–3463.

Liu H-N., Wu H., Chen Y-Z., Chen Y-J., Shen, X-Z., Liu T-T. (2017). Altered molecular signature of intestinal microbiota in irritable bowel syndrome patients compared with healthy controls, a systematic review and meta-analysis. *Digestive and Liver Disease*, 49(4), 331–337.

London L. E. E., Price N. P. J., Ryan P., Wang L., Auty M. A. E., Fitzgerald G. F., Stanton C., Ross R. P. (2014). Characterization of a bovine isolate *Lactobacillus mucosae* DPC 6426 which produces an exopolysaccharide composed predominantly of mannose residues. *Journal of Applied Microbiology*, 117(2), 509–517.

London L. E.E., Chaurin V., Auty M. A. E., Fenelon M. A., Fitzgerald G. F., Ross R. P., Stanton C. (2015). Use of *Lactobacillus mucosae* DPC 6426, an exopolysaccharide-producing strain, positively influences the techno-functional properties of yoghurt. *International Dairy Journal*, 40, 33–38.

López P., Monteserín D. C., Gueimonde M., de los Reyes-Gavilán C. G., Margolles A., Suárez A., Ruas-Madiedo P. (2012). Exopolysaccharide-producing *Bifidobacterium* strains elicit different in vitro responses upon interaction with human cells. *Food Research International*, 46(1), 99–107.

Louis P., Scott K. P., Duncan S. H., Flint H. J. (2007). Understanding the effects of diet on bacterial metabolism in the large intestine. *Journal of Applied Microbiology* 102(5), 1197–1208.

Lynch K. M., Zannini E., Coffey A., Arendt E. K. (2018). Lactic acid bacteria exopolysaccharides in foods and beverages: Isolation, properties, characterization, and health benefits. *Annual Review of Food Science and Technology*, 9(1), 155–176.

Maeda H., Zhu X., Omura K., Suzuki S., Kitamura S. (2004). Effects of an exopolysaccharide (kefiran) on lipids, blood pressure, blood glucose, and constipation. *BioFactors*, 22(1–4), 197-200.

Malang S. K., Maina N. H., Schwab C., Tenkanen M., Lacroix C. (2015). Characterization of exopolysaccharide and ropy capsular polysaccharide formation by *Weissella*. *Food Microbiology*, 46, 418–427.

Malik A., Radji M., Kralj S., Dijkhuizen L. (2009). Screening of lactic acid bacteria from indonesia reveals glucansucrase and fructansucrase genes in two different Weissella confusa strains from soya. *FEMS Microbiology Letters*, 300(1), 131–38.

Meijerink M., Ferrando M. L., Lammers G., Taverne N., Smith H. E., Wells J. M. (2012). Immunomodulatory effects of *Streptococcus suis* capsule type on human dendritic cell responses, phagocytosis and intracellular survival. *PLoS ONE,* 7(4), e35849.

Mende S., Peter M., Bartels K., Rohm H., Jaros D. (2013). Addition of purified exopolysaccharide isolates from *S. thermophilus* to milk and their impact on the rheology of acid gels. *Food Hydrocolloids,* 32(1), 178–85.

Meulen R. Van der, Grosu-Tudor S., Mozzi F., Vaningelgem F., Zamfir M., Font de Valdez G., De Vuyst L. (2007). Screening of lactic acid bacteria isolates from dairy and cereal products for exopolysaccharide production and genes involved. *International Journal of Food Microbiology,* 118(3), 250–258.

Mozzi F, Savoy de Giory G., Oliver G., Font de Valdez G. (1995). Exopolysaccharide production by *Lactobacillus casei.* II. Influence of the carbon source. *Milchwissenschaft,* 50, 307–309.

Mozzi F., Torino M. I., Font de Valdez G. 2004. Identification of exopolysaccharide-producing, lactic acid bacteria a method for the isolation of polysaccharides in milk cultures. In *Food Microbiology Protocols*, 183–190. New Jersey, Humana Press.

Nagai T., Makino S., Ikegami S., Itoh H., Yamada H. (2011). Effects of oral administration of yogurt fermented with *Lactobacillus delbrueckii* ssp. *bulgaricus* OLL1073R-1 and its exopolysaccharides against influenza virus infection in mice. *International Immunopharmacology,* 11(12), 2246–2250.

Nikolic M., López P., Strahinic I., Suárez A., Kojic M., Fernández-García M., Topisirovicb L., Golicb N., Ruas-Madiedo P. (2012). Characterisation of the exopolysaccharide (EPS)-producing *Lactobacillus paraplantarum* BGCG11 and its non-EPS producing derivative strains as potential probiotics. *International Journal of Food Microbiology,* 158(2), 155–162.

Nishimura-Uemura J., Kitazawa H., Kawai Y., Itoh T., Oda M. and Saito T. (2003). Functional alteration of mucrine macrophages stimulated with extracellular polysaccharides from *Lactobacillus delbrueckii* ssp. bulgaricus OLL1073R-1. *Food Microbiology,* 20, 267–273.

Nowak B., Ciszek-Lenda M., Śróttek M., Gamian A., Kontny E., Górska-Frączek S., Marcinkiewicz J., Sabina Górska-Frączek. (2012). *Lactobacillus rhamnosus* exopolysaccharide ameliorates arthritis induced by the systemic injection of collagen and lipopolysaccharide in DBA/1 mice. *Archivum Immunologiae et Therapiae Experimentalis,* 60(3), 211–220.

Oberg E. N. N, Oberg C. J. J., Motawee M. M. M., Martini S., McMahon D. J. J. (2015). Increasing stringiness of low-fat mozzarella string cheese using polysaccharides. *Journal of Dairy Science* 98(7), 4243–4254.

Owusu-Kwarteng J., Tano-Debrah K., Akabanda F., Jespersen L. (2015). Technological properties and probiotic potential of Lactobacillus fermentum strains isolated from west African fermented millet dough. *BMC Microbiology,* 15(1), 261.

Paiva I. M. de, Steinberg R., Lula I. S., de Souza-Fagundes E. M., de Oliveira Mendes T., Valenzuela Bell M. J., Nicoli J. R., Cantini Nunes A., Neumann E. (2016). *Lactobacillus kefiranofaciens* and *Lactobacillus satsumensis* isolated from Brazilian kefir grains produce alpha-glucans that are potentially suitable for food applications. *LWT - Food Science and Technology,* 72, 390–398.

Patten D. A., and Collett A. (2013). Exploring the immunomodulatory potential of microbial-associated molecular patterns derived from the enteric bacterial microbiota. *Microbiology,* 159(8), 1535–1544.

Patten D. A., Leivers S., Chadha M. J., Maqsood M., Humphreys P. N., Laws A. P., Collett A. (2014). The structure and immunomodulatory activity on intestinal epithelial cells of the EPSs isolated from *Lactobacillus helveticus* sp. r*osyjski* and *Lactobacillus acidophilus* sp. 5E2. *Carbohydrate Research,* 384, 119–127.

Péant B., LaPointe G., Gilbert C., Atlan D., Ward P., Roy D. (2005). Comparative analysis of the exopolysaccharide biosynthesis gene clusters from four strains of *Lactobacillus rhamnosus*. *Microbiology,* 151(6), 1839–1851.

Perry D. B., McMahon D. J., Oberg C. J. (1998). Manufacture of low fat mozzarella cheese using exopolysaccharide-producing starter cultures. *Journal of Dairy Science,* 81(2), 563–566.

Petry S., Furlan S., Crepeau M-J., Cerning J., Desmazeaud M. (2000). Factors affecting exocellular polysaccharide production by *Lactobacillus delbrueckii* subsp. *bulgaricus* grown in a chemically defined medium. *Applied and Environmental Microbiology,* 66(8), 3427–3431.

Picton L., Bataille I., Muller G. (2000). Analysis of a complex polysaccharide (gum arabic) by multi-angle laser light scattering coupled on-line to size exclusion chromatography and flow field flow fractionation. *Carbohydrate Polymers,* 42(1), 23–31.

Polak-Berecka M., Waśko A., Skrzypek H., Kreft A. (2013). Production of exopolysaccharides by a probiotic strain of *Lactobacillus rhamnosus*: ‚Biosynthesis and purification methods. *Acta Alimentaria*, 42(2), 220–28.

Polak-Berecka M., Waśko A., Szwajgier D., Chomaz A. (2013). Bifidogenic and antioxidant activity of exopolysaccharides produced by *Lactobacillus rhamnosus* E/N cultivated on different carbon sources. *Polish Journal of Microbiology,* 62(2), 181–188.

Riaz R., Sahid M., Jin M., Haobin Z., Li Q., Shao D., Jiang C., Huang Q., Yang H., Shi J., Hussain N. (2018). Functional characterization and biotechnological potential of exopolysaccharide produced by *Lactobacillus rhamnosus* strains isolated from human breast milk. *LWT- Food Science and Technology,* 89, 638–47.

Ricciardi A., Parente E., Clementi F. (1997). A simple method for the screening of lactic acid bacteria for the production of exopolysaccharides in liquid media. *Biotechnology Techniques,* 11(5), 271–275.

Ruas-Madiedo P., de los Reyes-Gavilán C. G. (2005). Invited review: Methods for the screening, isolation, and characterization of exopolysaccharides produced by lactic acid bacteria. *Journal of Dairy Science,* 88(3), 843–56.

Ruas-Madiedo P., Abraham A., Mozzi F., Mayo B., López P., Pérez-Martínez G. (2008). Functionality of exopolysaccharides produced by lactic acid bacteria. *Molecular Aspects of Lactic Acid Bacteria for Traditional and New Applications,* 661(2), 137–166.

Ruas-Madiedo P., Hugenholtz J., Zoon P. (2002). An overview of the functionality of exopolysaccharides produced by lactic acid bacteria. *International Dairy Journal,* 12(2–3), 163–171.

Ruas-Madiedo P., Medrano M., Salazar N., De Los Reyes-Gavilán C. G., Pérez P. F., Abraham A. G. (2010). Exopolysaccharides produced by *Lactobacillus* and *Bifidobacterium* strains abrogate in vitro the cytotoxic effect of bacterial toxins on eukaryotic cells. *Journal of Applied Microbiology,* 109(6), 2079–2086.

Ruas-Madiedo P., Salazar N., de los Reyes-Gavilan C. G. (2009.) Biosyntesis and chemical composition of exopolysaccharides produced by lactic acid bacteria. In, *Bacterial Polysaccharides, Current Innovations and Future Trends,* Matthias Ullrich (Ed.), Caister Academic Press, Norfolk, UK, pp. 279-310.

Rühmann B., Schmid J., Sieber V. (2016). Automated modular high throughput exopolysaccharide screening platform coupled with highly sensitive carbohydrate fingerprint analysis. *Journal of Visualized Experiments,* no. 110.

Ryan P.M., Burdíková Z., Beresford T., Auty M. A. E., Fitzgerald G. F., Ross R. P., Sheehan J. J., Stanton C. (2015). Reduced-fat Cheddar and swiss-type cheeses harboring exopolysaccharide-producing probiotic *Lactobacillus mucosae* DPC 6426. *Journal of Dairy Science,* 98(12), 8531–8544.

Salazar N., López P., Garrido P., Moran J., Cabello E., Gueimonde M., Suárez A., González C., de Los Reyes-Gavilán C. G., Ruas-Madiedo P. (2014). Immune modulating capability of two exopolysaccharide-producing *Bifidobacterium* strains in a wistar rat model. *BioMed Research International,* 2014, 106290.

Salvetti E., Harris H. M. B., Felis G. E., O'Toole P. W. (2018). Comparative genomics of the genus *Lactobacillus* reveals robust phylogroups that provide the basis for reclassification. *Applied and Environmental Microbiology,* 84(17), AEM.00993-18.

Sanz M. L., Martínez-Castro I. (2007). Recent developments in sample preparation for chromatographic analysis of carbohydrates. *Journal of Chromatography* A 1153(1–2), 74–89.

Sasikumar K., Kozhummal Vaikkath D., Devendra L., Nampoothiri K. M. (2017). An exopolysaccharide (EPS) from a *Lactobacillus plantarum* BR2 with potential benefits for making functional foods. *Bioresource Technology,* 241, 1152–1156.

Schmid J. (2018). Recent insights in microbial exopolysaccharide biosynthesis and engineering strategies. *Current Opinion in Biotechnology,* 53, 130–136.

Schwab C., Walter J., Tannock G. W., Vogel R. F., Gänzle M. G. (2007). Sucrose utilization and impact of sucrose on glycosyltransferase expression in *Lactobacillus reuteri. Systematic and Applied Microbiology,* 30(6), 433–43.

Şengül N., Aslím B., Uçar G., Yücel N., Işk S., Bozkurt H., Sakaoğullarí Z., Atalay, F. (2006). Effects of exopolysaccharide-producing probiotic strains on experimental colitis in rats. *Diseases of the Colon & Rectum,* 49(2), 250–58.

Shao L., Wu Z., Tian F., Zhang H., Liu Z., Chen W., Guo B. (2015). Molecular characteristics of an exopolysaccharide from *Lactobacillus rhamnosus* KF5 in solution. *International Journal of Biological Macromolecules,* 72, 1429–1434.

Shao L., Wu Z., Zhang H., Chen W., Ai L., Guo B. (2014). Partial characterization and immunostimulatory activity of exopolysaccharides from *Lactobacillus rhamnosus* KF5. *Carbohydrate Polymers,* 107(1), 51–56.

Shi T., Aryantini N. P. D, Uchida K., Urashima T., Fukuda K. (2014). Enhancement of exopolysaccharide production of *Lactobacillus fermentum* TDS030603 by modifying culture conditions. *Bioscience of Microbiota, Food and Health,* 33(2), 85–90.

Siezen R. J. Tzeneva V. A., Castioni A., Wels M., Phan H. T. K., Rademaker J. L. W., Starrenburg M. J. C., Kleerebezem M., Molenaar D., Hylckama Vlieg J. E. T. van. (2010). Phenotypic and genomic diversity of *Lactobacillus plantarum* strains isolated from various environmental niches. *Environmental Microbiology,* 12(3), 758–773.

Silver R. P., Prior K., Nsahlai C., Wright L. F. (2001). ABC transporters and the export of capsular polysaccharides from gram-negative bacteria. *Research in Microbiology,* 152(3–4), 357–364.

Sönmez Ş., Darilmaz D. O., Beyatli Y. (2018). Determination of the relationship between oxalate degradation and exopolysaccharide production by different *Lactobacillus* probiotic strains. *International Journal of Dairy Technology,* 71(3), 741–752.

Staudacher H. M., Lomer M. C. E., Farquharson F. M., Louis P., Fava F., Franciosi E., Scholz M., Tuohy K. M., Lindsay J. O., Irving P. M., Whelan K. (2017). A diet low in fodmaps reduces symptoms in patients with irritable bowel syndrome and a probiotic restores *Bifidobacterium* species, a randomized controlled trial. *Gastroenterology,* 153(4), 936–947.

Stingele F., Neeser J-R., Mollet, B. (1996). Identification and characterization of the EPS (exopolysaccharide) gene cluster from *Streptococcus thermophilus* SFI6. *Journal of Bacteriology* 178(6), 1680-1690.

Tahoun A., Masutani H., El-Sharkawy H., Gillespie J., Honda R. P., Kuwata K., Inagaki M., Yabe T., Nomura I., Suzuki T. (2017). Capsular polysaccharide inhibits adhesion of *Bifidobacterium longum* 105-A to enterocyte-like caco-2 cells and phagocytosis by macrophages. *Gut Pathogens,* 9(1), 27.

Takeda K., Akira S. (2003). Toll receptors and pathogen resistance. *Cellular Microbiology,* 5(3), 143–53.

Tallon R., Bressollier P., Urdaci M. C. (2003). Isolation and characterization of two exopolysaccharides produced by *Lactobacillus plantarum* EP56. *Research in Microbiology,* 154(10), 705–712.

Tedelind S., Westberg F., Kjerrulf M., Vidal A. (2007). Anti-inflammatory properties of the short-chain fatty acids acetate and propionate, a study with relevance to inflammatory bowel disease. *World Journal of Gastroenterology,* 13(20), 2826.

Thapa D., Zhang H. (2009). *Lactobacillus rhamnosus* exopolysaccharide reduces mutagenic potential of genotoxins. *International Journal of Probiotics and Prebiotics,* 4(2), 79–82.

Tieking M., Gänzle M. G. (2005). Exopolysaccharides from cereal-associated lactobacilli. *Trends in Food Science and Technology,* 16(1–3), 79–84.

Tieking M., Ehrmann M. A., Vogel R. F., Gänzle M. G. (2005). Molecular and functional characterization of a levansucrase from the sourdough isolate *Lactobacillus sanfranciscensis* TMW 1.392. *Applied Microbiology and Biotechnology,* 66(6), 655–663.

Toba T., Kotani T., Adachi S. (1991). Capsular polysaccharide of a slime-forming *Lactococcus lactis* ssp. *cremoris* LAPT 3001 isolated from Swedish fermented milk 'långfil.' *International Journal of Food Microbiology,* 12(2–3), 167–171.

Tok E., Aslim B. (2010). Cholesterol removal by some lactic acid bacteria that can be used as probiotic. *Microbiology and Immunology,* 54(5), 257-264.

Torino M. I., Font de Valdez G., Mozzi F. (2015). Biopolymers from lactic acid bacteria. Novel applications in foods and beverages. *Frontiers in Microbiology,* 6, 1–16.

Tsuda H., Miyamoto T. (2010). Production of exopolysaccharide by *Lactobacillus plantarum* and the prebiotic activity of the exopolysaccharide. *Food Science and Technology Research,* 16(1), 87–92.

Tsuda H., Hara K., Miyamoto T. (2008). Binding of mutagens to exopolysaccharide produced by Lactobacillus plantarum mutant strain 301102S. *Journal of Dairy Science,* 91(8), 2960–66.

Tuinier R., van Casteren W. H. M., Looijesteijn P. J., Schols H. A., Voragen A. G. J., Zoon P. (2001). Effects of structural modifications on some physical characteristics of exopolysaccharides from Lactococcus lactis. *Biopolymers,* 59(3), 160–166.

Uemura J., Itoh T., Kasneko T., Noda K. (1998). Chemical characterization of extracellular polysaccharide from *Lactobacillus belbrueckii* subsp. *bulgaricus* OLL1073R-1. *Milchwissenschaft,* 53, 443–446.

Vedamuthu E. R., Neville J. M. (1986). Involvement of a plasmid in production of ropiness (mucoidness) in milk cultures by Streptococcus cremoris MS. *Applied and Environmental Microbiology,* 51(4), 677–682.

Vinderola G., Perdigón G., Duarte J., Farnworth E., Matar C. (2006). Effects of the oral administration of the exopolysaccharide produced by *Lactobacillus kefiranofaciens* on the gut mucosal immunity. *Cytokine*, 36(5–6), 254–260.

Vitlic A., Sadiq S., Ahmed H. I., Ale E. C., Binetti A. G., Collett A., Humphreys P., Laws A. (2018). Evidence for the modulation of the immune response in peripheral blood mononuclear cells after stimulation with a high molecular weight β-glucan isolated from Lactobacillus fermentum Lf2. *BioRXiv*, https.//doi.org/10.1101/400267.

Wachi S., Kanmani P., Tomosada Y., Kobayashi H., Yuri T., Egusa S., Shimazu T., Suda Y., Aso H., Sugawara M., Saito T., Mishima T., Villena J., Kitazawa H. (2014). *Lactobacillus delbrueckii* TUA4408l and its extracellular polysaccharides attenuate enterotoxigenic *Escherichia coli*-induced inflammatory response in porcine intestinal epitheliocytes via Toll-like receptor-2 and 4. *Molecular Nutrition & Food Research*, 58(10), 2080–2093.

Walker A. W., Duncan S. H., McWilliam E. C., Child L. M. W., Flint H. J. 2005. pH and peptide supply can radically alter bacterial populations and short-chain fatty acid ratios within microbial communities from the human colon. *Applied and Environmental Microbiology*, 71(7), 3692–3700.

Walter J. 2008. Ecological role of lactobacilli in the gastrointestinal tract implications for fundamental and biomedical research. *Applied and Environmental Microbiology*, 74(16), 4985–4996.

Werning M. L., Corrales M. A., Prieto A., Palencia P. F. de., Navas J., Lopez P. (2008). Heterologous expression of a position 2-substituted (1→3)-β-D-glucan in *Lactococcus lactis*. *Applied and Environmental Microbiology*, 74(16), 5259–5262.

Werning M. L., Ibarburu I., Dueñas M. T., Irastorza A., Navas J., López P. (2006). *Pediococcus parvulus* gtf gene encoding the GTF glycosyltransferase and its application for specific PCR detection of β-D-glucan-producing bacteria in foods and beverages. *Journal of Food Protection*, 69 (1), 161–169.

Werning M. L., Notararigo S., Nácher M., Fernández de Palencia P., Aznar R., López P. (2012). *Biosynthesis, purification and biotechnological use of exopolysaccharides produced by lactic acid bacteria*. InTech, 2012.

Yasuda E., Serata M., Sako T. (2008). Suppressive effect on activation of macrophages by *Lactobacillus casei* strain Shirota genes determining the synthesis of cell wall-associated polysaccharides. *Applied and Environmental Microbiology*, 74(15), 4746–4755.

Yilmaz M. T., Dertli M., Toker O. S., Tatlisu N. B., Sagdic O., Arici M. (2015). Effect of in situ exopolysaccharide production on physicochemical, rheological, sensory, and microstructural properties of the yogurt drink ayran, an optimization study based on fermentation kinetics. *Journal of Dairy Science*, 98(3), 1604–1624.

Zannini E., Waters D. M. Coffey A., Arendt E. K. (2016). Production, properties, and industrial food application of lactic acid bacteria-derived exopolysaccharides. *Applied Microbiology and Biotechnology,* 100(3), 1121–1135.

Zarour K., Goretti Llamas M., Prieto A., Rúas-Madiedo P., Dueñas M. T., Fernández de Palencia P., Aznar R., Kihal M., López P. (2017). Rheology and bioactivity of high molecular weight dextrans synthesised by lactic acid bacteria. *Carbohydrate Polymers,* 174, 646–657.

Zeidan A. A., Poulsen V. K., Janzen T., Buldo P., Derkx P. M. F., Øregaard G., Neves A. N. (2017). Polysaccharide production by lactic acid bacteria, from genes to industrial applications. *FEMS Microbiology Reviews,* 41(1), 168–200.

Zhang J., Zhao X., Jiang Y., Zhao W., Guo T., Cao Y., Teng J., Hao X., Zhao J., Yang Z. (2017). Antioxidant status and gut microbiota change in an aging mouse model as influenced by exopolysaccharide produced by *Lactobacillus plantarum* YW11 isolated from tibetan kefir. *Journal of Dairy Science,* 100(8), 6025–6041.

Zhang L., Folkenberg D. M., Amigo J. M., Ipsen R. (2016). Effect of exopolysaccharide-producing starter cultures and post-fermentation mechanical treatment on textural properties and microstructure of low fat yoghurt. *International Dairy Journal,* 53, 10–19.

Zhang L., Li X., Ren H., Liu L., Ma L., Li M., Bi W. (2015). Impact of using exopolysaccharides (EPS)-producing strain on qualities of half-fat Cheddar cheese. *International Journal of Food Properties,* 18(7), 1546–1559.

Zhang L., Liu C., Li D., Zhao Y., Zhang X., Zeng X., Yang Z., Li S. (2013). Antioxidant activity of an exopolysaccharide isolated from *Lactobacillus plantarum* C88. *International Journal of Biological Macromolecules,* 54, 270–275.

Zhang Y., Li S., Zhang C., Luo Y., Zhang H., Yang Z. (2011). Growth and exopolysaccharide production by *Lactobacillus fermentum* F6 in skim milk. *African Journal of Biotechnology,* 10(11), 2080–2091.

Zhang Y., Zhou L., Bao Y. L., Wu Y., Yu C. L., Huang Y. X., Sun Y., Zheng L. H., Li Y. X. (2010). Butyrate induces cell apoptosis through activation of JNK MAP kinase pathway in human colon cancer RKO cells. *Chemico-Biological Interactions,* 185(3), 174–181.

Zisu B., Shah N. P. (2003). Effects of pH, temperature, supplementation with whey protein concentrate, and adjunct cultures on the production of exopolysaccharides by Streptococcus thermophilus 1275. *Journal of Dairy Science,* 86(11), 3405–3415.

Zisu B., Shah N. P. (2005). Textural and functional changes in low-fat Mozzarella cheeses in relation to proteolysis and microstructure as influenced by the use of fat replacers, pre-acidification and EPS starter. *International Dairy Journal,* 15(6–9), 957–972.

Zisu B., Shah N. P. (2007). Texture characteristics and pizza bake properties of low-fat Mozzarella cheese as influenced by pre-acidification with citric acid and use of encapsulated and ropy exopolysaccharide producing cultures. *International Dairy Journal,* 17(8), 985–997.

Živković M., Miljković M. S., Ruas-Madiedo P., Markelić M. B., Veljović K., Tolinački M., Soković S., Korać A., Golić N. (2016). EPS-SJ exopolisaccharide produced by the strain *Lactobacillus paracasei* subsp. *paracasei* BGSJ2-8 is involved in adhesion to epithelial intestinal cells and decrease on *E. coli* association to Caco-2 cells. *Frontiers in Microbiology,* 7.

In: The Many Benefits of Lactic Acid Bacteria
Editors: J. G. LeBlanc and A. de Moreno

ISBN: 978-1-53615-388-0
© 2019 Nova Science Publishers, Inc.

Chapter 13

SPORE-FORMING LACTIC ACID BACTERIA AND THEIR ABILITIES FOR UTILIZATION

Sitanan Thitiprasert[1],, Nuttha Thongchul[1] and Somboon Tanasupawat[2],†*

[1]Institute of Biotechnology and Genetic Engineering,
Chulalongkorn University, Bangkok, Thailand
[2]Department of Biochemistry and Microbiology,
Faculty of Pharmaceutical Sciences,
Chulalongkorn University, Bangkok, Thailand

ABSTRACT

Nowadays, many studies have reported the potential spore-forming lactic acid bacteria (SFLAB) with the high performance of lactic acid production and high specificity of lactic acid isomers. This group of bacteria belongs to the genera *Bacillus* and *Sporolactobacillus*. Most of the SFLAB are homofermentative bacteria which can only produce lactic acid from sugar. Additionally, these SFLAB strains are dominant to produce an optically pure D(-)- or L(+)-lactic acid which is preferred to the specific application, especially in the manufacturing of biodegradable and biocompatible polylactic acid (PLA) which can be a replacement of synthetic plastics derived from petroleum resources. Considering the general characteristics of SFLAB, both genera are endospore-forming, rod-shaped, and gram-positive bacteria. The members of the genus *Bacillus* include *B. coagulans*, *B. stearothermophilus*, *B. licheniformis*, *B. subtilis*, and *Bacillus* sp. All of these *Bacillus* species could grow and form endospore under aerobic condition. On the other hand, the genus *Sporolactobacillus* differs from strains of

---

*Corresponding Author's Email: Sitanan.T@chula.ac.th.
†Corresponding Author's Email: Somboon.T@chula.ac.th.

Bacillus species in its response to oxygen, because the optimal growth has been found in microaerophilic or anaerobic condition. At the present, increasing the *Sporolactobacillus* species has been reported, including *S. inulinus*, *S. laevolacticus*, *S. nakayamae*, *S. racemicus*, *S. terrae*, *S. kufoensis*, *S. lactosus*, *S. putidus*, *S. vineae*, *S. pectinovorans*, *S. shoreicoticis*, *S. shoreae* and *S. spathodeae*. A few years ago, the novel genus *Terrilactibacillus* was discovered and reported as a potential spore-forming lactic acid bacterium. The type species is *Terrilacticbacillus laevilacticus* which has individual characteristics compared to the others. Currently, SFLAB have been applied in the wide range of applications, including biochemical compounds, probiotics, and commercial enzymes. This chapter describes the recent status of SFLAB strains and their advantages which could be found from the studies and their applications.

INTRODUCTION

Generally, lactic acid production has been well recognized from the biological conversion of carbohydrate by lactic acid bacteria (LAB). Plenty information of this bacteria group can been easily found. The phenotypic characterization of this group is Gram-positive, fastidious, non-spore forming rods or cocci which mainly produce lactic acid from carbohydrates under anaerobic environment. The group members of LAB comprise of genera *Lactobacillus*, *Leuconostoc*, *Pediococcus*, and *Streptococcus* (Lui et al., 2014). At the present, spore-forming lactic acid bacteria (SFLAB) are one of the potential lactic acid producers, including the genera *Bacillus* and *Sporolactobacillus*. Historically, the definition of a spore-forming organism was originally referred to the genus *Bacillus* that are able to form endospores which are heat-resistant bodies that were not easily killed by mere boiling (Cohn, 1876; Koch, 1876). The ability to form endospores of the bacteria results in resistance to unfavorable growth conditions. Later, the first classification of endospore-forming bacteria was proposed by de Bary (1884) and by Hüppe (1886) but the generic name *Bacillus* was more used for various rod-shaped bacteria (Fritze & Clausaccepted, 1995). However, to define as SFLAB, the genus *Bacillus* was respected as comprising spore-bearing rod producing lactic acid, which is facultative anaerobic or aerobic catalase positive, following the 8th edition of Bergey's Manual Determinative Bacteriology (Abriouel et al., 2011). Another SFLAB member is defined to the genus *Sporolactobacillus*. These bacteria strains were classified using the phylogenetic data between *Lactobacillaceae* and *Bacillaceae* (Nakayama & Sakaguchi, 1950). The phenotypic characterization was described as Gram-positive, microaerophilic, mesophilic rods, and catalase negative. Comparing to LAB, the group members of SFLAB were smaller discovered. Nowadays, some *Bacillus* and *Sporolactobacillus* species have been reported as lactic acid producer, including *B. coagulans*, *B. stearothermophilus*, *B. licheniformis*, *B. subtilis*, *Bacillus* sp., *S. inulinus*, *S. laevolacticus*, *S. nakayamae*, *S. terrae*, *S. kofuensis*, *S. lactosus*, *S. putidus*, and *Sporolactobacillus* sp. (Abdel-Rahman et al., 2013; Doores, 2014). Besides, most of SFLAB strains produced an optically pure lactic acid isomer (D(-) or L(+)-lactic acid) with a high lactic acid titer, depending on their cultivation conditions. Additionally, the recent novel genus *Terrilactibacillus* has been found as a promising D-lactic acid producer. This finding indicates that the novel strains could be a candidate for D-lactic acid production (Prasirtsak et al., 2016; 2017). Besides, the several performances of SFLAB to produce a wide range of products using the alternative substrates are now being developed.

Recently, the spore-forming lactic acid bacteria are being explored for the variety of applications. Besides their ability to produce lactic acid, the probiotics aspect is being widely studied. Additionally, another feature of this SFLAB group is the source of enzymes which can be applied in food and non-food industries. This information indicates that SFLAB could be a candidate for a variety of industry. This chapter updates the research regarding the spore-forming lactic acid strains and highlights the utilization of SFLAB strains in various applications.

SPORE-FORMING LACTIC ACID BACTERIA

SFLAB are microorganisms which mainly produce lactic acid and relates to lactic acid bacteria (LAB) strains. In particular, due to the remarkable and resistant dormancy of their spores and response to oxygen these characteristics are used to differentiate between SFLAB and LAB. In order to obtain interesting strains, several environment resources, including soil, foods, and feed were used for screening and isolation (Doores, 2014; Chang & Stackebrandt, 2014; Abriouel et al., 2011). Previously, the genus *Bacillus* and *Sporolactobacillus* were classified as a member of SFLAB. Besides, some basic characterizations of these 2 genera were different. The genus *Bacillus* is facultatively anaerobic or aerobic and catalase positive. Most of the *Bacillus* strains are able to grow and ferment the hexoses and pentoses to L(+)- or DL(-)-lactic acid. Moreover, they can produce lactic acid in the thermal fermentation (Abdel-Rahman et al., 2013). On contrary, *Sporolactobacillus* strains were described as the mesophilic, catalase negative, microaerophilic, and produce D(-)-lactic acid by a homofermentative pathway (Doores, 2014; Chang & Stackebrandt, 2014). Recently, the novel genus *Terrilactibacillus* has been discovered and published as a Gram-positive, catalase-positive, facultatively anaerobic, spore-forming, rod-shaped bacterium. It was found as the homofermentative D(-)-lactic acid producer (Prasirtsak et al., 2016). Additionally, their differentiating characteristics were also clarified (Table 1).

The Genus *Bacillus*

The genus *Bacillus* has long been known to be spore-forming bacteria due to its endospore formation in adverse environments. The generic name *Bacillus* was established by a German Botanist, Ferdinand Cohn, who has discovered rod-shaped bacteria which grow in filaments (Cohn, 1872). Later, in this genus the resting stages resistant to various adverse environments were detected; it was not killed by heat (Cohn, 1876; Koch, 1876). According to its heat-resistant behavior, it was found that the genus *Bacillus* was able to form endospores that are more resistant than vegetative cells to heat, drying, and other destructive agents (Gordon et al., 1973). The classification of *Bacillus* species was originally proposed by de Bary (1884) and Hüppe (1886). However, the *Bacillus* lactic acid producing strains were officially published in the Bergey's Manual of Determinative Bacteriology. According to the publication, the genus *Bacillus* was defined as the spore-forming rods capable of lactic acid production. Additionally, the general characterizations of the strains are Gram-positive, facultatively anaerobic or aerobic, and catalase positive (Abriouel et al., 2011; Zeigler and Perkins, 2008).

Bacillus strains have been utilized in a wide range of the applications, including enzyme technology, probiotics, food industry, medicine, cosmetics, agriculture, and chemistry (Poudel et al., 2016). In order to obtain the new functions and improve existing applications, the novel *Bacillus* strains have been continuously screened. Recently, the *Bacillus* strains have been recognized as a potential L(+)-lactic acid producing strain. The list of *Bacillus* lactic acid producing strains includes *B. coagulans*, *B. stearothermophilus*, *B. licheniformis*, *B. subtilis*, and new isolated *Bacillus* species. All these *Bacillus* strains produced an optically pure L(+)-lactic acid (97 - 100%) (Poudel et al., 2016). Interestingly, *Bacillus* species are able to grow and produce L(+)-lactic acid by using a simple mineral salts medium with a small amount of yeast extract (Poudel et al., 2015; Wang et al., 2011; Heriban, 1993). Furthermore, this species can tolerate high temperatures (45 - 60°C) and pH (5.0 - 9.0) (Thitiprasert et al., 2017; Poudel et al., 2015; Ma et al., 2014; Ye et al., 2013; Ou et al., 2009; Ramos et al., 2000). These characteristics indicate the several advantages can be gained in the L(+)-lactic acid fermentation by these *Bacillus* species. First, to operate the fermentation at high temperature could reduce the cooling cost, comparing to mesophiles which require the temperature control during fermentation (Abdel-Rahman et al., 2013). Second, to control the high-temperature fermentation would prevent contamination problems (Ye et al., 2013). Third, there are some researches that have claimed that the rate of L(+)-lactic acid production could be improved by operating the fermentation at high temperature (Heriban et al., 1993). Fourth, to use the thermophilic *Bacillus* species could provide the benefit on the simultaneous saccharification and fermentation (SSF) of lignocellulosic biomass with hydrolytic enzymes (cellulase, amylase) at an optimum temperature (Budhavaram & Fan, 2009; Maas et al., 2008a; Ou et al., 2011; Patel et al., 2006). In addition, the thermotolerant *Bacillus* species, including *B. coagulans* and *B. licheniformis* have been used for open fermentation using nonsterile media at high temperature (Wang et al., 2011; Wang et al., 2019).

In sugar metabolism, the *Bacillus* species are able to metabolite hexose and pentose sugars via the Embden-Meyerhof-Parnas and the pentose-phosphate pathway, respectively. The maximum lactic acid yield that could be achieved was up to 1 g/g-consumed sugar (Patel et al., 2006; Ye et al., 2013, 2014). Currently, the alternative carbohydrate resources have been extensively determined for lactic acid production using the *Bacillus* strains. A variety of crops have been being investigated for lactic acid production, including lignocellulosic biomass (e.g., crop residues), sucrose (e.g., sugar cane and beet molasses), starchy materials (e.g., cassava, potato, cereal-based products), and food-wastes (Abdel-Rahman et al., 2013; Poudel et al., 2016). However, the use of these alternative carbohydrates is still limited; in particular these materials require pretreatments that form the inhibitor compounds (e.g., furfural, 5-hydroxymethylfurfural, and some organic acids) resulting in decreasing the performance of microbial lactic acid production (Jönsson and Martín, 2016). On the other hand, there are some reports that have investigated the fermentation of lactic acid in lignocelluloses containing inhibitors medium. It was found that the *Bacillus* strains could tolerate to the pretreatment inhibitors and convert them to less-toxic substance. Especially, *B. coagulans* has been reported the resistant performance of the inhibitors at certain levels (Peng et al., 2013; Ye et al., 2013; Bischoff et al., 2010; Zhang et al., 2014).

Table 1. Characteristics of *Bacilus*, *Sporolactobacillus* and *Terrilactibacillus*

Characteristic	*Bacillus*	*Sporolactobacillus*	*Terrilactibacillus*
Cell form	Rods	Rods	Rods
Endospore position	Terminal	Terminal/subterminal	Terminal
Oxygen requirement	Aerobe or facultative anaerobe	Facultative anaerobe	Aerobe or facultative anaerobe
Growth temperature (°C)	Wide range (psychrophilic to thermophilic species)	15-45	20-45
pH range	Wide range (acidophilic to alkaliphilic species)	5-8	5-8.5
Maximum NaCl (%) for growth	Some are halophilic, some are salt tolerant	10	3
Catalase production	+	-	+
Oxidase	+/- depended on species	-	-
Nitrate reduction	+/- depended on species	-	+
Homolactic acid fermentation	Depended on species and cultivation conditions	+	+
Isomer of lactic acid	Most optically pure L(+)-lactic acid	D(-)-lactic acid	D(-)-lactic acid
G+C mol %	32-69	43-50	42.6

Commonly, lactic acid bacteria producing strains employ a facultative anaerobic environment to produce lactic acid because of their lack of an oxygen electron acceptor, leading to a small amount of lactic acid production although they are able to grow in an environment presenting oxygen (Hofvendahl & Hahn-Hägerdal, 2000; Bobillo et al.,

1991). Considering to the *Bacillus* strains, they can produce lactic acid as a main metabolite either under strict anaerobic or aerobic condition (Thongpim et al., 2014; Thitiprasert et al., 2017).

In view of the aforementioned studies, it should be noted that the *Bacillus* lactic acid producing strains have been gained popularity as an interesting lactic acid producer. Particularly, *B. coagulans*, *B. subtilis*, *B. licheniformis*, and *B. thermoamylovorans* have been recognized as the promising L(+)-lactic acid strain (Poudel et al., 2016). Furthermore, the *Bacillus* species are recently determined for other applications, including probiotics, thermostable enzymes, and antimicrobial activity (Konuray & Erginkaya, 2018; Elshaghabee et al., 2017; Pandey 2017; Tolieng et al., 2018).

The Genus *Sporolactobacillus*

A spore-forming and homofermentative lactic acid bacteria in the genus *Sporolactobacillus* was originally isolated from chicken feed by Kitahara and Suzuki (1963). *Sporolactobacillus inulinus*, the type species was a Gram-positive, catalase negative, microaerophilic, mesophilic rods, producing exclusively D(-)-lactic acid, motile by a small number of peritrichous flagella and formed oval endospores in the terminal position (Kitahara and Suzuki, 1963; Suzuki & Kitahara 1964). Later, another species of spore forming lactic acid bacteria were isolated from the soil of the plant rhizospheres (Nakayama & Yanoshi, 1967a, 1967b). The species showed different catalase activity and an enantiomeric form of lactic acid production. In the group of catalase-positive which consisted of *B. laevolacticus* and *B. racemilacticus*, optically pure D(-)-lactic acid and DL-lactic acid were produced, respectively. For the catalase negative strains which related to *S. inulinus* the production of DL-lactic acid were found. However, the bacterial names proposed by Nakayama & Yanoshi (1967a) were not included in the Approved Lists of Bacterial names (Skerman et al., 1980). According to the proposal of Andersch et al. (1994), they determined the strains of *B. laevolacticus* genetically and phenotypically distinct from other lactic acid-producing, aerobic, spore-forming bacteria. Then the name *B. laevolacticus* was revived and assigned to the same taxon as it was originally located, following the International Code of Nomenclature of Bacteria. Based on the properties of the single strain resulted in the name *B. racemilacticus* could not be validated (Andersch et al., 1994). In the year 2006, from the results of phylogenetic analyses, as well as chemotaxonomic and phenotypic characterization revealed that *B. laevolacticus* was not included in the cluster of the genus *Bacillus* but should be reclassified into the genus *Sporolactobacillus*. Therefore, *B. laevolacticus* was reclassified as *S. laevolacticus* comb. Nov. (Hatayama et al., 2006). Initially, the species were considered to be similar to *Bacillus* and *Lactobacillus* (Kitahara & Lai, 1967). Furthermore, there are more studies on the phenotypic and genotypic characterizations of the strains. Studies have indicated that the *Sporolactobacillus* species show a separate phylogenetic position and differ in phenotypic properties from the genera *Bacillus* and *Lactobacillus* (Fox et al., 1980; Ash et al., 1991; Farrow et al., 1994; Suzuki & Yamasato, 1994; Yanagida et al., 1997). According to the previous findings, the members of the genus *Sporolactobacillus* could be divided into 2 major clusters, including catalase positive and catalase negative cluster. *S. laevolacticus* is a member in the catalase positive cluster which referred to mesophilic, gram-positive, an optically pure D(-)-lactic acid-producing, motile, and spore-forming organism (Andesch et al., 1994; Chang & Stackerbrandt, 2014). In the cluster of catalase-negative bacteria, the *Sporolactobacillus* species are included *S. inulinus*, *S. nakayamae* subsp. *nakayamae*, *S. nakayamae* subsp. *racemicus*, *S. terrae*, *S. kufoensis*, and *S. lactosus* (Yanagida et al., 1997). Recently, more novel *Sporolactobacillus* species have been discovered, that include *S. putidus* (Fujita et al., 2010), *S. vineae* (Chang et al., 2008), *S. pectinovorans* (Lan et al., 2016), *S. shoreicoticis* (Toleing et al., 2017), *S. shoreae*, and *S. spathodeae* (Thamacharoensuk et al., 2015).

Since these studies, the screening and isolation of the new *Sporolactobacillus* species have been successfully done. It was found that simple techniques can be applied. From various environmental resources, e.g., natural samples (soil and tree barks), spoiled foods, and feed can be heated at high temperature (~ 80°C) to eliminate the asporogeneous cells (Yanagida & Suzuki, 2009). In the enrichment step, the GYP medium containing glucose, yeast extract, peptone, and other minerals have been used for the cultivation of the isolates. For the selection, both MRS and GYP medium have been used. However, the previous study reported that the *Sporolactobacillus* species require a non-complex medium in particular GYP containing $CaCO_3$ medium (Chang & Stackebrandt, 2014). Until now, the phenotypic and molecular analysis of all *Sporolactobacillus* species have been clarified (Table 2). All of them are homofermentative D(-) or DL-lactic acid producer with using a variety of carbohydrates, e.g., glucose, fructose, and sucrose. Additionally, other medium components are required for growth and product formation. Most of the essential nitrogen sources, e.g., yeast extract, peptone, and inorganic nitrogen sources (NH_4Cl, NaH_2PO_4), play an important role on the concentration and productivity of lactic acid. While the free amino acids and peptides also influence in the fermentation of *Sporolactobacillus* strains, the lactic acid production might be limited by the concentration of free amino acid in the medium. The trace elements could affect lactic acid production in particular phosphate is necessary for the phosphorylation in a biological reaction which causes a significant product formation. Additionally, lactate dehydrogenase activity is induced by Mn (Klotz et al., 2017). At present, many studies have investigated using the alternative nutrition resources, e.g., protein hydrolysate from agricultural wastes for replacing the complex nutritional resources; especially yeast extract. However, lower product yield and productivity were found, compared with using yeast extract supplemented medium (Lu et al., 2009; Altaf et al., 2005).

Table 2. Differential characteristics of the *Sporolactobacillus* species*

Characteristic	1	2	3	4	5	6	7	8	9	10	11	12
Optimum temp. (°C) for growth	30	30	37	37	30	30	30	30	30	30	37	30
Optimum pH for growth	6.0	6.0 - 7.0	5.0	5.0	6.0 - 7.0	6.0 - 7.0	6.0 - 7.0	6.0 - 7.0	6.0 - 7.0	6.0 - 7.0	6.0	ND
Growth in 2% NaCl	w	-	-	+	+	+	+	+	+	w	w	+
Growth in 3% NaCl	-	-	-	+	+	+	+	+	+	-	-	+
Acid production from:												
D-Cellobiose	+	-	-	-	-	-	-	-	+	w	-	ND
D-Galactose	+	+	+	-	-	-	-	-	+	-	-	+
D-Lactose	-	-	-	-	-	-	-	-	-	-	ND	-
D-Maltose	-	-	w	+	-	-	+	-	+	w	-	w
D-Sorbitol	-	-	-	w	+	-	+	-	-	+	-	-
Starch	-	-	-	-	-	-	-	-	-	w	-	-
Sucrose	+	-	+	+	+	+	+	+	+	+	+	+
D-Trehalose	-	-	+	w	+	+	+	+	+	+	-	+
Enzyme activity												
Acid phosphatase	w	w	+	w	w	w	-	-	w	-	ND	+
Cystine arylamidase	w	-	-	-	+	-	-	-	w	-	ND	-
Esterase (C4)	-	-	-	-	-	-	-	-	w	-	ND	w
α-Galactosidase	-	-	-	-	-	-	-	-	+	-	ND	+
β-Galactosidase	-	-	-	-	-	-	-	-	+	w	ND	+
α-Glucosidase	-	-	+	w	-	-	-	-	w	-	ND	-
Leucine arylamidase	+	+	w	+	+	+	+	-	+	+	ND	ND
Naphthol-AS-BI-phosphohydrolase	-	-	-	-	+	-	+	-	w	-	ND	w
Valine arylamidase	+	+	-	+	+	+	+	-	+	-	ND	w

Strains: 1, *S. shoreae* BK92[T]; 2, *S. spathodeae* BK117-1[T]; 3, *S. putidus* JCM 15325[T]; 4, *S. vineae* JCM 14637[T]; 5, *S. inulinus* JCM 6014[T]; 6, *S. terrae* JCM 3516T; 7, *S. nakayamae* subsp. *nakayamae* JCM 3514[T]; 8, *S. kofuensis* JCM 3419[T]; 9, *S nakayamae* subsp. *racemicus* JCM 3417[T]; 10, *S. laevolacticus* JCM 2513[T]; 11, *S. pectinivorans* GD201205[T]; 12, *S. shoreicorticis* MK21-7[T]. +, Positive; w, weakly positive; -, negative reaction; ND, no data.

*Data from Thamacharoensuk et al. (2015), Lan et al. (2016), and Tolieng et al. (2017).

Recently, the members of genus *Sporolactobacillus* have been extensively studied their performance on another application, including probiotics. Furthermore, the high efficient D(-)-lactic acid-producing *Sporalctobacillus* strains are still being explored in order to obtain the strains are able to produce high lactic acid concentration and optically pure D(-)-lactic acid for supporting the synthesis of polylactic acid (PLA) application.

The Novel Genus *Terrilactibacillus*

The novel genus *Terrilactibacillus* was firstly reported by Prasirtsak et al. (2016) in their screening and isolation of lactic acid-producing bacteria from soil samples in Thailand. The strain NK26-11$^\text{T}$ was isolated and defined as a Gram-positive, catalase-positive, facultatively anaerobic, endospore-forming, rod-shaped, and motile by means of peritrichous flagella. The isolate showed the performance of homofermentative D(-)-lactic acid production from glucose medium. Based on its 16S rRNA gene sequence analysis, it was found that strain NK26-11$^\text{T}$ was located between the genera *Bacillus* and *Sporolactobacillus*. *Terrilactibacillus laevilacticus* is a type species for the genus *Terrilacticbacillus*. Later, during more screening and isolation on D(-)-lactic acid-producing bacteria conducted by Prasirtsak et al. (2017), *T. laevilacticus* SK5-6 as a potential D(-)-lactic acid producer was also isolated. The strain was studied the fermentation optimization in 5 L stirred fermenter using glucose medium. The results indicated that this strain could be used in industrial fermentation. Furthermore, the 2-phase fermentation approach revealed that the use of catalase-positive bacteria provides the advantage of homofermentative lactic acid production (Prasirtsak et al., 2017).

PERFORMANCE OF SPORE-FORMING LACTIC ACID BACTERIA IN VARIOUS APPLICATIONS

Lactic Acid Production

Lactic acid has long been known as a utility organic acid which can be applied in a variety of applications such as in food and non-food industries. In the field of food industry, lactic acid is well known to be used as food acidulant, preservative, flavoring, pH regulation, and mineral fortification. Lactic acid can be used as active ingredients in cosmetics industry which has the functionality as moisturizers, skin-lightening agents, humectants, anti-acne agents, and skin-rejuvenating agents (Wee et al., 2006). Considering its chemical functional groups containing a carboxylic and a hydroxy group; lactic acid can be utilized as precursors in the formulation of cleaning reagents, green solvents, and descaling agents. Moreover, lactic acid is able to be raw materials for synthesizing the useful chemicals such as propionic acid, acetic acid, acrylic acid, etc. (Dimerci et at., 1993; John et al., 2007). Currently, lactic acid is an important chemical which used as a monomer for the production of biodegradable and biocompatible polylactic acid (PLA). From the distinctive feature of PLA, it has received wide attention for utilizing in several applications, including pharmaceutical (prosthetic devices, sutures), textile (protective clothing), packaging and labeling (Wee et al., 2006; Martinez et al., 2013). Naturally, lactic acid has a chiral carbon atom which exists in two enantiomeric forms, including L(+)-lactic acid and D(-)-lactic acid. Because its optically pure isomer affects a polymer grade of PLA; therefore, the use of optically pure isomer is more preferred than a racemic mixture of lactic acid. The recent lactic acid production is manufactured from the biotechnological process. Lactic acid bacteria (LAB) are a mandatory group as the lactic acid producer in particularly the genus *Lactobacillus*. Several advantages could be obtained from the microbial lactic acid fermentation such as high yield and productivity of lactic acid production. Moreover, the biotechnological lactic acid production provides high specificity to produce an optically pure L(+)- or D(-)-lactic acid (Pandey et al., 2001). However, the limitations and challenges of lactic acid fermentation are being continuously discussed. This includes the use of renewable resources (cellulosic sugar, food carbohydrates) as substrates, low pH inhibition, contamination due to the mesophilic characteristic of LAB strains, and decrease optical purity (Abdel-Rahman et al., 2016). Therefore, numerous researchers have studied the isolation of new microbes from natural resources and also the improvement of existing microbes by molecular technology.

Currently, SFLAB are gained attractive attention for high performance of lactic acid production with optically pure lactic acid isomer (Table 3). The genus *Bacillus* is the most used as a L(+)-lactic acid producer due to its special features compared to other genera. There are several L(+)-lactic acid fermentations using *Bacillus* strains that have been developed. Strains of *Bacillus coagulans* are reported for their remarkable L(+)-lactic acid production features, including low nutrient requirements, tolerance to high temperature and pH, strong ability to ferment both hexose and pentose sugars to high optically pure L(+)-lactic acid via pentose phosphate pathway (PPP) with 98 - 100% theoretical yield (Wang et al., 2013, 2015; Patel et al., 2006; Ye et al., 2013, 2014). The novel homofermentative *Bacillus* sp. BC-001 was firstly reported as the potent industrial L-lactate-producing strain with a high lactic acid yield (0.96 and 0.87 g/g) and productivity (2.8 and 2.6 g/L·h) using glucose and sucrose as substrates. Interestingly, the pH during L-lactic acid fermentation by the novel BC-001 could be simply controlled using monovalent bases (NH_4OH, NaOH, and KOH) (Thitiprasert et al., 2017).

Table 3. Lactic acid production by the recent strains of spore-forming bacteria

Strain	Substrate	Fermentation mode	Lactic acid				Reference
			Titer (g/L)	Yield (g/g)	Productivity (g/L·h)	%ee (Isomer)	
B. coagulans IPE22	Non-sterile soluble starch	SLSF[a] Batch	68.72	0.99	1.72	N/D[b]	Wang et al., 2019
B. coagulans arr4	Granulated sugar and yeast extract	Exponential fed-batch	206.81	0.97	5.30	99.11(L)	Coelhe et al., 2018
B. coagulans NRBC 12714	Glucose	Open batch	99.84	0.99	3.99	99.9 (L)	Ma et al., 2016
B. coagulans NRBC 12714	Xylose	Open batch	92.82	0.93	2.90	99.8 (L)	Ma et al., 2016
B. coagulans NRBC 12714	Corn stover hydrolysate	Open batch with	98.25	0.93	2.90	> 99.5 (L)	Ma et al., 2016
B. coagulans NRBC 12714	Corn stover hydrolysate	Open cell-recycle and repeated batch	93.76	0.93	3.80	> 99.5 (L)	Ma et al., 2016
B. coagulans NRBC 12714	Corn stover hydrolysate	Open cell-recycle and continuous	92	0.92	13.80	> 99.5 (L)	Ma et al., 2016
Bacillus sp. BC-001	Glucose	3000 L Batch	104.11	0.87	6.31	100 (L)	Thitiprasert et al., 2017
B. licheniformis BL1	Glucose	2 L Batch	Up to 130	0.86	Up to 7.80	> 99 (L)	Wang et al. 2011
S. nakayamae	sugar, yeast extract, mineral salts	Multipulse fed-batch	122.41	N/D	3.65	N/D[b] (D)	Beitel et al., 2017
S. nakayamae	Sucrose and peanut flour	Batch	112.93	0.98	1.57	98.75 (D)	Beitel et al., 2016
Sporolactobacillus sp. CASD	Glucose and peanut meal	30-L Pulse fed-batch	207	0.93	3.80	99.3 (D)	Wang et al., 2011
S. laevolacticus JCM 2513	Raw cane sugar	Continuous	67.3	0.98	11.20	N/D[b] (D)	Mimitsuka et al., 2012
T. laevilacticus SK5-6	Glucose	5-L Batch	92.60	0.84	1.93	99.56 (D)	Prasirtsak et al., 2017

Note: [a]One-step simultaneous liquefaction, saccharification and fermentation.

[b]There is no report for the quantity of isomeric lactic acid purity.

According to D-lactic acid production, the wild-type *Sporolactobacillus* strains have recent been described as potential D-lactic acid producer which could produce D-lactic acid from a simple medium composition and provide a high optically pure D-lactate isomer. Additionally, the *Sporolactobacillus* species is being explored for their ability to utilize the alternative resources as substrate for D-lactic acid production. The efficient D-lactic acid production by *S. nakayamae* utilizing cheap peanut flour as a nitrogen source and commercial sucrose as a carbon source was determined. The results indicated that *S.*

According to D-lactic acid production, the wild-type *Sporolactobacillus* strains have recent been described as potential D-lactic acid producer which could produce D-lactic acid from a simple medium composition and provide a high optically pure D-lactate isomer. Additionally, the *Sporolactobacillus* species is being explored for their ability to utilize the alternative resources as substrate for D-lactic acid production. The efficient D-lactic acid production by *S. nakayamae* utilizing cheap peanut flour as a nitrogen source and commercial sucrose as a carbon source was determined. The results indicated that *S. nakayamae* is a promising new D-lactic acid producer with a high D(-)-lactic acid concentration of 112.93 g/L by fermentation using 110 g/L of sucrose, 150.40 g/L of peanut flour, and 0.16 mL/L of tween 80 (Beitel et al., 2016). Wang et al. (2011) have demonstrated the performance of *Sporolactobacillus* sp. strain CASD using various alternative nitrogen sources. It was found that a very high D-lactic acid production was achieved at a concentration of 207 g/L, the average productivity of 3.8 g/L·h and optical purity of 99.3% using 40 g/L of peanut meal in 30-L fed-batch fermentation. Compared to L-lactate yield, the production of D-lactate needs to be improved, leading to many studies that have discovered the potential strains and explored more technologies to improve D-lactate production by the strains of *Sporolactobacillus* (Nakano et al., 2018; Tolieng et al., 2017; Klotz et al., 2017; Zheng et al., 2017; Thamacharoensuk et al., 2016; Sun et al., 2015). As above mentioned, the novel genus *Terrilactibacillus* has been just discovered in a few years ago. It was found as another promising D-lactate-producing bacterium for use in industrial fermentation. *Terrilactibacillus laevilacticus* SK5-6 was screened and determined its performance of D-lactic acid fermentation using glucose medium. In this study identified *T. laevilacticus* SK5-6 as an industrial strain with 96.6 g/L final lactate titer, 0.84 g/g yield, 1.93 g/L·h, and 99.6% optical purity of D-lactate (Prasirtsak et al., 2017). The use of SFLAB members as lactic acid producers has been increasing. Because of their several advantages on providing an optically pure lactic acid isomer and high efficiency of lactic acid production could lead SFLAB to be the lactic acid-producing candidate in future approaches.

Probiotics

The recent trend of probiotics has been extensively increasing. Probiotics are live microorganisms which identified as the beneficial microbes providing the several applications in health and clinical scenarios (Elshaghabee et al., 2017). The spore-forming lactic acid bacteria have long been explored and commercially produced as probiotics for human and animal health, in particular, *Bacillus* species (Table 4) (Khatri et al., 2016; Elshaghabee et al., 2017; Cutting, 2011; Hong et al., 2005), although a number of SFLAB information is not compared to LAB species. The findings have presented that the strains of *Bacillus* are remarked as probiotics due to their inherent ability of secretory proteins, enzymes, antimicrobial compounds, and vitamins. Various *Bacillus* lactic acid-producing species have been studied to examine their probiotic attributes. Among the *Bacillus* species, the candidate probiotic of *Bacillus* lactic acid-producing strains includes *B. subtilis*, *B. coagulans*, and *B. licheniformis*. Furthermore, the genus *Bacillus*, spore-forming lactic acid strains, is mostly recognized as the Generally Regarded As Safe (GRAS) microorganisms (Khatri et al., 2016). The distinctive endospores characteristic of the *Bacillus* species is able to resist in extreme conditions, including nutrient deprivation and environment stress (low pH, high temperatures, UV irradiation, and solvents). By their spore-forming character, the strains of *Bacillus* play a crucial role as the potential probiotic functionalities by mean of bile and acid tolerance, mucin binding, immune stimulation ability, and clinical efficacy (Khatri et al., 2016; Elshaghabee et al., 2017). Recently, several studies have reported the beneficial effect of *Bacillus* strains on the gut metabolism and physiology. Different *Bacillus* strains have shown their ability as antimicrobial, anti-oxidative and immune-modulatory in the host (Elshaghabee et al., 2017). Additionally, the *Bacillus* strains exhibit the antagonist activity which suppresses the pathogenic activities. *B. subtilis* strains were reported their antagonist activity by exhibiting aninocoumacin A activity which could attribute to the anti *H. pylori* activity (Pinchuk et al., 2001). In 2016, Gobi et al. reported the impact of *B. licheniformis* Dahb1 on reducing the formation of biofilm in the virulent *V. parahaemolyticus* Dahv2, an Asian catfish. Bacteriocin-producing strains of probiotic *B. coagulans* have been also identified. The bacteriocins of *B. subtilis* strains have a great antimicrobial activity against food-born and related clinically relevant pathogens, besides having lower cytotoxicity (Abdhul et al., 2015). Also, *B. coagulans* has been evaluated because of beneficial effect on the gut metabolism. It was indicated that feeding of *B. coagulans* to cholic acid fed rat, improved intestinal permeability and reduced the bactericidal effect of bile acid (Lee et al., 2016). Furthermore, the positive effect of *Bacillus* strains to distant cells, beyond GIT has also been established by numerous studies (Foligné et al., 2012; Abhari et al., 2016). At present, the potential probiotic *Bacillus* species are being explored for more applications. Lee et al. (2012) studied the effect of *B. subtilis* CSY191 that secretes a surfactin like compound which could inhibit the growth of anti-human breast cancer cells. Additionally, the direct effect of active compounds, which are produced by the *Bacillus* species have been determined as the alternative and effective therapy for metabolic syndromes (Choi et al., 2016; Yang et al., 2014; Ghoneim et al., 2016; Zouari et al., 2015). Commercially, the probiotic products are under regulated, depending on the regulatory system of the country. In terms of the *Bacillus* lactic acid-producing strains, these are one of the mostly commonly used as probiotic based products in the range of human health supplements, animal feed, and aquaculture (Elshaghabee et al., 2017).

The spore former *Sporolactobacillus* species have been also studied as probiotics. However, only a few studies have worked on the function of probiotic *Sporolactobacillus* species. Hyrominus et al. (2000) studied the performance of various *Sporolactobacillus* strains for being as probiotics by testing their acid and bile tolerance in different cultivations. The study reported that the examined *Sporolactobacillus* strains could survive in acidic conditions (pH 2.5 - 3) and some are weakly tolerant to 0.3% bile.

Table 4. Probiotics of spore-forming LAB in the worldwide market

Strain and description	Probiotics	Target
B. licheniformis (NCTC 13123) at 10^9 - 10^{10} spores/kg, non-bacteriocin strain and not licensed in the EU.	AlCare™	Swine
B. licheniformis (1.6×10^9 CFU/g) and *B. subtilis* (1.6×10^9 CFU/g).	BioGrow®	Poultry, calves and swine
Mixture (1/1) of *B. licheniformis* (DSM 5749) and *B. subtilis* (DSM 5750) at (1.6×10^9 CFU/g) of each bacterium. EU approved.	BioPlus 2B®	Piglets, chickens, turkeys for fattening
Carries 4 strains of *B. subtilis*	Promarine®	Aquaculture shrimps
B. coagulans (1.0×10^9 spores/dose)	Lactopure	Poultry, swine, calves
B. coagulans and *B. subtilis*	NutriCommit	Human
B. coagulans (1.0×10^9 spores/dose)	Lactospore	Human
B. coagulans GBI-30	Sustenex	Human
B. coagulans	Neolactoflorene	Human
B. coagulans 15B and fructooligosaccharide	Nutrition essentials Probiotic	Human
Labelled as dietary supplement contained a stable *B. coagulans* that survives in stomach acid and retains its potency in the intestines.	THORNE®	Human
B. coagulans (*B. coagulans* GBI-30, 6086)	GanedenBC^{30}	Human and animal

*Information was modified from Cutting (2011), Elshaghabee et al. (2017), Konuray and Erginkaya (2018), and Hong et al. (2005).

Moreover, the study suggested that other parameters such as the spore-forming behavior in the gastrointestinal tract should be considered. The basic characteristics of *S. inulinus* BCRC 14647 were determined for its potential probiotic properties. Several factors, including acid and bile tolerance, adhesiveness, and antagonistic effects on pathogenic *Salmonella enteritidis* BCRC 10744 were tested. The results revealed that the vegetative cell form of *S. inulinus* exhibits probiotic features (Huang et al., 2007). The draft genome sequence of *S. vineae* SL153^T was analyzed to determine its probiotic characteristics. The study found that *S. vinae* SL153^T presents a high level of cell adhesion activity and growth inhibitory effect on pathogenic bacteria, including *Vibrio cholerea*, *Vibrio alginolyticus*, *Vibrio fluvialis*, *Vibrio parahaemolyticus*, *Aeromonas bivalvium*, *Listonella anguillarum* which improves the effectiveness of the prevention and treatment of disease infected by these pathogenic microorganisms (Kim et al., 2012).

Several numbers of researches have been developing in both phenotypic and genotypic probiotics in order to achieve the greatly beneficial human and animal health. Thereby, in order to promote health benefits, suitable probiotic formulations should be considered. Furthermore, a little information on probiotics spore-forming bacteria for clinical trial testing has been revealed. It is noted that further studies on the functionality and safety of probiotic spore-forming bacteria are required.

Other Applications and Future Perspectives

SFLAB, especially the *Bacillus* strains, have long been considered as the cell factories for the production of significant substances. Enzyme technology is one of the most important application that could be obtained from SFLAB. Various extracellular enzymes are commercially produced by *B. subtilis, B. lichniformis, and B. coagulans.* Moreover, most of enzymes produced by *Bacillus* strains have been used as thermostable enzymes (Frizte & Claus, 1995). Bacterial enzymes including amylase, pullulanase, xylanase, lipase, proteinase, α-glucosidase, cyclodextrin glycosyltransferase, glucose isomerase, 1,4-β-galctosidase, protease, DNA polymerase and restrict endonucleases have been used by food, pharmaceutical, textile, cosmetic, and household detergents industry (Meima et al., 2004; Kumar et al., 2005; Kanwar et al., 2006).

As previously mentioned for probiotics, SFLAB strains have the ability to produce antimicrobial substances against different food-borne pathogens, such as *Salmonella typhimurium, Clostidium perfringens*, and *Escherichia coli* (Shivaramaiah et al., 2011; Tactacan et al., 2013; La et al., 2001). Additionally, *Bacillus* strains are also the commercial product in the field of agriculture. *Bacillus subtilis* has been found as a promotor of plant growth by improving the survival and robustness of the seedlings (Cavaglieri et al., 2005; Ugoji et al., 2005).

Nowadays, the trend of healthy and functional food is widely interesting. Probiotic SFLAB have been extensively explored. However, compared with LAB, the lactic acid spore formers have not gained high popularity yet. Thereby, there is a need to elucidate and evaluate the features of spore former probiotic strain. Moreover, the association of safety and licensing issues that influence the use of bacteria spore former should be clarified. In addition to improve the polymer grade, SFLAB is presented as a good lactic acid producer which provide a high product yield and the desired stereoisomer, optically pure L(+)- or D(-)-lactic acid. So far, a stable and highly productive lactic acid fermentation is required. Therefore, the potential of SFLAB should be further explored. Furthermore, the cost reduction of biotechnological lactic acid production is now being developed.

CONCLUSION

SFLAB have been gaining attractiveness as microorganisms because of their many efficacies utilized in many applications. The endospore-forming characteristic provides an excellent ability of SFLAB to tolerate extreme conditions (like high temperature, pH, salts, and reagents). Interestingly, SFLAB has resilient features, for instance, the SFLAB strains could grow and produce product either under aerobic or facultative anaerobic condition, including the use of alternative resources as a substrate. Thereby, SFLAB have become an attractive outcome for being lactic acid-producing candidate and probiotics. Additionally, SFLAB are able to be the enzymes factory which produce variety of enzymes. Furthermore, most of these SFLAB enzymes are now applied in various industries. Nonetheless, there is a little information regarding SFLAB; therefore, it is necessary to develop more studies for understanding SFLAB features.

REFERENCES

Abdel-Rahman, M. A. & Sonomoto, K. (2016). Opportunities to overcome the current limitations and challenges for efficient microbial production of optically pure lactic acid. *Journal of Biotechnology*, *236*, 176-192. https://doi.org/10.1016/j.jbiotec.2016.08.008.

Abdel-Rahman, M. A., Tashiro, Y. & Sonomoto, K. (2013). Recent advances in lactic acid production by microbial fermentation processes. *Biotechnology Advances*, *31(6)*, 877-902. https://doi.org/10.1016/j.biotechadv.2013.04.002.

Abdhul, K., Ganesh, M., Shanmughapriya, S., Vanithamani, S., Kanagavel, M., Anbarasu, K. & Natarajaseenivasan, K. (2015). Bacteriocinogenic potential of a probiotic strain *Bacillus coagulans* [BDU3] from Ngari. *International Journal of Biological Macromolecules*, *79*, 800-806. https://doi.org/https://doi.org/10.1016/j.ijbiomac. 2015.06.005.

Abhari, K., Shekarforoush, S. S., Hosseinzadeh, S., Nazifi, S., Sajedianfard, J. & Eskandari, M. H. (2016). The effects of orally administered *Bacillus coagulans* and inulin on prevention and progression of rheumatoid arthritis in rats. *Food and Nutrition Research*, *60*, 30876. https://doi.org/10.3402/fnr.v60.30876.

Abriouel, H., Franz, C. M., Ben Omar, N. & Gálvez, A. (2011). Diversity and applications of *Bacillus* bacteriocins. *FEMS Microbiology Reviews*, *35*, 201-232.

Andersch, I., Pianka, S., Fritze, D. & Claus, D. (1994). Description of *Bacillus laevolacticus* (Ex Nakayama and Yanoshi 1967) sp. nov., nom. rev. *International Journal of Systematic and Evolutionary Microbiology*, *44*, 659664.

Ash, C., Farrow, J. A. E., Wallbanks, S. & Collins, M. D. (1991). Phylogenetic heterogeneity of the Genus Bacillus revealed by comparative analysis of small-subunit-ribosomal RNA sequences. *Letters in Applied Microbiology*, *13(4)*, 202-206. https://doi.org/10.1111/j.1472-765X.1991.tb00608.x.

Beitel, S. M., Sass, D. C., Fontes Coelho, L. & Contiero, J. (2016). High D (−) lactic acid levels production by *Sporolactobacillus nakayamae* and an efficient purification. *Annals of Microbiology*, *66(4)*, 1367-1376. https://doi.org/10.1007/s13213-016-1224-4.

Beitel, S. M., Fontes Coelho, L., Sass, D. C. & Contiero, J. (2017). Environmentally friendly production of D(-)Lactic acid by *Sporolactobacillus nakayamae*: Investigation of fermentation parameters and fed-batch strategies. *International Journal of Microbiology*, *2017*, 1367-1376. https://doi.org/10.1155/2017/4851612.

Bischoff, K. M., Liu, S., Hughes, S. R. & Rich, J. O. (2010). Fermentation of corn fiber hydrolysate to lactic acid by the moderate thermophile *Bacillus coagulans*. *Biotechnology Letters*, *32(6)*, 823-828. https://doi.org/10.1007/s10529-010-0222-z.

Bobillo, M. & Marshall, V. M. (1991). Effect of salt and culture aeration on lactate and acetate production by *Lactobacillus plantarum*. *Food Microbiology*, *8(2)*, 153-160. https://doi.org/https://doi.org/10.1016/0740-0020(91)90008-P.

Budhavaram, N. K. & Fan, Z. (2009). Production of lactic acid from paper sludge using acid-tolerant, thermophilic *Bacillus coagulan* strains." [In Eng]. *Bioresource Technology*, *100(23)*, 5966-5972. https://doi.org/10.1016/j.biortech.2009.01.080.

Cavaglieri, L., Orlando, J., Rodríguez, M. I., Chulze, S. & Etcheverry, M. (2005). Biocontrol of *Bacillus subtilis* against *Fusarium verticillioides* in vitro and at the maize root level. *Research in Microbiology*, *156(5)*, 748-754. https://doi.org/ https://doi.org/10.1016/j.resmic.2005.03.001.

Chang, Y. H. & Stackebrandt, E. (2014). The Family *Sporolactobacillaceae*. In: *The prokaryotes: Firmicutes and tenericutes*, E. Rosenberg et al. (Eds), Verlag Berlin Heidelberg, Gemany: Springer.

Chang, Y. H., Jung, M. Y., Park, I. S. & Oh, H. M. (2008). *Sporolactobacillus vineae* sp. nov., a spore-forming lactic acid bacterium isolated from vineyard soil. *International Journal of Systematic and Evolutionary Microbiology*, *58(10)*, 2316-2320. https:// doi.org/10.1099/ijs.0.65608-0.

Choi, J. H., Pichiah, P. B., Kim, M. J. & Cha, Y. S. (2016). Cheonggukjang, a soybean paste fermented with *B. licheniformis*-67 prevents weight gain and improves glycemic control in high fat diet induced obese mice. *Journal of Clinical Biochemistry and Nutrition*, *59(1)*, 31-38. https://doi.org/10.3164/jcbn.15-30.

Coelho, L. F., Beitel, S. M., Sass, D. C., Neto, P. M. A. & Contiero, J. (2018). High-titer and productivity of L-(+)-lactic acid using exponential fed-batch fermentation with *Bacillus coagulans* arr4, a new thermotolerant bacterial strain. *3 Biotech*, *8(4)*, 213. https://doi.org/10.1007/s13205-018-1232-0.

Cohn, F. (1872). Untersuchungen Aber Bakterien [Investigations on Bacteria]. *Beitriige zur Biologie der Pflanzen*, *1*, 127-224.

Cutting, S. M. (2011). *Bacillus* probiotics. *Food Microbiology*, *28(2)*, 214-220. https://doi.org/10.1016/j.fm.2010.03.007.

de Bary, A. (1884). Vergleichende Morphologie Und Biologie Der Pilze. *Mycetozoen Und Bakterien* [Comparative Morphology and Biology of Mushrooms. *Mycetozoa and bacteria*]. Leipzig, Germany: Wilhelm Engelmann, 1884.

Demirci, A. Pometto A. L. & Johnson, k. E. (1993). Lactic acid production in a mixed-culture biofilm reactor. *Applied and Environmental Microbiology*, *59(1)*, 203.

Doores, S. (2014). The genus *Sporolactobacillus*. In: *Lactic acid bacteria biodiversity and taxonomy*, Brian J.B. Wood Wilhelm H. Holzapfel (Eds). Oxford, UK: John Wiley & Sons, Ltd.

Elshaghabee, F. M. F., Rokana, N., Gulhane, R. D., Sharma, C. & Panwar, H. (2017). *Bacillus* as potential probiotics: Status, concerns, and future perspectives." *Frontiers in Microbiology*, 8, 1490. https://doi.org/10.3389/fmicb.2017.01490.

Farrow, J. A., Wallbanks, S. & Collins, M. D. (1994). Phylogenetic interrelationships of round-spore-forming bacilli containing cell walls based on lysine and the non-spore-forming genera *Caryophanon*, *Exiguobacterium*, *Kurthia*, and *Planococcus*. *International Journal of Systematic and Evolutionary Microbiology*, 44, 74-82.

Foligné, B., Peys, E., Vandenkerckhove, J., Van Hemel, J., Dewulf, J., Breton, J. & Pot, B. (2012). Spores from two distinct colony types of the strain *Bacillus subtilis* PB6 substantiate anti-inflammatory probiotic effects in mice. *Clinical Nutrition*, *31(6)*, 987-994. https://doi.org/https://doi.org/10.1016/j.clnu.2012.05.016.

Fox, G. E., Stackebrandt, E., Hespell, R. B., Gibson, J., Maniloff, J., et al. (1980). The phylogeny of prokaryotes. *Science*, 209, 457-463.

Fritze, D. D. (1995). Spore-forming, lactic acid producing bacteria of the genera *Bacillus* and *Sporolactobacillus*. In *The genera of lactic acid bacteria*, B.J.B. & Holzapfel (Eds), Boston, MA: Springer, 368-391.

Fujita, R., Mochida, K., Kato, Y. & Goto, K. (2010). *Sporolactobacillus putidus* sp. nov., an endospore-forming lactic acid bacterium isolated from spoiled orange juice. *International Journal of Systematic and Evolutionary Microbiology*, *60(7)*, 1499-1503. https://doi.org/10.1099/ijs.0.002048-0.

Gobi, N., Malaikozhundan, B., Sekar, V., Shanthi, S., Vaseeharan, B., Jayakumar, R. & Nazar, A. K. (2016). Gfp tagged *Vibrio parahaemolyticus* Dahv2 infection and the protective effects of the probiotic *Bacillus licheniformis* Dahb1 on the growth, immune and antioxidant responses in *Pangasius hypophthalmus*. *Fish & Shellfish Immunology*, 52, 230-238. https://doi.org/https://doi.org/10.1016/j.fsi.2016.03.006.

Gordon, R. E., Haynes, W. C. & Pang, C. H. N. (1973). The Genus *Bacillus*. Chap. 116-119, In *USDA Agriculture Handbooks*.

Hatayama, K., Shoun, H., Ueda, Y. & Nakamura, A. (2006). *Tuberibacillus calidus* gen. nov., sp. nov., isolated from a compost pile and reclassification of *Bacillus naganoensis* Tomimura et al. 1990 as *Pullulanibacillus naganoensis* gen. nov., comb. nov. and *Bacillus laevolacticus* Andersch et al. 1994 as *Sporolactobacillus laevolacticus* comb. nov. *International Journal of Systematic and Evolutionary Microbiology*, *56(11)*, 2545-2551. https://doi.org/doi:10.1099/ijs.0.64303-0.

Heriban, V., ŠTurdík, E., Zalibera, Ľ. & Matuš, P. (1993). Process and metabolic characteristics of *Bacillus coagulans* as a lactic acid producer. *Letters in Applied Microbiology*, *16*, 243-246. https://doi.org/10.1111/j.1472-765X.1993.tb01409.x.

Hiippe, F. (1886). *Die Formen Der Bakterien Und Ihre Beziehungen Zu Den Gattungen Und Arten* [*The Forms of Bacteria and Their Relationships to Genera and Species*]. Verlag C. W. Kreidel VIII, Germany: Wiesbaden.

Hofvendahl, K. & Hahn–Hägerdal, B. (2000). Factors affecting the fermentative lactic acid production from renewable resources. *Enzyme and Microbial Technology*, *26(2)*, 87-107. https://doi.org/https://doi.org/10.1016/S0141-0229(99)00155-6.

Hong, H. A., Duc, L. H. & Cutting, S. M. (2005). The use of bacterial spore formers as probiotics. *FEMS Microbiology Reviews*, *29(4)*, 813-835. https://doi.org/https://doi:10.1016/j.femsre.2004.12.001.

Huang, H. Y., Huang, S. Y., Chen, P. Y., An-Erl King, V., Lin, Y. P. & Tsen, J. H. (2007). Basic characteristics of *Sporolactobacillus inulinus* BCRC 14647 for potential probiotic properties. *Current Microbiology*, *54(5)*, 396-404. https://doi.org/10.1007/s00284-006-0496-5.

John, R. P., Nampoothiri, K. M. & Pandey, A. (2007). Fermentative production of lactic acid from biomass: an overview on process developments and future perspectives. *Applied Microbiology and Biotechnology*, *74(3)*, 524-534. https://doi.org/10.1007/s00253-006-0779-6.

Jönsson, L. J. & Martín, C. (2016). Pretreatment of lignocellulose: formation of inhibitory by-products and strategies for minimizing their effects. *Bioresource Technology*, *199*, 103-112. https://doi.org/https://doi.org/10.1016/ j.biortech.2015. 10.009.

Khatri, I., Sharma, S., Ramya, T. N. C. & Subramanian, S. (2016). Complete genomes of *Bacillus coagulans* S-Lac and *Bacillus subtilis* TO-A JPC, two phylogenetically distinct probiotics. *PLOS One*, *11(6)*, 1-25. https://doi.org/10.1371/journal. pone.0156745.

Kim, D. S., Sin, Y., Kim, D. W., Paek, J., Kim, R. N., Jung, M. Y., Park, I. S., Kim, A., Kang, A., Park, H. S., Choi, S. H. & Chang, Y. H. (2012). Genome sequence of the probiotic bacterium *Sporolactobacillus vineae* SL153T. *Journal of Bacteriology*, *194(11)*, 3015-3016. https://doi.org/10.1128/JB.00452-12.

Kitahara, K. & Lai, C. L. (1967). On the spore formation of *Sporolactobacillus inulinus*. *The Journal of General and Apply Microbiology*, *13*, 97-203.

Kitahara, K. & Suzuki, J. (1963). *Sporolactobacillus* nov. subgen. *Journal of General and Apply Microbiology*, *9*, 59-71.

Klotz, S., Kuenz, A. & Prüße, U. (2017). Nutritional requirements and the impact of yeast extract on the D-lactic acid production by *Sporolactobacillus inulinus*. *Green Chemistry*, *19(19)*, 4633-4641. https://doi.org/10.1039/c7gc01796k.

Koch, R. (1876). Die Aetiologie Der Milzbrandkrankheit [The etiology of anthrax]. *Beitriige zur Biologie der Pflanzen, 2*, 277-310.

Konuray, G. & Erginkaya, Zerrin. (2018). Potential use of *Bacillus coagulans* in the food industry. *Foods, 7(6)*. https://doi.org/10.3390/foods7060092.

La Ragione, R. M., Casula, G., Cutting, S. M. & Woodward, M. J. (2001). *Bacillus subtilis* spores competitively exclude *Escherichia coli* O78:K80 in poultry. *Veterinary Microbiology.*, *79(2)*, 133-142. https://doi.org/https://doi.org/10.1016/S0378-1135 (00)00350-3.

Lan, Q. X., Chen, J., Lin, L., Ye, X. L., Yan, Q. Y., et al. (2016). *Sporolactobacillus pectinivorans* sp. nov., an anaerobic bacterium isolated from spoiled jelly. *International Journal of Systematic and Evolutionary Microbiology, 66(11)*, 4323-4328. https://doi.org/10.1099/ijsem.0.001351.

Lee, J. H., Nam, S. H., Seo, W. T., Yun, H. D., Hong, S. Y., Kim, M. K. & Cho, K. M. (2012). The production of surfactin during the fermentation of cheonggukjang by potential probiotic *Bacillus subtilis* CSY191 and the resultant growth suppression of MCF-7 human breast cancer cells. *Food Chemistry, 131(4)*, 1347-1354. https://doi.org/https://doi.org/10.1016/j.foodchem.2011.09.133.

Lee, Y., Yoshitsugu, R., Kikuchi, K., Joe, G. H., Tsuji, M., Nose, T., Shimizu, H., Hara, H., Minamida, K., Miwa, K. & Ishizuka, S. (2016). Combination of soya pulp and *Bacillus coagulans* lilac-01 improves intestinal bile acid metabolism without impairing the effects of prebiotics in rats fed a cholic acid-supplemented diet. *British Journal of Nutrition, 116(4)*, 603-610. https://doi.org/10.1017/S0007 114516002270.

Lu, Z., Lu, M., He, F. & Yu, L. (2009). An economical approach for D-Lactic acid production utilizing unpolished rice from aging paddy as major nutrient source. *Bioresource Technology, 100(6)*, 2026-2031. https://doi.org/https://doi.org/ 10.1016/j.biortech.2008.10.015.

Lui, W., Pang, H., Zhang, H. & Cai, Y. (2014). *Biodiversity of lactic acid bacteria.* In *Lactic acid bacteria fundamental and practice*, H. Zhang and Y. Cai (Eds), Germany: Springer Science+Business Media Dordrecht, 103-203.

Ma, K., Maeda, T., You, H. & Shirai, Y. (2014). Open fermentative production of L-lactic acid with high optical purity by thermophilic *Bacillus coagulans* using excess sludge as nutrient. *Bioresource Technology.*, *151*, 28-35. https://doi.org/10.1016/j.biortech.2013.10.022.

Ma, Kedong., Guoquan, Hu., Liwei, Pan., Zichao, Wang., Yi, Zhou., et al. Highly efficient production of optically pure L-Lactic Acid from corn stover hydrolysate by thermophilic *Bacillus coagulans*. *Bioresource Technology, 219(2016)*, 114-122. http://dx.doi.org/10.1016/j.biortech.2016.07.100.

Maas, R. H. W., Springer, J., Eggink, G. & Weusthuis, R. A. (2008). Xylose metabolism in the fungus *Rhizopus oryzae*: Effect of growth and respiration on L+-lactic acid production. *Journal of Industrial Microbiology and Biotechnology, 35(6)*, 569-578. https://doi.org/10.1007/s10295-008-0318-9.

Martinez Castillo, F. A., Balciunas, E. M., Salgado, J. M., Domínguez González, J. M., Converti, A. & Pinheiro de Souza Oliveira, R. (2013). Lactic acid properties, applications and production: A review. *Trends in Food Science and Technology, 30(1)*, 70-83. https://doi.org/10.1016/j.tifs.2012.11.007.

Mimitsuka, T., Na, K., Morita, K., Sawai, H., Minegishi, S., Henmi, M., Yamada, K., Shimizu, S. & Yonehara, T. (2012). A membrane-integrated fermentation reactor system: its effects in reducing the amount of sub-raw materials for D-lactic acid continuous fermentation by *Sporolactobacillus laevolacticus*. *Bioscience Biotechnology and Biochemistry, 76(1)*, 67-72. https://doi.org/10.1271/bbb. 110499.

Mohammad, A., Naveena, B. J. & Reddy, G. (2005). Screening of inexpensive nitrogen sources for production of L(+) lactic acid from starch by amylolytic *Lactobacillus amylophilus* GV6 in single step fermentation. *Food Technology and Biotechnology, 43*, 235-239.

Nakano. Sawada S., Yamada, R., Mimitsuka, T. & Ogino, H. (2018). Enhancement of the catalytic activity of D-lactate dehydrogenase from *Sporolactobacillus laevolacticus* by site-directed mutagenesis. *Biochemical Engineering Journal, 133*, 214-218. https://doi.org/10.1016/j.bej.2018.02.015.

Nakayama, O. & i Yanoshi, M. (1967a). Spore-bearing lactic acid bacteria isolated From rhizosphere. I. Taxonomic studies on *Bacillus laevolacticus* nov. sp. and *Bacillus racemilacticus* nov. sp. *The Journal of General and Applied Microbiology, 13*, 139-153.

Nakayama, O. & Sakaguchi, K. (1950). Studies on the spore-bearing lactic acid-forming bacilli. Part 1. *Journal of the Agricultural Chemical Society of Japan, 23*, 513-517.

Nakayama, O. & i Yanoshi, M. (1967b). Spore-bearing lactic acid bacteria isolated from rhizosphere. II. taxonomic studies on the catalase-negative strains. *The Journal of General and Applied Microbiology, 13*, 155-165.

Ou, M. S., Mohammed, N., Ingram, L. O. & Shanmugam, K. T. (2009). Thermophilic *Bacillus coagulans* requires less cellulases for simultaneous saccharification and fermentation of cellulose to products than mesophilic microbial biocatalysts. *Applied Biochemistry and Biotechnology, 155(1)*, 76-82. https://doi.org/10.1007/s12010-008-8509-4.

Pandey, A., Soccol, C. R., Rodriguez-Leon, J. A. & Nigam, P. (2001). *Solid state fermentation in biotechnology: Fundamentals and applications*. New Delhi, India: Asiatech.

Pandey, K. R. (2017). Mini review on the dynamic probiotic *Bacillus coagulans*. *Journal of Dairy and Veterinary Sciences, 3(3)*. https://doi.org/10.19080/jdvs.2017. 03.555614.

Patel, M. A., Ou, M. S., Harbrucker, R., Aldrich, H. C., Buszko, M. L., Ingram, L. O. & Shanmugam, K. T. (2006). Isolation and characterization of acid-tolerant, thermophilic bacteria for effective fermentation of biomass-derived sugars to lactic acid. *Applied and Environmental Microbiology, 72(5)*, 3228-3235. https://doi.org/ 10.1128/AEM.72.5.3228-3235.2006.

Peng, L., Wang, L., Che, C., Yang, G., Yu, B. & Ma, Y. (2013). *Bacillus* sp. strain P38: An efficient producer of L-lactate from cellulosic hydrolysate, with high tolerance for 2-furfural. *Bioresource Technology, 149*, 169-176. https://doi.org/https://doi.org/ 10.1016/j.biortech.2013.09.047.

Pinchuk, I. V., Bressollier, P., Verneuil, B., Fenet, B., Sorokulova, I. B., Mégraud, F. & Urdaci, M. C. (2001). *In vitro* anti-helicobacter pylori activity of the probiotic strain *Bacillus subtilis* 3 is due to secretion of antibiotics. *Antimicrobial Agents and Chemotherapy, 45(11)*, 3156-3161. https://doi.org/10.1128/AAC.45.11.3156-3161.2001.

Poudel, P., Tashiro, Y., Miyamoto, H., Miyamoto, H., Okugawa, Y. & Sakai, K. (2015). Direct starch fermentation to L-lactic acid by a newly isolated thermophilic strain, *Bacillus* sp. MC-07. *Journal of Industrial Microbiology and Biotechnology, 42*, 143-149.

Poudel, P., Tashiro, Y. & Saka, K. (2016). New application of *Bacillus* strains for optically pure L-lactic acid production: General overview and future prospects. *Bioscience, Biotechnology, and Biochemistry, 80*(4), 642-654. https://doi.org/ 10.1080/09168451.2015.1095069.

Prasirtsak, B., Thitiprasert, S., Tolieng, V., Assabumrungrat, S., Tanasupawat, S. & Thongchul, N. (2017). Characterization of D-lactic acid, spore-forming bacteria and *Terrilactibacillus laevilacticus* SK5-6 as potential industrial strains. *Annals of Microbiology, 67(11)*, 763-778. https://doi.org/10.1007/s13213-017-1306-y.

Prasirtsak, B., Thongchul, N., Tolieng, V. & Tanasupawat, S. (2016). *Terrilactibacillus laevilacticus* gen. nov., sp. nov., isolated from soil. *International Journal of Systematic and Evolution Microbiology, 66(3)*, 1311-1316. https://doi.org/ 10.1099/ijsem.0.000954.

Ramos, C. H., Hoffmann, T., Marino, M., Nedjari, H., Presecan-Siedel, E., Dreesen, O., Glaser, P. & Jahn, D. (2000). Fermentative metabolism of *Bacillus subtilis*: Physiology and regulation of gene expression. *Journal of Bacteriology, 182(11)*, 3072-3080.

Skerman, V. B. D., McGowan, V. & Sneath, P. H. A. (1980). Approved lists of bacterial names. *International Journal of Systematic and Evolution Microbiology, 30*, 225-420.

Sun, J., Wang, Y., Wu, B., Bai, Z. & He, B. (2015). Enhanced production of D-lactic acid by *Sporolactobacillus* sp. Y2-8 mutant generated by atmospheric and room temperature plasma. *Biotechnology and Applied Biochemistry*, *62(2)*, 287-292. https://doi.org/10.1002/bab.1267.

Suzuki, J. & Kitahara, K. (1964). Base compositions of deoxyribonucleic acid in *Sporolactobacillus inulinus* and other lactic acid bacteria. *The Journal of General and Applied Microbiology.*, *10*, 305-311. https://doi.org/10.2323/jgam.10.305.

Suzuki, T. & Yamasato, K. (1994). Phylogeny of spore-forming lactic acid bacteria based on 16S rRNA gene sequences. *FEMS Microbiology Letters*, *115*, 13-17. https://doi.org/10.1111/j.1574-6968.1994.tb06607.x.

Tactacan, G. B., Schmidt, J. K., Miille, M. J. & Jimenez, D. R. (2013). A *Bacillus subtilis* (QST 713) spore-based probiotic for necrotic enteritis control in broiler chickens. *The Journal of Applied Poultry Research*, *22(4)*, 825-831. https://doi.org/10.3382/japr.2013-00730.

Thamacharoensuk, T., Kitahara, M., Ohkuma, M., Thongchul, N. & Tanasupawat, S. (2015). *Sporolactobacillus shoreae* sp. nov. and *Sporolactobacillus spathodeae* sp. nov., two spore-forming lactic acid bacteria isolated from tree barks in Thailand. *International Journal of Systematic and Evolution Microbiology*, *65*, 1220-1226. https://doi.org/10.1099/ijs.0.000084.

Thamacharoensuk, T., Kitahara, M., Ohkuma, M., Thongchul, N. & Tanasupawat, S. (2016). Characterisation of lactic acid producing *Sporolactobacillus* strains from tree barks in Thailand. *Annals of Microbiology*, *67(2)*, 215-218. https://doi.org/10.1007/s13213-016-1248-9.

Thitiprasert, S., Kodama, K., Tanasupawat, S., Prasitchoke, P., Rampai, T., Prasirtsak, B., Tolieng, V., Piluk, J., Assabumrungrat, S. & Thongchul, N. (2017). A homofermentative *Bacillus* sp. BC-001 and its performance as a potential L-lactate industrial strain. *Bioprocess and Biosystems Engineering*, *40(12)*, 1787-1799. https://doi.org/10.1007/ s00449-017-1833-8.

Tolieng, V., Prasirtsak, B., Miyashita, M., Shibata, C., Tanaka, N., Thongchul, N. & Tanasupawat, S. (2017). *Sporolactobacillus shoreicorticis* sp. nov., a lactic acid-producing bacterium isolated from tree bark. *International Journal of Systematic and Evolution Microbiology*, *67(7)*, 2363-2369. https://doi.org/10.1099/ ijsem.0.001959.

Tongpim, S., Meidong, R., Poudel, P., Yoshino, S., Okugawa, Y., Tashiro, Y., Taniguchi, M. & Sakai, K. (2014). Isolation of thermophilic L-lactic acid producing bacteria showing homo-fermentative manner under high aeration condition. *Journal of Bioscience and Bioengineering*, *117(3)*, 318-324. https://doi.org/ https://doi.org/10.1016/j.jbiosc. 2013.08.017.

Ugoji, E. O., Laing, M. D. & Hunter, C. H. (2005). Colonization of *Bacillus* spp. on seeds and in plant rhizoplane. *Journal of Environmental Biology*, *26*, 459-466.

Wang, L., Xue, Z., Zhao, B., Yu, B., Xu, P. & Ma, Y. (2013). Jerusalem artichoke powder: A useful material in producing high optical purity L-lactate using an efficient sugar-utilizing thermophilic *Bacillus coagulans* strain. *Bioresource*

Technology, *130*, 174-180. https://doi.org/https://doi.org/10.1016/j.biortech.2012. 11.144.

Wang, L., Zhao, B., Li, F., Xu, K., Ma, C., Tao, F., Li, Q. & Xu, P. (2011). Highly efficient production of D-lactate by *Sporolactobacillus* sp. CASD with simultaneous enzymatic hydrolysis of peanut meal. *Applied Microbiology and Biotechnology*, *89(4)*, 1009-1017. https://doi.org/10.1007/s00253-010-2904-9.

Wang, Q., Zhao, X., Chamu, J. & Shanmugam, K. T. (2011). Isolation, characterization and evolution of a new thermophilic *Bacillus licheniformis* for lactic acid production in mineral salts medium. *Bioresource Technology*, *102(17)*, 8152-8158. https:// doi.org/10.1016/j.biortech.2011.06.003.

Wang, Y., Tashiro, Y. & Sonomoto, K. (2015). Fermentative production of lactic acid from renewable materials: Recent achievements, prospects, and limits. *Journal of Bioscience and Bioengineering*, *119(1)*, 10-18. https://doi.org/10.1016/j.jbiosc. 2014.06.003.

Wang, Y., Cao, W., Luo, J., Qi, B. & Wan, Y. (2019). One step open fermentation for lactic acid production from inedible starchy biomass by thermophilic *Bacillus coagulans* IPE22. *Bioresource Technology*, *272*, 398-406. https://doi.org/10.1016/ j.biortech.2018.10.043.

Wee, Y. J., Kim, J. N. & Ryu, H. W. (2006). Biotechnological production of lactic acid and its recent applications. *Food Technology and Biotechnology*, *44*, no. 2, 163-172.

Yanagida, F., Suzuki, K. I.., Kozaki, M. & Komagata, K. (1997). Proposal of *Sporolactobacillus nakayamae* subsp. *nakayamae* sp. nov., subsp. nov., *Sporolactobacillus nakayamae* subsp. *racemicus* subsp. nov., *Sporolactobacillus* terrae sp. nov., Sporolactobacillus *kofuensis* sp. nov., and *Sporolactobacillus lactosus* sp. nov. *International Journal of Systematic Bacteriology, 47(2)*, 499-504. https://doi.org/doi:10.1099/00207713-47-2-499.

Yanagida, F. & Suzuki, K. I. (2009). Genus I. *Sporolactobacillus* Kitahara and Suzuki 1963, 69[AL]. In De Vos P, Garrity Gm, Jones D, Krieg Nr, Ludwig W, Rainey Fa, Schleifer K-H, Whitman Wb (Eds) *Bergey's Manual of Systematic Bacteriology*, Dordrecht: Springer, 386-391.

Yang, H. J., Kwon, D. Y., Kim, H. J., Kim, M. J., Jung, D. Y., Kang, H. J., Kim, D. S., Kang, S., Moon, N. R., Shin, B. K. & Park, S. (2015). Fermenting soybeans with *Bacillus licheniformis* potentiates their capacity to improve cognitive function and glucose homeostaisis in diabetic rats with experimental Alzheimer's type dementia. *European Journal of Nutrition*, *54(1)*, 77-88. https://doi.org/10.1007/s00394-014-0687-y.

Ye, L., Hudari, M. S., Li, Z. & Wu, J. C. (2014). Simultaneous detoxification, saccharification and co-fermentation of oil palm empty fruit bunch hydrolysate for l-lactic acid production by *Bacillus coagulans* JI12. *Biochemical Engineering Journal*, *83*, 16-21. https://doi.org/https://doi.org/10.1016/j.bej.2013.12.005.

Ye, L., Zhou, X., Hudari, M. S., Li, Z. & Wu, J. C. (2013). Highly efficient production of L-lactic acid from xylose by newly isolated *Bacillus coagulans* C106. *Bioresource Technology*, *132*, 38-44. https://doi.org/10.1016/j.biortech.2013.01.011.

Zeigler, D. R. (2001). The Genus *Geobacillus*. In *Bacillus genetic stock center catalog of strains*, Ohio: Ohio State University, USA, 1-25.

Zhang, Y., Chen, X., Qi, B., Luo, J., Shen, F., Su, Y., Khan, R. & Wan, Y. (2014). Improving lactic acid productivity from wheat straw hydrolysates by membrane integrated repeated batch fermentation under non-sterilized conditions. *Bioresource Technology*, *163*, 160-166. https://doi.org/https://doi.org/10.1016/j. biortech. 2014.04.038.

Zheng, L., Liu, M., Sun, J., Wu, B. & He, B. (2017). Sodium ions activated phosphofructokinase leading to enhanced D-lactic acid production by *Sporolactobacillus inulinus* using sodium hydroxide as a neutralizing agent. *Applied Microbiology and Biotechnology*, *101(9)*, 3677-3687. https://doi.org/10.1007/ s00253-017-8120-0.

Zouari R., Ben Abdallah-Kolsi R., Hamden K., El Feki A., Chaabouni K., Makni-Ayadi F., Sallemi F., Ellouze-Chaabouni S. & Ghribi-Aydi D. (2015). Assessment of the antidiabetic and antilipidemic properties of *Bacillus subtilis* SPB1 biosurfactant in alloxan-induced diabetic rats. *Peptide Science*, *104(6)*, 764-774. https://doi.org/doi:10.1002/bip.22705.

ABOUT THE EDITORS

Jean Guy LeBlanc, PhD

Centro de Referencia para Lactobacilos (CERELA-CONICET),
San Miguel de Tucumán, Argentina

Dr. Jean Guy LeBlanc, PhD in Biochemistry (U. Nacional de Tucuman, Argentina), MSc and BSc in Biochemistry (U. Moncton, Canada), is a Principal Researcher (CERELA-CONICET, Argentina) and a post-graduate teacher (Faculty of Biochemistry, Chemistry and Pharmacy, U. Nacional Tucumán, Argentina). His principle areas of study include the use of lactic acid bacteria to increase bioactive compounds (vitamins, digestive enzymes, antioxidants, etc.) for the fermentation of foods or as bio-pharmaceuticals to treat and prevent vitamin deficiencies, inflammatory diseases and some types of cancer. He has edited 6 books, published 91 peer-reviewed articles and 28 book chapters and has participated in 135 works in scientific meetings (March, 2019).

Alejandra de Moreno de LeBlanc, PhD

Centro de Referencia para Lactobacilos (CERELA-CONICET),
San Miguel de Tucumán, Argentina

Dr. Alejandra de Moreno de LeBlanc, PhD in Biochemistry (U. Nacional de Tucuman, Argentina), is an Independent Researcher (CERELA-CONICET, Argentina) and a post-graduate teacher (Faculty of Biochemistry, Chemistry and Pharmacy, U. Nacional Tucumán, Argentina). Her principle areas of study include the evaluation of lactic acid bacteria as modulators of the host immune response in healthy conditions and in hosts suffering certain pathologies (cancer, inflammatory diseases, and neurodegenerative diseases, among others). She has edited 2 books, published 80 peer-reviewed articles and 23 book chapters and has participated in 113 works in scientific meetings (February, 2019)

INDEX

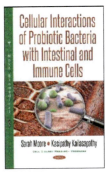
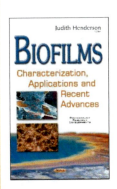

Related Nova Publications

BACTERIOPHAGES: AN OVERVIEW AND SYNTHESIS OF A RE-EMERGING FIELD

EDITOR: Daniel Harrington

SERIES: Bacteriology Research Developments

BOOK DESCRIPTION: This current book provides an overview of current research on bacteriophages.

SOFTCOVER ISBN: 978-1-63485-455-9
RETAIL PRICE: $95

To see a complete list of Nova publications, please visit our website at www.novapublishers.com